An Applied Approach

TIME SERIES AND FORECASTING

BRUCE L. BOWERMAN
and RICHARD T. O'CONNELL

Miami University

DUXBURY PRESS | *North Scituate Massachusetts*

Time Series and Forecasting: An Applied Approach
was edited and prepared for composition by M. N. Lewis
Interior design was provided by Dorothy Booth.
The cover was designed by Michael Fender.

Duxbury Press
A Division of Wadsworth, Inc.

Library of Congress Cataloging in Publication Data
Bowerman, Bruce L
  Time series and forecasting.

  Includes bibliographical references and index.
   1. Time-series analysis. 2. Forecasting.
I. O'Connell, Richard T., joint author.
II. Title.
QA280.B66   519.2′32   78-20869
ISBN 0-87872-218-1

Printed in the United States of America
1 2 3 4 5 6 7 8 9 — 83 82 81 80 79

To our parents

Edgar G. Bowerman
Esther U. Bowerman
and
Thomas C. O'Connell
Grace B. O'Connell

Contents

Preface

The objective of this book is to give a complete, applied, and easy to understand presentation of many of the statistical techniques that are useful in the short-term forecasting of time series. This book can be understood by a reader familiar with only high school algebra and basic statistics. Yet, this modest prerequisite has not been achieved at a cost of merely "talking about and around" forecasting techniques. A detailed presentation of each of the techniques discussed is given, and understanding is achieved through thorough explanation and detailed examples. Each forecasting technique is illustrated in an example, which applies the technique to a data set. In each case, an appropriate model is "fit" to the data set and is then used to forecast future values of the time series.

Some "starred sections" and problems in this book employ elementary matrix algebra. The matrix notation is used to demonstrate how various multiple regression calculations are done. These starred sections may be omitted without loss of continuity. However, Appendix A gives a very simple presentation of matrix algebra. Upon reading Appendix A, the reader can easily understand the starred sections and work the starred problems.

This book is designed to be used as a textbook for applied courses in time series analysis and forecasting. Specifically, it is written for advanced undergraduate and beginning graduate students in business, engineering, and the sciences (including mathematics, statistics, operations research, and computer science). Although very few proofs are presented, this book should also be useful as a supplementary textbook, giving detailed examples and an overview, for more theoretical courses in time series analysis and forecasting. Finally, this book should be valuable as a reference book for practitioners who must forecast real-world time series.

Since different teachers and practitioners feel that different forecasting techniques are important, this book has been organized for maximum flexibility, so that the reader may, without loss of continuity, omit those techniques considered less important and concentrate on those techniques considered more important.

This book is organized into four parts, each part containing several chapters.

Part I introduces basic concepts in time series analysis and forecasting and studies in some detail forecasting by regression analysis. One feature of Part I is Chapter 2, which studies using historical data to build an appropriate regression model.

Part II discusses the forecasting of time series described by trend and irregular components. Both the regression and exponential smoothing approaches to forecasting such time series are presented. A feature of Part II is emphasis on using historical data to determine appropriate smoothing constants for exponential smoothing models.

Part III describes the use of decomposition techniques, regression analysis, and exponential smoothing to forecast time series described by trend, seasonal, and irregular components. In this part we concentrate on forecasting two time series. Each of these time series is forecasted using several of the techniques presented. The forecasts obtained using the techniques are then compared.

Part IV studies the Box-Jenkins methodology of time series analysis. The approach that we take in our discussion of this topic differs from the traditional approach taken in most textbooks. The traditional approach begins with a lengthy, detailed discussion of the theoretical properties of the many models used by the Box-Jenkins methodology and then progresses to a discussion of how the models are used. We have found that readers often get bogged down in these preliminary theoretical discussions, because they do not understand why the properties are being discussed. Hence, in order to motivate the reader, we begin by combining discussion of the properties of one model with discussion of how the properties of this model relate to the steps that are taken in building and then forecasting with an appropriate Box-Jenkins model. We then proceed to a full discussion of the Box-Jenkins models. Although our approach is somewhat nonstandard, we have found it to be very effective both in the classroom and in written presentation of the Box-Jenkins methodology. An important feature of Part IV is a systematic, intuitive approach to the general modeling of seasonal time series by the Box-Jenkins methodology. We discuss a general strategy that can be used to identify an appropriate Box-Jenkins seasonal model, rather than merely presenting one example of seasonal modeling as many other books do. Another feature of Part IV is a case study of Box-Jenkins seasonal modeling that is presented in Chapter 12. This case study was written by Professor William Q. Meeker of the Department of Statistics at Iowa State University. It illustrates the modeling of a fairly complicated seasonal time series and also includes output from the TSERIES computer package, a computer program, written by Professor Meeker, that is designed to analyze and forecast time series using the Box-Jenkins methodology.

Many people have contributed to this book. We would especially like to thank Professor William Q. Meeker for writing the Box-Jenkins case study in Chapter 12. Professor Meeker also provided several extremely helpful reviews of this book during various stages of the writing. We would like to express our appreciation to Daniel Barnathan, Deborah Hanesworth, and Garen Wisner who, as our students, made many contributions to this book, including much of the computer work that had to be done in preparing examples. We would further like to thank our many friends and colleagues at Miami University for their many helpful suggestions and encouragement. We also wish to thank Kathleen Prescott for her superb typing of the manuscript and Hazel Spencer for the work she did duplicating the manuscript. We are deeply indebted to our parents, Esther and Edgar Bowerman and Grace and

Thomas O'Connell for the help, love, and encouragement they have given us over the years. Finally, we would like to thank our wives, Kathy and Jean, for their constant love and understanding during this writing endeavor.

Bruce L. Bowerman
Richard T. O'Connell

Oxford, Ohio, 1979

PART
I

Forecasting and Multiple Regression Analysis

Predictions of future events and conditions are called forecasts, and the act of making such predictions is called forecasting.

There are many techniques available to forecast future values of time series. Multiple regression, exponential smoothing, and Box-Jenkins techniques are useful quantitative forecasting methods.

If multiple regression is to be used as a forecasting technique, we must use historical data to build a multiple regression model that is likely to provide accurate forecasts.

CONTENTS

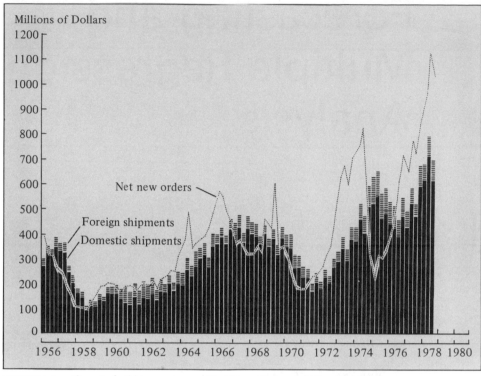

FIGURE I-1   *Quarterly Shipments and Net New Orders of Machine Tools*
Source: National Machine Tool Builders' Association. Reprinted by permission.

OBJECTIVES FOR PART I

In this book we consider the problem of predicting what will happen in the future. This problem is extremely important because many of the decisions we make in our everyday lives are based on predictions of conditions that are likely to exist in the future. For example, the clothing a person decides to wear is often influenced by the weather forecast he or she heard the night before. Similarly, the house that a family decides to buy often depends on predictions of future family earnings and expenses, since these predictions help the family decide whether they can afford the mortgage payments for a house they might like to purchase. As another example, the career path a college student chooses depends on his or her prediction of how marketable he or she will be several years in the future. Thus predictions of future conditions and events influence our lives in many ways. The objectives of Part I are to introduce the topic of forecasting and illustrate some of the ideas we will use throughout this book.

INTRODUCTION TO PART I

This book is about forecasting and some of the statistical techniques that can be used to produce forecasts. We will begin by making the following definition.

> Predictions of future events and conditions are called *forecasts,*
> and the act of making such predictions is called *forecasting.*

Clearly, forecasting is very important in many types of organizations since predictions of the future must be incorporated into the decision-making process. The government of a country must be able to forecast such things as air quality, water quality, unemployment rate, inflation rate, and welfare payments in order to formulate its policies. A university must be able to forecast student enrollment in order to make decisions concerning faculty resources, housing availability, and the like. The university might also wish to forecast daily mean temperature so that it can plan its fuel purchases for the coming months. A local school board must be able to forecast the number of school children of elementary school age who will be living in their school district years in the future in order to decide whether a new school should be built. Any organization must be able to make forecasts in order to make intelligent decisions.

In particular, business firms require forecasts of many events and conditions in all phases of their operations. Some examples of situations in which business forecasts are required are given below.

In marketing departments, reliable forecasts of demand must be available so that sales strategies can be planned. For example, total demand for products must be forecasted in order to plan total promotional effort. Besides this, demand in various market regions and among various consumer groups must be predicted in order to plan effective advertising strategies.

In finance, interest rates must be predicted so that new capital acquisitions can be planned and financed. Financial planners must also forecast receipts and expenditures in order to predict cash flows and maintain company liquidity.

In personnel management, forecasts of the number of workers required in different job categories are required in order to plan job recruiting and training programs. In addition, personnel managers need predictions of the supply of labor in various areas and of the amount of absenteeism and the rate of labor turnover to be expected.

In production scheduling, predictions of demand for each product line are needed. Such predictions are made for specific time periods, for example, for specific weeks and months. These forecasts allow the firm to plan production schedules and inventory maintenance. Forecasts of demand for individual products can be translated into forecasts of raw material requirements so that purchases can be planned.

The planning of resource purchases also requires predictions about resource availabilities and prices.

Process control requires forecasts of the behavior of the process in the future. For example, an industrial process may begin to produce increasing numbers of defective items as the process operates over time. If the behavior of this process can be predicted accurately, it will be possible to determine when the process should be shut down and overhauled so that the number of defective items produced can be minimized.

Strategic management requires forecasts of general economic conditions, price and cost changes, technological change, market growth, and the like in order to plan the long-term future of the company. For example, such forecasts might be used to determine whether investment in new plant and equipment will be needed in the future.

Since accurate forecasts are required in so many management decision-making situations, forecasting is one of the most important activities carried on in an organization. In Part I we will begin our discussion of forecasting and the techniques used to produce forecasts. In Chapter 1 we will introduce the concept of a time series, discuss errors in forecasting, preview some of the methods that can be used to make forecasts, and discuss how to choose a forecasting method. We will illustrate these concepts by introducing a forecasting technique called multiple regression analysis. Chapter 2 presents a more complete discussion of multiple regression analysis. In particular, we will discuss how to use historical data to build a multiple regression model.

CHAPTER 1

An Introduction to Time Series, Forecasting, and Multiple Regression Analysis

1–1 INTRODUCTION

In this chapter we introduce some basic concepts in time series, forecasting, and multiple regression analysis. In Section 1–2 we will begin our discussion of time series, and in Section 1–3 we will discuss the fact that forecasts of future values of time series are not likely to be perfectly accurate. In Section 1–4 we will introduce forecasting methods, and in Section 1–5 we will begin our discussion of multiple regression analysis, which is a very useful forecasting method. In Section 1–6 we will present an overview of quantitative forecasting techniques, including multiple regression, exponential smoothing, and the Box-Jenkins techniques.

1–2 TIME SERIES

In generating forecasts of events that will occur in the future, a forecaster must rely on information concerning events that have occurred in the past. That is, in order to prepare a forecast, the forecaster must analyze past data and must base the forecast on the results of this analysis. Forecasters use past data in the following way. First, *the forecaster analyzes this data in order to identify a pattern* that can be used to describe it. Then *this pattern is extrapolated, or extended, into the future in order to prepare a forecast*. This basic strategy is employed in most forecasting techniques and *rests on the assumption that the pattern that has been identified will continue* in the future. It should be noted that a forecasting technique cannot be expected to give good predictions unless this assumption is valid. If the data pattern that has been identified does not persist in the future, the forecasting technique being used will likely produce inaccurate predictions. A forecaster should not be surprised by such a situation. Rather, the forecaster must try to anticipate when such a change in pattern will take place so that appropriate changes in the forecasting system being employed can be made before the predictions become too inaccurate. We will return to a discussion of this situation in Section 1–3.

In this book the past data that will be used in order to prepare forecasts will be time series data.

> A *time series* is a chronological sequence of observations on a particular variable.

As an example, the data in Table 1.1 is a time series which gives the quarterly total value of time deposits held by the Baarth National Bank during 1977 and 1978. Notice that the value of time deposits was observed at equally spaced time points (quarterly). Equally spaced time points are used in most time series studies, although they are not always used. Business time series often involve yearly, quarterly, or monthly observations; but any other time period may be used. There are many, many examples of time series data; some of these are listed below.

Sales of a particular product over time

Total sales for a company over time

Number of unemployed or unemployment rate over time

Production of a product over time

Air or water quality over time

Inventory level for a product over time

Population changes over time

Daily mean temperature over time

TABLE 1.1   *Time Series Data: Quarterly Values of Time Deposits*

Year	Quarter	Value of Time Deposits (in millions of dollars)
1977	1	35.3
	2	37.6
	3	38.1
	4	39.5
1978	1	37.9
	2	39.9
	3	40.1
	4	41.2

University enrollment over time

Total earnings for a firm over time

Time series data are often examined in hopes of discovering a historical pattern that can be exploited in the preparation of a forecast. In order to identify this pattern, it is often convenient to think of a time series as consisting of several components.

The components of a time series are:

1. Trend
2. Cycle
3. Seasonal variations
4. Irregular fluctuations

We will consider each of these components in turn.

Trend refers to the upward or downward movement that characterizes a time series over a period of time. Thus, trend reflects the long-run growth or decline in the time series.

Trend movements can represent a variety of factors. For example, long-run movements in the sales of a particular industry might be determined by one, some, or all of the factors listed below.

1. Technological change in the industry
2. Changes in consumer tastes
3. Increases in per capita income
4. Increases in total population
5. Market growth
6. Inflation or deflation (price changes)

Cycle refers to recurring up and down movements around trend levels. These fluctuations can have a duration of anywhere from 2 to 10 years or even longer measured from peak to peak or trough to trough.

One of the common cyclical fluctuations found in time series data is the "business cycle." The business cycle is represented by fluctuations in the time series caused by recurrent periods of prosperity and recession. Economists have identified several phases in the business cycle. A period of *expansion* in economic or business activity (boom) ends at the *peak* or upper turning point of the business cycle. This peak is followed by a period of *contraction* in economic activity (bust) during which economic activity diminishes. This contraction ends at the lower turning point or *trough* of activity and is then followed by a renewed period of expansion or increase in economic activity.

Cyclical fluctuations need not be caused by changes in economic factors, however. For example, cyclical fluctuations in agricultural yields might reflect changes in weather cycles; the cyclical fluctuations in sales of a particular item of clothing might reflect changes in clothing styles, which are determined by the whims of Paris fashion designers who are bored with the current length of hem lines.

Because there is no single explanation for cyclical fluctuations, they vary greatly in both length and magnitude.

Seasonal variations are periodic patterns in a time series that complete themselves within the period of a calendar year and are then repeated on a yearly basis. Seasonal variations are usually caused by factors such as weather and customs. For example, the average monthly temperature clearly is seasonal in nature since it directly measures changes in the weather. Similarly, the number of monthly housing starts might have a seasonal pattern due to changes in the weather. There might be a high level of housing starts in spring and early summer because of good weather in future months. Housing starts might then decline through late summer and fall, reaching a low point during the coldest months of winter, and then increase rapidly again in early spring. Another time series that might contain a seasonal component is the monthly sales volume in a department store. Here seasonal variation might be caused by the observance of various holidays. Thus department store sales volume might reach high points in December and April because of shopping for the Christmas and Easter holidays.

Ordinarily, series of monthly or quarterly data are used to examine seasonal variations. Clearly, one single yearly observation would not reveal variations that occur during the year.

Irregular fluctuations are erratic movements in a time series that follow no recognizable or regular pattern. Such movements represent what is "left over" in a time series after trend, cycle, and seasonal variations have been accounted for. Many irregular fluctuations in time series are caused by "unusual" events that cannot be forecasted—earthquakes, accidents, hurricanes, wars, wildcat strikes, and the like. Irregular fluctuations can also be caused by errors on the part of the time series analyst.

Time series that exhibit trend, seasonal, and cyclical components are illustrated in Figure 1.1. In Figure 1.1a a time series of sales observations that has an essentially straight line or linear trend is plotted. Figure 1.1b portrays a time series of sales observations that contains a seasonal pattern that repeats annually. Figure 1.1c exhibits a time series of agricultural yields that is cyclical in nature, repeating a cycle about once every ten years.

It should be pointed out that the time series components we have discussed do not always occur alone; they can occur in any combination or can occur all together. For this reason, *no single best forecasting technique exists.* A forecasting technique that can be used to forecast a time series characterized by trend alone may not be appropriate in forecasting a time series characterized by a combination of trend and seasonal variations. Thus one of the most important problems to be solved in forecasting is that of *trying to match the appropriate forecasting technique to the pattern of the available time series data.* Once an appropriate technique has been selected, the methodology usually involves analyzing the time series data in such a way that the different components that are present can be estimated.

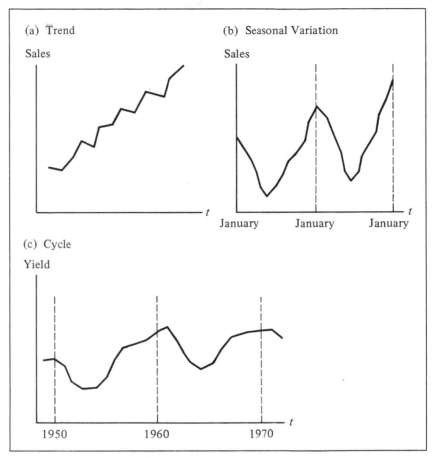

FIGURE 1.1   *Time Series Exhibiting Trend, Seasonal, and Cyclical Components*

The different estimates obtained are then combined in order to produce a forecast. For example, if a time series is characterized by a combination of trend and seasonal components, the appropriate forecasting technique would first estimate these two components. Forecasts would then be obtained by combining the estimate of the trend component with the estimate of the seasonal component. Again, however, it should be emphasized that the key to this methodology is finding a technique that matches the pattern of the historical data that is available.

1–3 ERRORS IN FORECASTING

Unfortunately, all forecasting situations involve some degree of uncertainty. We recognize this fact by including an irregular component in the description of a time series. The presence of this irregular component, which represents unexplained or unpredictable fluctuations in the data, means that some error in forecasting must be expected. If the effect of the irregular component is substantial, our ability to forecast accurately will be limited. If, however, the effect of the irregular component is small, determination of the appropriate trend, seasonal, or cyclical patterns should allow us to forecast with more accuracy.

The irregular component is not the only source of errors in forecasting, however. The accuracy with which we can predict each of the other components of a time series also influences the magnitude of error in our forecasts. Since these components cannot be perfectly predicted in a practical situation, the errors in forecasting represent the combined effects of the irregular component and the accuracy with which the forecasting technique can predict trend, seasonal, or cyclical patterns. Hence, large forecasting errors may indicate that the irregular component is so large that no forecasting technique will produce accurate forecasts, or they may indicate that the forecasting technique being used is not capable of accurately predicting the trend, seasonal, or cyclical components and, therefore, that the technique being used is inappropriate.

1–3.1   TYPES OF FORECASTS

The fact that forecasting techniques often produce predictions that are somewhat in error has a bearing on the form of the forecasts we require. In this book we will consider two types of forecasts: (1) the *point forecast* and (2) the *confidence interval forecast.*

A point forecast is a number that represents our best prediction of the value of the variable of interest at a given point in time. It is essentially our "best guess" for the future value of the variable being forecast. For example, the Blendo Manufacturing Company, a company that produces the Stir-Crazy blender, might forecast that sales for the coming month will be 40,000 units. This forecast, which is simply one number or "point," is a point forecast. However, point forecasts are often in error. Because of this fact, a point forecast alone is often not adequate.

The Blendo Manufacturing Company might also require some estimate of how wrong its point forecast might be. Such an estimate is supplied by a confidence interval forecast. A confidence interval is an interval or range of values that is calculated so that we are "quite sure," say 95 percent sure, that the actual value of the variable being forecast will be contained in that interval (range). If the interval is constructed so that we are 95 percent sure that the actual value will be in the interval, we call this interval, a "95 percent confidence interval" and say that the level of confidence is 95 percent. Although confidence intervals can be constructed with any desired level of confidence, it is customary to construct 95 percent confidence intervals, and we will follow this practice throughout this book. Thus, for example, the Blendo Manufacturing Company might forecast its sales for next month by calculating a 95 percent confidence interval of [38,000 units, 42,000 units]. This confidence interval says that the firm is 95 percent sure that actual sales next month will be no less than 38,000 units and no more than 42,000 units.

Confidence interval forecasts are often very helpful from many planning standpoints. For example, the Blendo Manufacturing Company might use the fact that it is quite sure that sales will not be more than 42,000 units to determine the level of inventory it should carry next month. It might also use the fact that sales are not likely to be lower than 38,000 units, to find the minimum amount of cash it will have available next month. This information will indicate whether the company must be prepared to borrow money next month to meet its cash obligations. Suppose that this firm is also trying to decide whether it should invest in more productive capacity in the coming year, and also suppose that a 95 percent confidence interval for monthly sales one year from now is [68,000 units, 72,000 units]. Then, if the present monthly productive capacity is more than 72,000 units (the upper limit of the confidence interval), the firm can be quite sure that it need not invest in additional capacity. However, if the present monthly productive capacity is less than 68,000 units (the lower limit of the confidence interval), the firm can be quite sure that investment in additional capacity will be needed in order to meet future demand for its product. In situations such as these, a confidence interval forecast can be a very valuable aid to the decision-making process. As another example, consider a situation in which State University wishes to predict the amount of coal it will need for heating purposes next January. Suppose that the university has determined that a 95 percent confidence interval for its coal requirement next January is [350 tons, 390 tons]. This confidence interval says that State University is 95 percent confident that its coal requirement next January will be no less than 350 tons and no more than 390 tons. Since running short of coal in January could well be disastrous, the university would like to order enough coal so that it will have a supply large enough to last through the month. If State University places an order for next January of at least 390 tons of coal (the upper limit of the confidence interval), it can be quite sure that its coal supply will be adequate.

1–3.2   MEASURING ERRORS IN FORECASTING

We now consider the problem of measuring forecasting errors. Let us denote the actual value of the variable of interest in time period t as y_t. Then, if we denote the predicted value of y_t as $\hat{y}_t$, we can subtract the predicted value of y_t from the actual value y_t to obtain the forecast error e_t. That is:

The forecast error for a particular forecast $\hat{y}_t$ is

$$e_t = y_t - \hat{y}_t$$

In Section 1–2 we noted the importance of matching a forecasting technique to the pattern of data characterizing a time series. An examination of forecast errors over time can often indicate whether the forecasting technique being used does or does not match this pattern. For example, if a forecasting technique is accurately forecasting the trend, seasonal, or cyclical components that are present in a time series, the forecast errors should reflect only the irregular component of the time

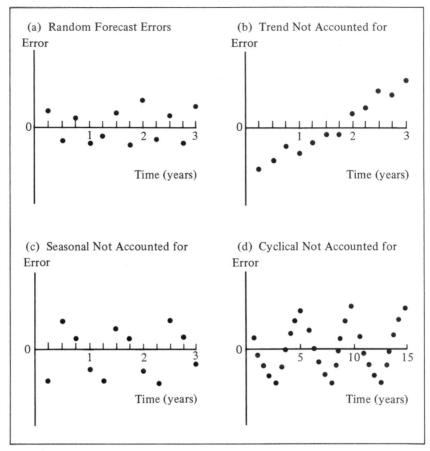

FIGURE 1.2   *Plots of Forecast Errors*

series. In such a case, the forecast errors should appear purely random. Figure 1.2a illustrates forecast errors that indicate that the forecasting technique being used appropriately accounts for the trend, seasonal, or cyclical components present in the time series being forecast. Sometimes, when the forecasting technique does not match the pattern of data, the forecasting errors themselves will exhibit a pattern over time. In Figure 1.2b, the forecast errors show an upward trend, which indicates that the forecasting methodology being used does not account for an upward trend in the time series. The forecast errors in Figure 1.2c indicate that a seasonal pattern in the data is not accounted for in the forecasting methodology being used; and the forecast errors in Figure 1.2d indicate that a cyclical pattern in the data is not being accounted for. Patterns of forecast errors such as those illustrated in Figures 1.2b, c, and d indicate that the forecasting technique being employed is not appropriate, that is, it does not match the pattern of the data characterizing the time series.

If the forecasting errors over time indicate that the forecasting methodology is appropriate (random distribution of errors), it is important to *measure the magnitude of the errors* so that we can determine whether accurate forecasting is possible. In order to do this, one might consider the sum of all forecast errors over time. That is, one might calculate

$$\sum_{t=1}^{n} (y_t - \hat{y}_t)$$

which reads "the summation of the differences between the predicted $(\hat{y}_t)$ and actual (y_t) values from time period $t = 1$ through time period $t = n$, where n is the total number of observed time periods." However, this quantity is not used because, if the errors display a random pattern, some errors will be positive while other errors will be negative; and the sum of the forecast errors will be near zero regardless of the size of the errors. That is, the positive and negative errors, no matter how large or small, will cancel each other out.

One way to remedy this problem is to consider the absolute values of the forecasting errors. These absolute values are called the *absolute deviations*. That is,

$$\text{Absolute Deviation} = |e_t| = |y_t - \hat{y}_t|$$

Given the absolute deviations, we can then define a measure known as the *mean absolute deviation* (MAD). This measure is simply the average of the absolute deviations for all forecasts. That is,

$$\text{Mean Absolute Deviation (MAD)} = \frac{\sum_{t=1}^{n} |e_t|}{n} = \frac{\sum_{t=1}^{n} |y_t - \hat{y}_t|}{n}$$

An example of the calculations involved in computing the MAD is given in Table 1.2. This measure can be used to determine the magnitude of the forecast errors generated by the forecasting methodology being used.

Another way to prevent positive and negative forecast errors from cancelling each other out is to square the forecast errors. These squares are called the *squared errors*:

$$\text{Squared Error} = (e_t)^2 = (y_t - \hat{y}_t)^2$$

TABLE 1.2   *Computation of Mean Absolute Deviation*

Actual Value y_t	Predicted Value $\hat{y}_t$	Error $e_t = y_t - \hat{y}_t$	Absolute Deviation $|e_t| = |y_t - \hat{y}_t|$
25	22	3	3
28	30	−2	2
30	29	1	1
			$\sum_{t=1}^{3} |e_t| = 6$

$$\text{MAD} = \frac{\sum_{t=1}^{3} |e_t|}{3} = \frac{6}{3} = 2$$

Given the squared errors, we can then define a measure known as the *mean squared error* (MSE). This measure is simply the average of the squared errors for all forecasts. That is

$$\text{Mean Squared Error (MSE)} = \frac{\sum_{t=1}^{n} (e_t)^2}{n} = \frac{\sum_{t=1}^{n} (y_t - \hat{y}_t)^2}{n}$$

An example of the calculations involved in computing the MSE is given in Table 1.3. This measure can also be used to determine the magnitude of the forecast errors generated by the forecasting technique being used.

Having shown that both the MAD and the MSE can be used to measure the magnitude of forecast errors, we now wish to discuss how these two measures

TABLE 1.3   *Computation of the Mean Squared Error*

Actual Value y_t	Predicted Value $\hat{y}_t$	Error $e_t = y_t - \hat{y}_t$	Squared Error $(e_t)^2 = (y_t - \hat{y}_t)^2$
25	22	3	9
28	30	−2	4
30	29	1	1
			$\sum_{t=1}^{3} (e_t)^2 = 14$

$$\text{MSE} = \frac{\sum_{t=1}^{3} (e_t)^2}{3} = \frac{14}{3} = 4.67$$

differ. The basic difference between these measures is that *the MSE, unlike the MAD, penalizes a forecasting technique much more for large errors than for small errors.* For example, an error of 2 produces a squared error of 4 while an error of 4 (an error twice as large) produces a squared error of 16 (a squared error four times as large). So, when using the MSE, the forecaster would prefer several smaller forecast errors to one large error. This situation is illustrated in Table 1.4, where we consider two different sets of forecasts generated by methods A and B. Forecasting method A has produced predictions yielding moderate forecast errors, while forecasting method B has produced predictions yielding two small errors along with one large error. Notice that forecasting method A has the larger MAD, while forecasting method B has the larger MSE. This is because, in the calculation of the MSE, forecasting method B is heavily penalized for its large error in forecasting the actual value 67.

Measures such as the MAD and MSE can be used in two different ways. First, they can be used to aid in the process of selecting a forecasting technique to be used in forecasting a time series. Suppose that we are trying to choose among several forecasting techniques in an attempt to determine which of them will likely produce the most accurate predictions of future values of some variable of interest. A common strategy used in making such a selection involves the simulation of historical data. In the simulation process we pretend that we do not know the values

TABLE 1.4   *Comparisons of the Errors Produced by Two Different Forecasting Methods*

	Actual y_t	Predicted $\hat{y}_t$	Error	Absolute Deviation	Squared Error
Forecasting	60	57	+3	3	9
Method A	64	61	+3	3	9
	67	70	−3	3	9
				9	27

$$\text{MAD} = \frac{9}{3} = 3$$

$$\text{MSE} = \frac{27}{3} = 9$$

Forecasting	60	59	+1	1	1
Method B	64	65	−1	1	1
	67	73	−6	6	36
				8	38

$$\text{MAD} = \frac{8}{3} = 2.67$$

$$\text{MSE} = \frac{38}{3} = 12.67$$

of the historical data. Then we use each of the forecasting techniques to produce "predictions" of the historical data. We next compare these "predictions" with the actual values of the historical data and measure their accuracy using the MAD or MSE. Now, in order to choose the technique we will use in actual forecasting, we compare the performance of the various techniques in "forecasting" the historical data to see which of them provided the most accurate simulated "predictions."

Second, the MAD or MSE can be incorporated in measures used to monitor a forecasting system in order to detect when something has "gone wrong" with the system. For example, in Section 1–2, we stated that the forecasts produced by forecasting techniques cannot be expected to be accurate unless the pattern that has been identified in the historical data continues in the future. Let us consider a situation in which a data pattern that has persisted for an extended period of time suddenly changes in some way. Any forecasting method we might have been using to forecast the variable of interest might now be expected to become inaccurate because of this change in the data pattern. In such a situation, we would like to discover the change in pattern as quickly as possible before the forecasts being generated by our forecasting system become very inaccurate. This can be done by using measures that incorporate the MAD or MSE to monitor the forecast errors and to "signal" us when these errors become "too large." We will return to this problem in Chapter 3.

1–4 FORECASTING METHODS

In Section 1–2 we pointed out that no single best forecasting technique exists. In fact, there are many forecasting methods that can be used to predict future events. These methods can be divided into two basic types—*qualitative* methods and *quantitative* methods.

1–4.1 QUALITATIVE FORECASTING METHODS

Qualitative forecasting methods generally use the opinions of experts to subjectively predict future events. Such methods are often required when historical data concerning the events to be predicted either are not available at all or are scarce. For example, consider a situation in which a new product is being introduced. In such a case, no historical sales data for the product are available. In order to forecast sales for the new product, a company must rely on expert opinion, which can be supplied by members of its sales force and market research team. Other situations in which historical data are not available might involve trying to predict if and when new technologies will be discovered and adopted. Qualitative forecasting techniques are also used to predict changes in historical data patterns. Since the use of historical data to predict future events is based on the assumption that the pattern of the historical data will persist, changes in the data pattern cannot be predicted on the basis of historical data. Thus qualitative methods are often used to predict such changes.

We will briefly describe several commonly used qualitative forecasting techniques. The first of these techniques involves *subjective curve fitting*. Consider a firm that is introducing a new product and wishes to forecast sales of this new product over the next several years so that it can determine the productive capacity needed to produce the product. In predicting sales of a new product, it is often convenient to consider what is known as the "product life cycle." This life cycle is usually thought of as consisting of several stages. During the first stage (growth), sales of the product start slowly, then increase rapidly, and then continue to increase at a slower rate. During the next stage (maturity), sales of the product stabilize, increasing slowly, reaching a plateau, and then decreasing slowly. During the last stage (decline), sales of the product decline at an increasing rate. This product life cycle is illustrated in Figure 1.3. In forecasting sales of the product during the growth stage, the company might use the expert opinion of its sales and marketing personnel to subjectively construct an S-curve, as illustrated in Figure 1.4. Such an S-curve, could then be used to forecast sales during this stage. In constructing this S-curve, the company must use its experience with other products and all its knowledge concerning the new product in order to predict how long it will take for the rapid increase in sales to begin, how long this rapid growth will continue, and when sales of the product will begin to stabilize. Note that the construction of this curve is done subjectively since there will be little or no sales data available for the new product. Estimating such a curve is an example of subjective curve fitting. Of course, one of the biggest problems in using this

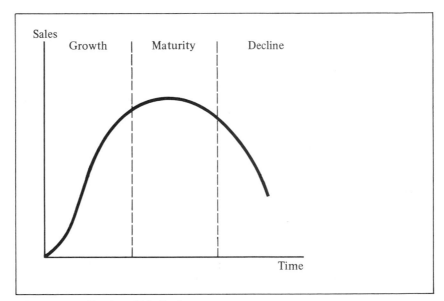

FIGURE 1.3   *Product Life Cycle*

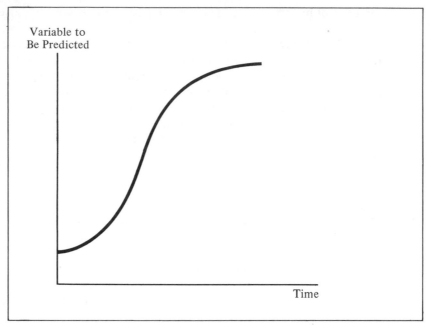

FIGURE 1.4   *S-Curve*

technique is deciding upon the form of the curve that should be used. In a product life cycle situation the use of an S-curve may be appropriate. But many other functional forms can be used. For example, an exponential curve, as illustrated in Figure 1.5, might be appropriate in some situations. Other situations might call for the use of a logarithmic curve. Thus, the forecaster must first subjectively determine the form of the curve to be used. The subjective construction of such curves is very difficult and requires a great deal of expertise and judgment.

Another common qualitative forecasting method is called the *Delphi Method*. This technique, which was developed by the RAND Corporation, involves using a panel of experts to produce predictions concerning a specific question, such as when a new development will occur in a particular field. The use of the Delphi Method assumes that the panel members are recognized experts in the field of interest, and it also assumes that the combined knowledge of the panel members will produce predictions at least as good as those that would be produced by any one member. When a panel of experts is called upon to make predictions, a panel discussion might seem to be appropriate. But such discussions are often dominated by one individual or by a small group of individuals. Also, the decisions made in panel discussions can be influenced by various kinds of social pressure. The Delphi Method attempts to avoid these problems by keeping the panel members physically separated. Each participant is asked to respond to a series of questionnaires and to return the completed questionnaire to a panel coordinator. After the first questionnaire is completed, subsequent questionnaires are accompanied by information concerning the opinions of the group as a whole. Thus, the participants can review their predictions relative to the group response. It is hoped that after several rounds of questionnaires the group response will converge on a consensus that can be used as a forecast. It should be noted, however, that the Delphi Method does not require that a consensus be reached. Instead, the method allows for justified differences of opinion rather than attempting to produce unanimity. We will not present a more detailed discussion of the Delphi Method here. The interested reader is referred to Brown [2] or Dalkey [4, 5].

The third qualitative forecasting technique we will discuss concerns the use of time independent *technological comparisons*. This method is often used in predicting technological change. The method involves predicting changes in one area by monitoring changes that take place in another area. That is, the forecaster tries to determine a pattern of change in one area, often called a *primary trend,* which he or she believes will result in new developments being made in some other area. A forecast of developments in the second area can then be made by monitoring developments in the first area. For example, consider the problem of trying to forecast when a new metal alloy of very high tensile strength will be used commercially. Suppose that the forecaster determines that metallurgical advances made in industry are related to metallurgical advances in the space program. Then, by following metallurgical advances made in the space program, the forecaster can predict when similar advances will take place in industry. Thus the development of high-tensile-strength alloys in the space program would allow the forecaster to predict when such alloys will be available for commercial use. This type of forecasting poses two basic problems. First, the forecaster must identify a primary trend that will reliably predict events in the area of interest. Second, the forecaster must use his expertise to determine the precise relationship between the primary trend and the events to be

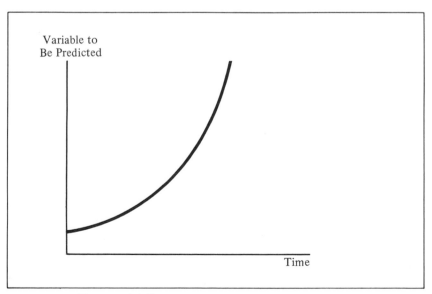

FIGURE 1.5   *Exponential Curve*

forecast. Once these determinations have been made, forecasts in the area of interest can be made by monitoring the primary trend. For a further discussion of this technique the reader is referred to Gerstenfeld [7].

The qualitative forecasting techniques we have discussed—subjective curve fitting, the Delphi Method, and time independent technological comparisons—represent only some of the subjective forecasting methods available. There are many other subjective methods used to generate predictions of future events—these include the cross-impact method, the relevance tree method, and the morphological research method. The interested reader is referred to, respectively, Gordon and Hayward [8], Sigford and Parvin [12], and Zwicky [13].

1–4.2   QUANTITATIVE FORECASTING METHODS

The rest of this book will be devoted to a discussion of *quantitative forecasting techniques*. These techniques involve the analysis of historical data in an attempt to predict future values of a variable of interest. Quantitative forecasting methods can be grouped into two kinds—*time series* and *causal*.

The most common quantitative forecasting methods are called *time series models*. In such models, historical data on the variable to be forecast is analyzed in an attempt to identify a data pattern. Then, assuming that it will continue in the future, this data pattern is extrapolated in order to produce forecasts. Note that time series models generate predictions that are based solely on the historical pattern of the variable to be forecast. Thus, any decisions management might implement in the future will not alter the predictions generated by a time series model. Time series forecasting models are, therefore, most useful when conditions are expected to remain the same; they are not very useful in forecasting the impact of changes in management policies. For example, while a time series model can be used to predict sales if a firm expects to continue using its present marketing strategy, such a model would not be useful in predicting the changes in sales that might result from a price increase, increase in advertising expenditures, or a new advertising campaign. In Section 1–6 we will outline the most commonly used time series models. Chapters 3 through 12 contain detailed presentations of these forecasting methods.

The use of *causal forecasting models* involves the identification of other variables that are related to the variable to be predicted. Once these related variables have been identified, a statistical model that describes the relationship between these variables and the variable to be forecast is developed. The statistical relationship derived is then used to forecast the variable of interest. For example, the sales of a product might be related to the price of the product, advertising expenditures to promote the product, competitors' prices charged for similar products, and so on. In such a case, sales would be referred to as the *dependent variable*, while the other variables are referred to as the *independent variables*. The forecaster's job is to statistically estimate the functional relationship between sales and the independent variables. Having determined this relationship, the forecaster would use predicted future values of the independent variables (price of the product, advertising expenditures, competitors' prices, etc.) to predict future values of sales (the dependent variable).

In the business world, causal models are advantageous because they allow management to evaluate the impact of various alternative policies. For example, management might wish to predict how various price structures and levels of advertising expenditures will affect sales. A causal model relating these variables could be used here. However, causal models have several disadvantages. First, they are quite difficult to develop. Also, they require historical data on all the variables included in the model, not only on the variable to be forecast. Besides this, the ability to predict the dependent variable depends on the ability of the forecaster to accurately predict future values of the independent variables. Despite these disadvantages, causal models are often used to generate predictions. We will consider causal models in Section 1–5 and in Chapter 2.

1–4.3   CHOOSING A FORECASTING TECHNIQUE

Before proceeding further, let us summarize our discussion of forecasting methods. Quantitative forecasting methods are used when historical data are available: Time series models predict future values of the variable of interest solely on the basis of the historical pattern of that variable, assuming that the historical pattern will continue; causal models predict future values of the variable of interest based on the relationship between that variable and other variables. Qualitative forecasting techniques are used when historical data are scarce or not available at all and depend on the opinions of experts who subjectively predict future events. It should also be noted that in actual practice most forecasting systems employ both quantitative and qualitative methods. For example, quantitative methods are used when the existing data pattern is expected to persist, while qualitative methods are used to predict when the existing data pattern might change. Thus, forecasts generated by quantitative methods are almost always subjectively evaluated by management. This evaluation may result in a modification of the forecast based on the manager's "expert opinion."

We now turn to the problem of choosing the particular method to be used in a forecasting situation. In choosing a forecasting technique, the forecaster must consider the following factors.

1.   The forecast form desired
2.   The time frame
3.   The pattern of data
4.   The cost of forecasting
5.   The accuracy desired
6.   The availability of data
7.   The ease of operation and understanding

The first factor to be considered in choosing a forecasting method is the form in which the forecast is desired. We have discussed the difference between a point forecast and a confidence interval forecast. In some situations a point forecast may be sufficient; in other situations a confidence interval forecast may be required. As we will see in later sections, the form of the forecast can influence the choice of a forecasting method because some techniques yield theoretically correct confidence intervals while others do not.

The second factor that can influence the choice of a forecasting method is the *time frame* of the forecasting situation. Forecasts are generated for points in time that may be a number of days, weeks, months, quarters, or years in the future. This length of time is called the time frame or time horizon. The length of the time frame is usually categorized as follows.

Immediate: less than one month

Short term: one to three months

Medium: more than three months to less than two years

Long term: two years or more

In general, the length of the time frame will influence the choice of the forecasting technique to be used. Typically, a longer time frame makes accurate forecasting more difficult, with qualitative forecasting techniques becoming more useful as the time frame increases.

As discussed in Section 1–2, the pattern of data must also be considered when choosing a forecasting method. Whether the data pattern that exists displays trend, seasonal, or cyclical components, or some combination of these components, often determines the forecasting technique that will be used. Thus, in any forecasting situation in which historical data are being used to generate a forecast, it is extremely important to identify the existing data pattern.

When choosing a forecasting technique, several costs are relevant. First, the cost of developing the method must be considered. We will see in later chapters that the development of a forecasting method requires that a set of procedures be followed. The complexity, and hence the cost, of these procedures vary from technique to technique. Second, the cost of storing the necessary data must be considered. Some forecasting methods require the storage of a relatively small amount of data, while other methods require the storage of large amounts of data. The cost of data storage often becomes an important consideration. Last, the cost of the actual operation of the forecasting technique is obviously very important. Some forecasting methods are operationally simple, while others are very complex. The degree of complexity can have a definite influence on the total cost of forecasting.

Another very important factor that has a bearing on the choice of a forecasting technique is the desired accuracy of the forecast. In some situations, a forecast that is in error by as much as 20 percent may be acceptable; in other situations, a forecast that is in error by 1 percent might be disastrous. The accuracy that can be obtained using any particular forecasting method is always an important consideration.

We have pointed out that historical data on the variable of interest are used when quantitative forecasting methods are employed. The availability of this information is a factor that may determine the forecasting method to be used. Since various forecasting methods require different amounts of historical data, the quantity of data available is important. Beyond this, the accuracy and timeliness of the data that are available must be examined, since the use of inaccurate or outdated historical data will obviously yield inaccurate predictions. If the needed historical data are not available, special data-collection procedures may have to be implemented. Such a situation would clearly have a bearing on the choice of the forecasting method to be used.

Last, the ease with which the forecasting method is operated and understood is important. Managers are held responsible for the decisions they make and if they are to be expected to base their decisions on predictions generated by forecasting techniques, they must be able to understand these techniques. A manager simply will not have confidence in the predictions obtained from a forecasting technique he or she does not understand, and if the manager does not have confidence in these predictions, they will not be used in the decision-making process. Thus, the manager's understanding of the forecasting system being used is of crucial importance.

Choosing the forecasting method to be used in a particular situation involves finding a technique that balances the factors just discussed. It is obvious that the "best" forecasting method for a given situation is not always the "most accurate." Instead, *the forecasting method that should be used is one that meets the needs of the situation at the least cost and inconvenience.*

Suppose that the Blendo Manufacturing Company wants to predict sales one month in advance. To accomplish this task, the firm develops a complicated forecasting system and finds that the mean absolute deviation for their technique is 2,000 units. The firm must then determine whether the cost and inconvenience of this forecasting system is justified. In order to make this decision, the mean absolute deviation of 2,000 units must be placed in perspective. If monthly sales average 5,000 units, then the mean absolute deviation of 2,000 units is very large and the forecasts will be quite inaccurate. In such a case, using such a complex forecasting system is probably not justified. If, however, monthly sales average 40,000 units, then the forecasts will be quite accurate. But the knowledge that the forecasting method produces accurate forecasts is not enough to tell the firm whether the method is *appropriate*. Suppose that, in the past, monthly sales have differed from 40,000 units by no more than 3,000 units. Then the firm could forecast sales of 40,000 units each month and predict sales nearly as well as with the complex forecasting system, and with much less cost and effort. Thus, if a forecast that is accurate within 3,000 units is adequate, the Blendo Manufacturing Company should simply forecast sales of 40,000 units each month rather than use the complicated forecasting system; if greater accuracy is required, the more complicated forecasting system would be justified. Thus, the cost and ease of using a simple forecasting method must be balanced against the greater accuracy but higher cost of a more complex forecasting technique.

1–5 FORECASTING USING MULTIPLE REGRESSION ANALYSIS

In this section we will introduce quantitative forecasting techniques by studying multiple regression analysis, which is an important tool often used in the analysis and forecasting of time series data. We will illustrate the forecasting process by showing how a causal multiple regression model can be used to forecast future values of a time series. We will also present some aspects of multiple regression analysis that will be used in future chapters.

This section and the greater portion of this book can be read without any loss of understanding by a reader unfamiliar with matrix algebra. However, the starred sections in this book use matrix algebra to study in greater detail how various multiple regression computations, including the calculation of confidence intervals, would be done. The reader who is already familiar with elementary matrix algebra or who reads Appendix A, which discusses the necessary matrix algebra, would benefit from reading these starred sections. However, it must be stressed that the reader who omits these starred sections can read this book without loss of continuity.

We will begin our study of multiple regression analysis with an example.

Example 1.1   Enterprise Industries produces Fresh, a brand of liquid laundry detergent. The company would like to develop a prediction model that can be used to predict the demand for the extra-large sized bottle of Fresh. With a reliable model, Enterprise Industries can more effectively plan its production schedule, plan its budget, and estimate requirements for producing and storing this product. The demand for Fresh (in hundreds of thousands of bottles) in "sales period t," where each sales period is defined to last 4 weeks, is denoted by the symbol y_t, is called the *dependent variable,* and is believed (subjectively) to be partially determined by one or more of the *independent* variables $x_{t1} =$ the price (in dollars) of Fresh in period t as offered by Enterprise Industries, $x_{t2} =$ the average industry price (in dollars) in period t of competitors' similar detergents, and $x_{t3} =$ the advertising expenditure (in hundreds of thousands of dollars) in period t of Enterprise Industries to promote Fresh. Assume the data in Table 1.5 have been observed over the past 30 sales periods.

Since the company can fairly easily determine x_{t1}, x_{t2}, and x_{t3} for a future period t, we can forecast demand y_t for the future period if we can use the data in Table 1.5 to develop a relationship that expresses y_t as a function of x_{t1}, x_{t2}, and x_{t3}. One possible relationship is the following *multiple regression model:*

$$y_t = \beta_0 + \beta_1 x_{t1} + \beta_2 x_{t2} + \beta_3 x_{t3} + \epsilon_t$$

where β_0, β_1, β_2, and β_3 are unknown constants, or numbers. These constants are often called parameters and relate the dependent variable y_t to the independent variables x_{t1}, x_{t2}, and x_{t3}. The term ϵ_t is an irregular component often called a random error component. This random error component describes the combined

TABLE 1.5   *Historical Data Relevant to the Demand for Fresh*

t	Demand y_t	Price x_{t1}	Average Industry Price x_{t2}	Advertising Expenditure x_{t3}
1	7.38	3.85	3.80	5.50
2	8.51	3.75	4.00	6.75
3	9.52	3.70	4.30	7.25
4	7.50	3.70	3.70	5.50
5	9.33	3.60	3.85	7.00
6	8.28	3.60	3.80	6.50
7	8.75	3.60	3.75	6.75
8	7.87	3.80	3.85	5.25
9	7.10	3.80	3.65	5.25
10	8.00	3.85	4.00	6.00
11	7.89	3.90	4.10	6.50
12	8.15	3.90	4.00	6.25
13	9.10	3.70	4.10	7.00
14	8.86	3.75	4.20	6.90
15	8.90	3.75	4.10	6.80
16	8.87	3.80	4.10	6.80
17	9.26	3.70	4.20	7.10
18	9.00	3.80	4.30	7.00
19	8.75	3.70	4.10	6.80
20	7.95	3.80	3.75	6.50
21	7.65	3.80	3.75	6.25
22	7.27	3.75	3.65	6.00
23	8.00	3.70	3.90	6.50
24	8.50	3.55	3.65	7.00
25	8.75	3.60	4.10	6.80
26	9.21	3.65	4.25	6.80
27	8.27	3.70	3.65	6.50
28	7.67	3.75	3.75	5.75
29	7.93	3.80	3.85	5.80
30	9.26	3.70	4.25	6.80

influence on y_t of factors other than the independent variables x_{t1}, x_{t2}, and x_{t3}. Two such factors are *measurement error,* which takes into account inaccurate measuring and/or reporting of the values y_t and *stochastic error,* which takes into account the effect on the dependent variable y_t of all independent variables other than the independent variables x_{t1}, x_{t2}, and x_{t3} explicitly included in the model. One such independent variable that is not explicitly included in the model and the effect of which is measured by ϵ_t is the competitors' average advertising expenditure in period t for promoting their detergents. One possible reason for this independent variable not being explicitly included in the model is that, while x_{t1}, x_{t2}, and x_{t3} would be fairly easy to determine for a future period, Enterprise Industries' competitors would probably not provide the information needed to determine their average advertising expenditure for the future period. It is assumed that the expected value of ϵ_t is 0, which says that in the long run the random errors average out to 0. For a particular period t, however, ϵ_t will probably not be 0. Consequently, the model

$$y_t = \beta_0 + \beta_1 x_{t1} + \beta_2 x_{t2} + \beta_3 x_{t3} + \epsilon_t$$

states that the time series y_t can be represented by an average level which changes over time according to the expression $\beta_0 + \beta_1 x_{t1} + \beta_2 x_{t2} + \beta_3 x_{t3}$ combined with random fluctuations (represented by ϵ_t) which cause the observations of the time series to deviate from the average level.

Using the above observed data it can be shown that "least squares" estimates of β_0, β_1, β_2, and β_3 are, respectively,

$$b_0 = 7.5891, \quad b_1 = -2.3577, \quad b_2 = 1.6122, \quad b_3 = .5012$$

The meaning of the term *least squares estimates* will be explained presently; for now, it is sufficient to know that these "estimates" are guesses, made on the basis of existing data, for values of the unknown parameters β_0, β_1, β_2, and β_3, which relate the dependent variable y_t to the independent variables x_{t1}, x_{t2}, and x_{t3}. Since it is assumed that the error component ϵ_t averages out to 0 in the long run it follows that 0 is a reasonable guess for any future value of ϵ_t. Thus, if x_{t1}, x_{t2}, and x_{t3} are known for period t, a prediction or forecast of y_t is

$$\hat{y}_t = b_0 + b_1 x_{t1} + b_2 x_{t2} + b_3 x_{t3}$$
$$= 7.5891 - 2.3577 x_{t1} + 1.6122 x_{t2} + .5012 x_{t3}$$

For example, if it is known that in sales period $t = 31$ the price of Fresh will be 3.80, the average competitors' price will be 3.90, and Enterprise Industries' advertising expenditure for Fresh will be 6.80, then a forecast of demand for Fresh detergent in period 31 is

$$\hat{y}_{31} = 7.5891 - 2.3577(3.80) + 1.6122(3.90) + .5012(6.80)$$
$$= 8.325$$

We have stated that the least squares estimates of β_0, β_1, β_2, and β_3 are, respectively, $b_0 = 7.5891$, $b_1 = -2.3577$, $b_2 = 1.6122$, and $b_3 = .5012$. We will now explain what is meant by the term "least squares estimates." Consider Table 1.6, where we have computed for each period t the squared *forecast error* $(y_t - \hat{y}_t)^2$, which is the squared difference between the actual observed demand y_t and the demand prediction $\hat{y}_t$ made using prediction equation

$$\hat{y}_t = 7.5891 - 2.3577 x_{t1} + 1.6122 x_{t2} + .5012 x_{t3}$$

From inspection of the table it seems that the forecast error $(y_t - \hat{y}_t)$ is reasonably small for almost every period t. Moreover, the sum of the squared forecast errors

TABLE 1.6   *Computation of the Squared Forecast Errors Using the Prediction Equation* $\hat{y}_t = 7.5891 - 2.3577x_{t1} + 1.6122x_{t2} + .5012x_{t3}$

t	x_{t1}	x_{t2}	x_{t3}	y_t	$\hat{y}_t$	$(y_t - \hat{y}_t)^2$
1	3.85	3.80	5.50	7.38	7.395	$(-.015)^2$
2	3.75	4.00	6.75	8.51	8.579	$(-.069)^2$
3	3.70	4.30	7.25	9.52	9.431	$(.089)^2$
4	3.70	3.70	5.50	7.50	7.587	$(-.087)^2$
5	3.60	3.85	7.00	9.33	8.816	$(.514)^2$
6	3.60	3.80	6.50	8.28	8.485	$(-.205)^2$
7	3.60	3.75	6.75	8.75	8.530	$(.220)^2$
8	3.80	3.85	5.25	7.87	8.648	$(-.222)^2$
9	3.80	3.65	5.25	7.10	7.145	$(-.045)^2$
10	3.85	4.00	6.00	8.00	8.411	$(-.411)^2$
11	3.90	4.10	6.50	7.89	8.262	$(-.372)^2$
12	3.90	4.00	6.25	8.15	7.975	$(.175)^2$
13	3.70	4.10	7.00	9.10	8.984	$(.116)^2$
14	3.75	4.20	6.90	8.86	8.977	$(-.117)^2$
15	3.75	4.10	6.80	8.90	8.766	$(.134)^2$
16	3.80	4.10	6.80	8.87	8.648	$(.222)^2$
17	3.70	4.20	7.10	9.26	9.195	$(.065)^2$
18	3.80	4.30	7.00	9.00	9.070	$(-.070)^2$
19	3.70	4.10	6.80	8.75	8.883	$(-.133)^2$
20	3.80	3.75	6.50	7.95	7.933	$(.017)^2$
21	3.80	3.75	6.25	7.65	7.808	$(-.158)^2$
22	3.75	3.65	6.00	7.27	7.639	$(-.369)^2$
23	3.70	3.90	6.50	8.00	7.968	$(.032)^2$
24	3.55	3.65	7.00	8.50	8.612	$(-.112)^2$
25	3.60	4.10	6.80	8.75	9.119	$(-.369)^2$
26	3.65	4.25	6.80	9.21	9.243	$(-.033)^2$
27	3.70	3.65	6.50	8.27	8.008	$(.262)^2$
28	3.75	3.75	5.75	7.67	7.675	$(-.005)^2$
29	3.80	3.85	5.80	7.93	7.743	$(.187)^2$
30	3.70	4.25	6.80	9.26	9.195	$(.075)^2$

$$\sum_{t=1}^{30}(y_t - \hat{y}_t)^2 = 1.432$$

$$\bar{y} = 8.38$$

equals 1.432. Note that the smallness of each forecast error $(y_t - \hat{y}_t)$ and hence the smallness of the sum of the squares of the errors (often designated as SSE),

$$\sum_{t=1}^{30}(y_t - \hat{y}_t)^2$$

depends upon the least squares estimates $b_0 = 7.589$, $b_1 = -2.3577$, $b_2 = 1.6122$, and $b_3 = .5012$ of the parameters β_0, β_1, β_2, and β_3. If we used a different set of estimates of β_0, β_1, β_2, and β_3, we would have a different prediction equation, a different forecast error $(y_t - \hat{y}_t)$, and hence a different value of $\Sigma_{t=1}^{30} (y_t - \hat{y}_t)^2$. We call $b_0 = 7.5891$, $b_1 = -2.3577$, $b_2 = 1.6122$, and $b_3 = .5012$ the *least squares estimates* of β_0, β_1, β_2, and β_3 because it can be shown by manipulations we need not go into here that these estimates give a value of $\Sigma_{t=1}^{30} (y_t - \hat{y}_t)^2$ that is smaller than any other estimates would give. For example, the estimates $b_0 = 6.421$, $b_1 = -2.4165$, $b_2 = 1.8214$, and $b_3 = .4474$ give the following prediction equation

$$\hat{y}_t = 6.421 - 2.4165x_{t1} + 1.8214x_{t2} + .4474x_{t3}$$

These estimates (and this prediction equation) are different from the least squares estimates (and prediction equation) used in Table 1.6, and they give larger forecast errors and, therefore, a larger value for the "sum of the squares" than do the least squares estimates (and prediction equation) used in Table 1.6. How the data in Table 1.5 are actually used to compute the least squares estimates is explained in Section *1–5. From the user's standpoint, it is sufficient to know that standard multiple regression computer packages will compute the least squares estimates when the data in Table 1.5 are provided as input.

The Fresh detergent example illustrates that a multiple regression model can be used to forecast a future value of a dependent variable by relating that variable to various independent variables. The general multiple regression model is

$$y_t = \beta_0 + \beta_1 x_{t1} + \beta_2 x_{t2} + \cdots + \beta_p x_{tp} + \epsilon_t$$

where y_t denotes the dependent variable in period t; p represents the number of independent variables used in the model; x_{t1}, x_{t2}, . . . , x_{tp} represent the values of those p independent variables in period t; β_0, β_1, . . . , β_p are unknown parameters relating the dependent variable y_t to the p independent variables x_{t1}, x_{t2}, . . . , x_{tp}; and ϵ_t is a random error component that describes the influence on y_t of all factors other than the p independent variables x_{t1}, x_{t2}, . . . , x_{tp}. (It is assumed that the expected value of ϵ_t is 0. For a particular period t, however, ϵ_t will probably not be 0.)

The general multiple regression model

$$y_t = \beta_0 + \beta_1 x_{t1} + \beta_2 x_{t2} + \cdots + \beta_p x_{tp} + \epsilon_t$$

states that time series y_t can be represented by an average level that changes over time according to the function $\beta_0 + \beta_1 x_{t1} + \beta_2 x_{t2} + \cdots + \beta_p x_{tp}$ combined with random fluctuations (represented by ϵ_t) which cause the observations of the time series to deviate from the average level. We will henceforth denote the average level by the symbol μ_t. That is,

$$\mu_t = \beta_0 + \beta_1 x_{t1} + \beta_2 x_{t2} + \cdots + \beta_p x_{tp}$$

The value μ_t represents the average of all of the values of the dependent variable y_t that could ever possibly be observed when the values of the independent variables are fixed at $x_{t1}, x_{t2}, \ldots, x_{tp}$. If $b_0, b_1, \ldots, b_p$ are the least squares estimates of $\beta_0, \beta_1, \ldots, \beta_p$, then the point estimate of μ_t is

$$\hat{y}_t = b_0 + b_1 x_{t1} + b_2 x_{t2} + \cdots + b_p x_{tp}$$

Then, since we assume the error component ϵ_t averages out to 0 in the long run and that 0 is therefore a reasonable guess for any future value of ϵ_t, it follows that $\hat{y}_t$ is the point forecast of the actual time series value

$$y_t = \mu_t + \epsilon_t$$
$$= \beta_0 + \beta_1 x_{t1} + \beta_2 x_{t2} + \cdots + \beta_p x_{tp} + \epsilon_t$$

In order to obtain $\hat{y}_t$, the least squares estimates are computed by gathering observations of the variables $y_t, x_{t1}, \ldots, x_{tp}$ for n periods. Note that the observed number of periods n must be greater than $(p + 1)$, the number of parameters in the model, in order to calculate meaningful estimates of $\beta_0, \beta_1, \ldots, \beta_p$. A formula is given in Section *1–5 for obtaining the least squares estimates $b_0, b_1, \ldots, b_p$, but it is sufficient for the user to know that least squares estimates will be computed by standard multiple regression computer packages.

Since the least squares estimates $b_0, b_1, \ldots, b_p$ are only guesses of the unknown parameters $\beta_0, \beta_1, \ldots, \beta_p$, it follows that the estimate

$$\hat{y}_t = b_0 + b_1 x_{t1} + \cdots + b_p x_{tp}$$

will be somewhat different from both

$$\mu_t = \beta_0 + \beta_1 x_{t1} + \cdots + \beta_p x_{tp} \quad \text{and} \quad y_t = \mu_t + \epsilon_t$$

Because of this, we wish to find out "how far" $\hat{y}_t$ might differ from μ_t and y_t. That is, we wish to find confidence intervals for μ_t and y_t. We will subsequently discuss the formulas for these confidence intervals. In order for these formulas to be valid, three assumptions concerning the random error component ϵ_t must be met. The first assumption is that for each and every period t the random error component ϵ_t follows a normal probability distribution. Since ϵ_t describes the influence on y_t of all factors other than the p independent variables $x_{t1}, x_{t2}, \ldots, x_{tp}$ and hence causes the time series value y_t to deviate from μ_t, the first assumption implies that all of the values of the dependent variable y_t that could ever possibly be observed when the values of the independent variables are fixed at $x_{t1}, x_{t2}, \ldots, x_{tp}$ are distributed in a normal curve around μ_t, as illustrated in Figure 1.6. For later reference, it should be noted that a table of normal curve areas is given in Appendix B. The second assumption is that the *variance* of y_t, which *measures the spread of all of the potential values of the dependent variable* y_t *around the average level* μ_t, is the same for each and every value of t. That is, for each and every combination of values of the independent variables $x_{t1}, x_{t2}, \ldots, x_{tp}$, the variance of y_t is the same. The third assumption is that the time series values $y_1, y_2, \ldots$ in different periods

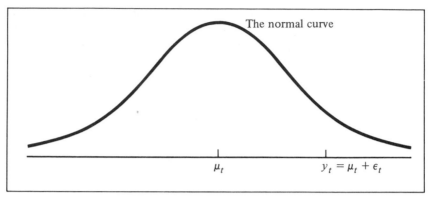

FIGURE 1.6   *All of the Potential Values of y_t Are Distributed in a Normal Curve Around μ_t*

are statistically independent of or not related to each other. We will henceforth call these three assumptions the *inference assumptions*. We use the word "inference" because they are the assumptions that must be met if statistical inferences concerning regression models—for example, calculations of confidence intervals for μ_t and y_t—are to be valid. It is probably true that the inference assumptions are not exactly met by any regression model in a real-world situation. Hence, the real question is whether, in an actual situation, the inference assumptions are closely enough met so that the confidence interval formulas are approximately valid. Without going into a lengthy discussion, we will simply say that one of the best ways to insure that a regression model meets the inference assumptions closely enough in a particular regression situation is to construct the model so that it contains all of the independent variables that *significantly* effect the dependent variable. Constructing such a model will be discussed in Chapter 2. In this book we will try to carefully construct the regression models that we use. Therefore, we can assume that the confidence interval formulas that we use are at least approximately valid. Finally, it should be noted that the method of residual plots is useful in determining whether the inference assumptions are at least approximately met by a regression model. The interested reader is referred to Draper and Smith [6].

Using statistical theory and the inference assumptions, it can be shown that an error $E_t(100 - \alpha)$ exists such that we are $(100 - \alpha)$ percent confident (for example, 95 percent confident, in which case $\alpha = 5$) that the forecast $\hat{y}_t$ will be within $E_t(100 - \alpha)$ of y_t; so we say that a $(100 - \alpha)$ percent *confidence interval* for y_t is

$$[\hat{y}_t - E_t(100 - \alpha), \hat{y}_t + E_t(100 - \alpha)]$$

This confidence interval says that we are $(100 - \alpha)$ percent confident that the actual time series value y_t will be no less than $(\hat{y}_t - E_t(100 - \alpha))$ and no more than

$(\hat{y}_t + E_t(100 - \alpha))$.[1] It should be apparent to the reader that $E_t(100 - \alpha)$ is a function of the time period t and the level of confidence, $100 - \alpha$, and that the larger the level of confidence is, the larger is the error, $E_t(100 - \alpha)$. The level of confidence $(100 - \alpha)$ can be set arbitrarily by the user. It is customary to set the level of confidence at $(100 - 5) = 95$, in which case we are computing 95 percent confidence intervals.

Using statistical theory and the inference assumptions, it also can be shown that an error $E_{\mu_t}(100 - \alpha)$ exists such that we are $(100 - \alpha)$ percent confident that the point estimate $\hat{y}_t$ is within $E_{\mu_t}(100 - \alpha)$ of μ_t, and hence we say that a $(100 - \alpha)$ percent confidence interval for μ_t is

$$[\hat{y}_t - E_{\mu_t}(100 - \alpha), \hat{y}_t + E_{\mu_t}(100 - \alpha)]$$

This confidence interval says that we are $(100 - \alpha)$ percent confident that the true average level μ_t is no less than $(\hat{y}_t - E_{\mu_t}(100 - \alpha))$ and no more than $(\hat{y}_t + E_{\mu_t}(100 - \alpha))$.[2]

As illustrated in Figure 1.7, the confidence intervals for μ_t and y_t are both centered at

$$\hat{y}_t = b_0 + b_1 x_{t1} + b_2 x_{t2} + \cdots + b_p x_{tp}$$

However, the error $E_{\mu_t}(100 - \alpha)$ in the confidence interval for

$$\mu_t = \beta_0 + \beta_1 x_{t1} + \beta_2 x_{t2} + \cdots + \beta_p x_{tp}$$

accounts for the uncertainty caused by the fact that the least squares estimates b_0, $b_1, \ldots, b_p$ differ from the true parameters $\beta_0, \beta_1, \ldots, \beta_p$, whereas the error $E_t(100 - \alpha)$ in the confidence interval for

$$y_t = \mu_t + \epsilon_t$$
$$= \beta_0 + \beta_1 x_{t1} + \beta_2 x_{t2} + \cdots + \beta_p x_{tp} + \epsilon_t$$

accounts for the uncertainty caused by both the fact that the least squares estimates $b_0, b_1, \ldots, b_p$ differ from the true parameters $\beta_0, \beta_1, \ldots, \beta_p$ and the fact that y_t differs from μ_t by the random error component ϵ_t. Although it is assumed that the expected value of ϵ_t is 0, it is nevertheless true that, for a particular period t, ϵ_t will probably not be 0. Thus, $E_t(100 - \alpha)$ takes into account more uncertainty than does

---

[1] The formula for $E_t(100 - \alpha)$ is given in Section *1–5. From the user's standpoint it is sufficient to know that standard multiple regression computer packages can be used to compute $E_t(100 - \alpha)$ and thus give a $(100 - \alpha)$ percent confidence interval for y_t.

[2] The formula for $E_{\mu_t}(100 - \alpha)$ is given in Section *1–5. From the user's standpoint, it is sufficient to know that standard multiple regression packages can be used to compute $E_{\mu_t}(100 - \alpha)$.

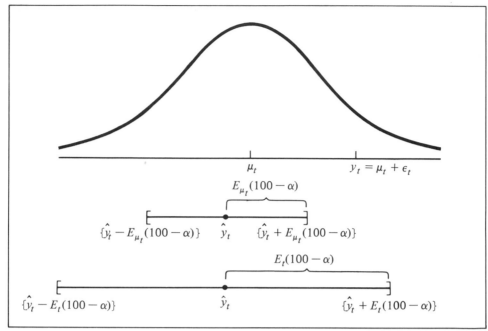

FIGURE 1.7   *Comparison of a Confidence Interval for μ_t with a Confidence Interval for*
$y_t = \mu_t + \epsilon_t$

$E_{\mu_t}(100 - \alpha)$ and therefore is substantially larger. This idea is illustrated in Figure 1.7. Note that the point estimate $\hat{y}_t$ of both μ_t and y_t differs from both μ_t and y_t. Note also that the confidence interval for μ_t contains the average level μ_t but is not "wide enough" to contain the actual time series value y_t. However, the confidence interval for y_t is "wide enough" and does contain y_t. Of course, it should be noted that the situation illustrated in Figure 1.7 is only typical. The relative positions of μ_t, y_t, and $\hat{y}_t$ may be different in different situations. To differentiate the confidence interval for μ_t from the confidence interval for y_t, some forecasters refer to the confidence interval for μ_t as a confidence interval and the confidence for y_t as a *prediction interval*. In this book we will not use the term "prediction interval."

Example 1.2   Consider once again the Fresh detergent situation in Example 1.1. Assume it is known that in sales period $t = 31$ the price of Fresh will be 3.80, the average competitors' price will be 3.90, and Enterprise Industries' advertising expenditure for Fresh will be 6.80. We wish to find a confidence interval for

$$\mu_{31} = \beta_0 + \beta_1(3.80) + \beta_2(3.90) + \beta_3(6.80)$$

which represents the average of all of the demands that could ever possibly be observed in a sales period during which the price of Fresh is 3.80, the average competitors' price is 3.90, and Enterprise Industries' advertising expenditure for Fresh is 6.80. We also wish to find a confidence interval for

$$y_{31} = \mu_{31} + \epsilon_{31}$$

which is the actual demand to be observed in period 31. The point estimate of μ_{31} and the point forecast of y_{31} is

$$\hat{y}_{31} = 7.5891 - 2.3577(3.80) + 1.6122(3.90) + .5012(6.80)$$

$$= 8.325$$

which has been computed using the least squares estimates of β_0, β_1, β_2, and β_3.

It can be shown that we are 95 percent confident that the point estimate $\hat{y}_{31} = 8.325$ is within $E_{\mu_{31}}(95) = .201$ of μ_{31}, that is, that a 95 percent confidence interval for μ_{31} is

$$[8.325 - .201, 8.325 + .201] \quad \text{or} \quad [8.124, 8.526]$$

This confidence interval says that we are 95 percent confident that the true *average demand* μ_{31} is no less than 8.124 and no more than 8.526. It also can be shown that we are 95 percent confident that the forecast $\hat{y}_{31} = 8.325$ will be within $E_{31}(95) = .523$ of y_{31}, that is, that a 95 percent confidence interval for y_{31} is

$$[8.325 - .523, 8.325 + .523] \quad \text{or} \quad [7.80, 8.85]$$

This confidence interval says that we are 95 percent confident that the *actual demand* in period 31 will be no less than 7.80 and no more than 8.85.

Enterprise Industries can now use these forecasts to plan its production schedule for the next sales period and to determine the quantity of raw materials it must purchase in meeting that schedule. Note that

$$[7.80, 8.85]$$

which is the confidence interval for the actual time series value y_{31}, is considerably wider than

$$[8.124, 8.526]$$

which is the confidence interval for the average level μ_{31}.

As another example, if it is known that in sales period $t = 32$ the price of Fresh will be 3.75, the average competitors' price will be 3.85, and Enterprise Industries' advertising expenditure for Fresh will be 6.85, then a forecast of demand for Fresh detergent in period 32 is

$$\hat{y}_{32} = 7.5891 - 2.3577(3.75) + 1.6122(3.85) + .5012(6.85)$$

$$= 8.388$$

And, since it can be shown that we are 95 percent confident that the forecast $\hat{y}_{32} = 8.388$ will be within $E_{32}(95) = .517$ of y_{32}, we say that a 95 percent confidence interval for y_{32} is

$$[8.388 - .517, 8.388 + .517] \quad \text{or} \quad [7.87, 8.90]$$

As yet another example, if it is known that in sales period $t = 33$ the price of Fresh will be 3.85, the average competitors' price will be 3.90, and Enterprise Industries' advertising expenditure for Fresh will be 6.90, then a forecast of demand for Fresh detergent in period 33 is

$$\hat{y}_{33} = 7.5891 - 2.3577(3.85) + 1.6122(3.90) + .5012(6.90)$$
$$= 8.257$$

and in this case it can be shown that we are 95 percent confident that the forecast $\hat{y}_{33}$ = 8.257 will be within $E_{33}(95) = .557$ of y_{33}, so we say that a 95 percent confidence interval for y_{33} is

$$[8.257 - .557, 8.257 + .557] \qquad \text{or} \qquad [7.70, 8.81]$$

Sometimes it is desirable to find a forecast for the sum of the values of a time series in periods l through $(l + q)$, that is

$$\sum_{t=l}^{l+q} y_t$$

Such a forecast would be the sum of the individual forecasts

$$\sum_{t=l}^{l+q} \hat{y}_t$$

Since, using statistical theory and the inference assumptions, it can be shown that an error $E_{(l,l+q)}(100 - \alpha)$ exists such that we are $(100 - \alpha)$ percent confident the sum of the individual forecasts will be within $E_{(l,l+q)}(100 - \alpha)$ of the sum of the actual values of the time series, we say that a $(100 - \alpha)$ percent confidence interval for $\sum_{t=l}^{l+q} y_t$ is

$$\left[\sum_{t=l}^{l+q} \hat{y}_t - E_{(l,l+q)}(100 - \alpha), \sum_{t=l}^{l+q} \hat{y}_t + E_{(l,l+q)}(100 - \alpha) \right]$$

This confidence interval says that we are $(100 - \alpha)$ percent confident that the actual sum of the values of the time series, will be no less than

$$\sum_{t=l}^{l+q} \hat{y}_t - E_{(l,l+q)}(100 - \alpha)$$

and no more than

$$\sum_{t=l}^{l+q} \hat{y}_t + E_{(l,l+q)}(100 - \alpha)$$

The formula for $E_{(l,l+q)}(100 - \alpha)$ is given in Section *1–5.

It should be noted that, in general for periods l through $(l + q)$, the "error of the sum" of the forecasts is less than "the sum of the errors." That is,

$$E_{(l,l+q)}(100 - \alpha) < \sum_{t=l}^{l+q} E_t(100 - \alpha)$$

This is because the forecasts of the individual values $y_l, \ldots, y_{l+q}$ are in error, some errors being positive, while others are negative. In forecasting the sum of these values, it is likely that these errors will tend to cancel each other out, resulting in a smaller error.

Example 1.3   Enterprise Industries often purchases the raw materials required for the production of Fresh detergent in large quantities so that it can take advantage of quantity discounts offered by its suppliers. In particular, the firm is interested in predicting the total demand for Fresh detergent in the next three sales periods so that it can place bulk orders for the raw materials it will require over this time span. A forecast of $(y_{31} + y_{32} + y_{33})$, the sum of the demands in periods 31, 32, and 33, is given by

$$\hat{y}_{31} + \hat{y}_{32} + \hat{y}_{33} = 8.325 + 8.388 + 8.257 = 24.97$$

It can be shown that we are 95 percent confident that the forecast $(\hat{y}_{31} + \hat{y}_{32} + \hat{y}_{33})$ = 24.97 will be within $E_{(31,33)}(95) = 1.062$ of $(y_{31} + y_{32} + y_{33})$, so we say that a 95 percent confidence interval for the cumulative demand $(y_{31} + y_{32} + y_{33})$, is

$$[24.97 - 1.062, 24.97 + 1.062] \quad \text{or} \quad [23.91, 26.03]$$

That is, we are 95 percent confident that the actual sum of the values $y_{31} + y_{32} + y_{33}$ will be no less than 23.91 and no more than 26.03. Note that, while the error in the confidence interval for $(y_{31} + y_{32} + y_{33})$ is 1.062, the sum of the errors in the confidence intervals for y_{31}, y_{32}, and y_{33} is .523 + .517 + .557 = 1.597. That is, the error in the confidence interval for $(y_{31} + y_{32} + y_{33})$ is less than the sum of the errors in the confidence intervals for y_{31}, y_{32}, and y_{33}. This illustrates the fact stated above that the "error of the sum" is less than "the sum of the errors."

*1–5   THE REGRESSION CALCULATIONS

This starred subsection[3] shows how the least squares estimates $b_0, b_1, \ldots, b_p$ are computed and how the errors $E_{\mu_t}(100 - \alpha)$, $E_t(100 - \alpha)$, and $E_{(l,l+q)}(100 - \alpha)$ in the

---

[3]   Starred sections in this book use matrix algebra and can be omitted without loss of continuity. Appendix A contains a discussion of the necessary matrix algebra for those readers who are interested.

confidence intervals for, respectively, μ_t, y_t, and $\Sigma_{t=l}^{l+q} y_t$ are computed. The general multiple regression model is

$$y_t = \mu_t + \epsilon_t = \beta_0 + \beta_1 x_{t1} + \beta_2 x_{t2} + \cdots + \beta_p x_{tp} + \epsilon_t$$

In order to obtain estimates (b_0, b_1, ... b_p) of the parameters β_0, β_1, ..., β_p, assume that we gather observations of y_t, x_{t1}, ..., x_{tp} for n periods. Using these data, computation of b_0, b_1, ..., b_p, $E_{\mu_t}(100 - \alpha)$, $E_t(100 - \alpha)$, and $E_{(l,l+q)}(100 - \alpha)$ can be accomplished by using the matrix $\mathbf{X}$ and column vector $\mathbf{y}$, where

$$\mathbf{y} = \begin{bmatrix} y_1 \\ y_2 \\ \cdot \\ \cdot \\ \cdot \\ y_n \end{bmatrix} \quad \text{and} \quad \mathbf{X} = \begin{bmatrix} 1 & x_{11} & x_{12} & \cdot & \cdot & \cdot & x_{1p} \\ 1 & x_{21} & x_{22} & \cdot & \cdot & \cdot & x_{2p} \\ \cdot & \cdot & \cdot & \cdot & \cdot & \cdot & \cdot \\ \cdot & \cdot & \cdot & \cdot & \cdot & \cdot & \cdot \\ \cdot & \cdot & \cdot & \cdot & \cdot & \cdot & \cdot \\ 1 & x_{n1} & x_{n2} & \cdot & \cdot & \cdot & x_{np} \end{bmatrix}$$

The least squares estimates b_0, b_1, ..., b_p of the parameters β_0, β_1, ..., β_p are given by the matrix algebra equation

$$\mathbf{b} = \begin{bmatrix} b_0 \\ b_1 \\ \cdot \\ \cdot \\ \cdot \\ b_p \end{bmatrix} = (\mathbf{X'X})^{-1}\mathbf{X'y}$$

Here, $\mathbf{X'}$ is the transpose of the matrix $\mathbf{X}$ and $(\mathbf{X'X})^{-1}$ is the inverse of the matrix $\mathbf{X'X}$.

Remember that

$$\hat{y}_t = b_0 + b_1 x_{t1} + b_2 x_{t2} + \cdots + b_p x_{tp}$$

is the estimate of μ_t and the forecast of y_t. Using the inference assumptions, it can be shown that a $(100 - \alpha)$ percent confidence interval for y_t is

$$[\hat{y}_t - E_t(100 - \alpha), \hat{y}_t + E_t(100 - \alpha)]$$

where the error $E_t(100 - \alpha)$ is given by the equation

$$E_t(100 - \alpha) = t_{\alpha/2}(n - (p + 1)) \cdot s \cdot f_t$$

Here, $t_{\alpha/2}(n - (p + 1))$ is the point on the scale of the t-distribution having $(n - (p + 1))$ degrees of freedom such that an area of $(100 - \alpha) \div 100$ exists under the curve of this t-distribution between $-t_{\alpha/2}(n - (p + 1))$ and $t_{\alpha/2}(n - (p + 1))$, while

$$s = \sqrt{\frac{\sum\limits_{t=1}^{n} (y_t - \hat{y}_t)^2}{n - (p + 1)}}$$

and

$$f_t = \sqrt{1 + \mathbf{x}_t'(\mathbf{X}'\mathbf{X})^{-1}\mathbf{x}_t}$$

where

$$\mathbf{x}_t' = [1 \; x_{t1} \; x_{t2} \ldots x_{tp}]$$

The derivation of the formula for $E_t(100 - \alpha)$ will be omitted here. Intuitively, however, the level of confidence is determined by $t_{\alpha/2}(n - (p + 1))$; s measures the variation of the observed values of the dependent variable y_t around the regression equation; and f_t is a measure of uncertainty that results from the lack of knowledge of the true values of the regression equation parameters $\beta_0, \beta_1, \ldots, \beta_p$. The product of these factors yields the error $E_t(100 - \alpha)$.

It is obvious that f_t is greater than 1. However, f_t can be assumed equal to 1 if the values of the parameters $\beta_0, \beta_1, \ldots, \beta_p$ are known; when these values are not known, f_t reflects this lack of knowledge. Sometimes, even if the values of $\beta_0, \beta_1, \ldots, \beta_p$ are not known, a confidence interval for y_t is computed by setting f_t equal to 1 and using the formula

$$E_t(100 - \alpha) = t_{\alpha/2}(n - (p + 1)) \cdot s$$

for the error in the confidence interval. This procedure is not theoretically correct, but is used by some because they desire an error in their confidence interval smaller than the error that would be obtained if f_t, which is greater than 1, were included in the calculations. These practitioners are willing to pay for a smaller error by a decreased level of confidence in the size of the error. In this book we will always include f_t in the calculations for $E_t(100 - \alpha)$.

Using the inference assumptions, it also can be shown that a $(100 - \alpha)$ percent confidence interval for μ_t is

$$[\hat{y}_t - E_{\mu_t}(100 - \alpha), \hat{y}_t + E_{\mu_t}(100 - \alpha)]$$

where the error $E_{\mu_t}(100 - \alpha)$ is given by the equation

$$E_{\mu_t}(100 - \alpha) = t_{\alpha/2}(n - (p + 1)) \cdot s \cdot \sqrt{\mathbf{x}_t(\mathbf{X}'\mathbf{X})^{-1}\mathbf{x}_t}$$

All quantities in the above equation are defined as in the formula for $E_t(100 - \alpha)$.

Finally, using the inference assumptions, it can be shown that a $(100 - \alpha)$ percent confidence interval for the sum of the values of a time series in periods l through $(l + q)$ is

$$\left[\sum_{t=l}^{l+q} \hat{y}_t - E_{(l,l+q)}(100 - \alpha), \sum_{t=l}^{l+q} \hat{y}_t + E_{(l,l+q)}(100 - \alpha) \right]$$

Here

$$E_{(l,l+q)}(100 - \alpha) = t_{\alpha/2}(n - (p + 1)) \cdot s \cdot \sqrt{(q + 1) + \left\{ \sum_{t=l}^{l+q} \mathbf{x}_t' \right\} (\mathbf{X}'\mathbf{X})^{-1} \left\{ \sum_{t=l}^{l+q} \mathbf{x}_t \right\}}$$

where all terms are defined as above.

Before proceeding to Example 1.4 it should be noted that

$$s = \sqrt{\frac{\sum_{t=1}^{n} (y_t - \hat{y}_t)^2}{n - (p + 1)}}$$

is an important component of each of the confidence intervals given above. Although

$$\sum_{t=1}^{n} (y_t - \hat{y}_t)^2$$

can be computed by first computing $\hat{y}_1, \hat{y}_2, \ldots, \hat{y}_n$, a computationally easier way to compute this value is as follows:

$$\sum_{t=1}^{n} (y_t - \hat{y}_t)^2 = \sum_{t=1}^{n} y_t^2 - \mathbf{b}'\mathbf{X}'\mathbf{y}$$

where $\mathbf{b}' = [b_0, b_1, \ldots, b_p]$ is a row vector containing the least squares estimates $b_0, b_1, \ldots, b_p$ and $\mathbf{X}'\mathbf{y}$ is a previously defined column vector used in the calculation of $b_0, b_1, \ldots, b_p$.

Example 1.4   Reconsider the Fresh detergent problem of Example 1.1 and the data of Table 1.5. We see that if we wish to find the least squares estimates of β_0, β_1, β_2, and β_3 in the regression model

$$y_t = \beta_0 + \beta_1 x_{t1} + \beta_2 x_{t2} + \beta_3 x_{t3} + \epsilon_t$$

we use

$$y = \begin{bmatrix} 7.38 \\ 8.51 \\ . \\ . \\ . \\ 9.26 \end{bmatrix} \qquad \text{and} \qquad X = \begin{bmatrix} 1 & 3.85 & 3.80 & 5.50 \\ 1 & 3.75 & 4.00 & 6.75 \\ . & . & . & . \\ . & . & . & . \\ . & . & . & . \\ 1 & 3.70 & 4.25 & 6.80 \end{bmatrix}$$

Thus,

$$(X'X)^{-1} = \begin{bmatrix} 108.556 & -26.7730 & 3.48877 & -3.45557 \\ -26.7730 & 7.39020 & -1.75917 & .947785 \\ 3.48877 & -1.75917 & 1.58405 & -.491626 \\ -3.45557 & .947785 & -.491626 & -.287712 \end{bmatrix}$$

and

$$X'y = \begin{bmatrix} 251.480 \\ 938.442 \\ 996.088 \\ 1632.781 \end{bmatrix}$$

and the least squares estimates of β_0, β_1, β_2, and β_3 are

$$b = \begin{bmatrix} b_0 \\ b_1 \\ b_2 \\ b_3 \end{bmatrix}$$

$$= (X'X)^{-1}X'y$$

$$= \begin{bmatrix} 108.556 & -26.7730 & 3.48877 & -3.45557 \\ -26.7730 & 7.39020 & -1.75917 & .947785 \\ 3.48877 & -1.75917 & 1.58405 & -.491626 \\ -3.45557 & .947785 & -.491626 & -.287712 \end{bmatrix} \begin{bmatrix} 251.480 \\ 938.442 \\ 996.088 \\ 1632.781 \end{bmatrix}$$

$$= \begin{bmatrix} 7.5891 \\ -2.3577 \\ 1.6122 \\ .5012 \end{bmatrix}$$

Moreover,

$$\sum_{t=1}^{30} y_t^2 = y_1^2 + y_2^2 + \cdots + y_{30}^2$$

$$= (7.38)^2 + (8.51)^2 + \cdots + (9.26)^2 = 2121.5486$$

and

$$\mathbf{b'X'y} = [7.5891 \;\; -2.3577 \;\; 1.6122 \;\; .5012] \begin{bmatrix} 251.480 \\ 938.442 \\ 996.088 \\ 1632.781 \end{bmatrix}$$

$$= 2120.1168$$

and therefore,

$$\sum_{t=1}^{30} (y_t - \hat{y}_t)^2 = \sum_{t=1}^{30} y_t^2 - \mathbf{b'X'y}$$

$$= 2121.5486 - 2120.1168 = 1.4318 \approx 1.432$$

A 95 percent confidence interval for μ_t is

$$[\hat{y}_t - E_{\mu_t}(95), \; \hat{y}_t + E_{\mu_t}(95)]$$

where

$$\hat{y}_t = 7.5891 - 2.3577x_{t1} + 1.6122x_{t2} + .5012x_{t3}$$

and

$$E_{\mu_t}(95) = t_{5/2}(30 - (3 + 1)) \cdot s \cdot \sqrt{\mathbf{x'_t(X'X)^{-1}x_t}}$$

Assuming that $\mathbf{x'_{31}} = [1 \;\; 3.80 \;\; 3.90 \;\; 6.80]$, and calculating

$$t_{5/2}(30 - (3 + 1)) = t_{5/2}(26) = 2.056$$

and

$$s = \sqrt{\frac{\sum_{t=1}^{30} (y_t - \hat{y}_t)^2}{30 - (3 + 1)}} = \sqrt{\frac{1.432}{26}} = .2347$$

we have that if $t = 31$, then

$$\hat{y}_t = 8.325$$

$$\mathbf{x'_t(X'X)^{-1}x_t} =$$

$$[1 \; 3.8 \; 3.9 \; 6.8] \begin{bmatrix} 108.556 & -26.7730 & 3.48877 & -3.45557 \\ -26.7730 & 7.39020 & -1.75917 & .947785 \\ 3.48877 & -1.75917 & 1.58405 & -.491626 \\ -3.45557 & .947785 & -.491626 & -.287712 \end{bmatrix} \begin{bmatrix} 1 \\ 3.8 \\ 3.9 \\ 6.8 \end{bmatrix}$$

$$= .173$$

and

$$E_{\mu_t}(95) = t_{5/2}(30 - (3 + 1)) \cdot s \cdot \sqrt{\mathbf{x}_t'(\mathbf{X}'\mathbf{X})^{-1}\mathbf{x}_t} = .201$$

The 95 percent confidence interval for y_t is

$$[\hat{y}_t - E_t(95), \hat{y}_t + E_t(95)]$$

where

$$\hat{y}_t = 7.5891 - 2.3577x_{t1} + 1.6122x_{t2} + .5012x_{t3}$$

and

$$E_t(95) = t_{5/2}(30 - (3 + 1)) \cdot s \cdot f_t$$

Assuming that

$$\mathbf{x}_{31}' = [1 \ 3.80 \ 3.90 \ 6.80]$$

$$\mathbf{x}_{32}' = [1 \ 3.75 \ 3.85 \ 6.85]$$

$$\mathbf{x}_{33}' = [1 \ 3.85 \ 3.90 \ 6.90]$$

and calculating

$$t_{5/2}(30 - (3 + 1)) = t_{5/2}(26) = 2.056$$

and

$$s = \sqrt{\frac{\sum_{t=1}^{30}(y_t - \hat{y}_t)^2}{30 - (3 + 1)}} = \sqrt{\frac{1.432}{26}} = .2347$$

we have

$$\hat{y}_t = \begin{cases} 8.325 & \text{if } t = 31 \\ 8.388 & \text{if } t = 32 \\ 8.257 & \text{if } t = 33 \end{cases}$$

$$f_t = \sqrt{1 + \mathbf{x}_t'(\mathbf{X}'\mathbf{X})^{-1}\mathbf{x}_t} = \begin{cases} 1.083 & \text{if } t = 31 \\ 1.071 & \text{if } t = 32 \\ 1.154 & \text{if } t = 33; \end{cases}$$

$$E_t(95) = t_{5/2}(30 - (3 + 1)) \cdot s \cdot f_t = \begin{cases} .523 & \text{if } t = 31 \\ .517 & \text{if } t = 32 \\ .557 & \text{if } t = 33 \end{cases}$$

The 95 percent confidence interval for

$$\sum_{t=l=31}^{t=l+q=31+2=33} y_t$$

is

$$\left[\sum_{t=31}^{33} \hat{y}_t - E_{(31,31+2)}(95), \ \sum_{t=31}^{33} \hat{y}_t + E_{(31,31+2)}(95) \right]$$

where

$$E_{(31,31+2)}(95) = t_{5/2}(26) \cdot s \cdot \sqrt{(2 + 1) + \left(\sum_{t=31}^{33} \mathbf{x}'_t \right)(\mathbf{X}'\mathbf{X})^{-1}\left(\sum_{t=31}^{33} \mathbf{x}_t \right)}$$

$$= 1.062$$

since

$$\sum_{t=31}^{33} \mathbf{x}'_t = [1 \ 3.80 \ 3.90 \ 6.80] + [1 \ 3.75 \ 3.85 \ 6.85] + [1 \ 3.85 \ 3.90 \ 6.90]$$

$$= [3 \ 11.40 \ 11.65 \ 20.55]$$

1-6 AN OVERVIEW OF QUANTITATIVE FORECASTING TECHNIQUES

In Section 1–5 we studied the basic multiple regression model

$$y_t = \beta_0 + \beta_1 x_{t1} + \beta_2 x_{t2} + \cdots + \beta_p x_{tp} + \epsilon_t$$

Each of the least squares estimates $b_0, b_1, b_2, \ldots, b_p$ of the unknown parameters $\beta_0, \beta_1, \beta_2, \ldots, \beta_p$ is a function of the data observed for n periods on $y_t, x_{t1}, x_{t2}, \ldots, x_{tp}$. Hence, the forecast

$$\hat{y}_T = b_0 + b_1 x_{T1} + b_2 x_{T2} + \cdots + b_p x_{Tp}$$

of the value of the time series in a future period T is a function of (1) the data previously observed for n periods—$y_1, x_{11}, \ldots, x_{1p}; y_2, x_{21}, \ldots, x_{2p}; \ldots; y_n, x_{n1}, \ldots, x_{np}$—and (2) the values of the independent variables in the future period— $x_{T1}, x_{T2}, \ldots, x_{Tp}$.

In Example 1.1, the independent variables are other time series. For example, x_{t1} = the price (in dollars) of Fresh in period t as offered by Enterprise Industries. The multiple regression model in Example 1.1 is therefore an example of a *causal model*, which exploits the relationship between various time series to forecast a time series of interest. We will further discuss causal multiple regression models in Chapter 2. Of course, as mentioned previously, one problem with the use of causal models is that the values of the time series used as the independent variables must be known for the time period for which the dependent variable is being forecast.

Chapters 3 through 12 will study the forecasting of time series by the use of *time series models,* which are generally of the form

$$y_t = f(\beta_0, \beta_1, \beta_2, \ldots, \beta_p; t) + \epsilon_t$$

where $f(\beta_0, \beta_1, \ldots, \beta_p; t)$ is a function of the unknown parameters $\beta_0, \beta_1, \ldots, \beta_p$ and of time t. This function is expressed in terms of trend and seasonal effects but not in terms of time series *other than* the time series y_t. Again, ϵ_t is a random error component assumed to have an expected value equal to 0. The model states that the time series y_t can be represented by an average level that changes over time according to the function $f(\beta_0, \beta_1, \beta_2, \ldots, \beta_p; t)$ combined with random fluctuations (represented by ϵ_t) that cause the observations of the time series to deviate from the average level. Many of the time series models we will study in this book are of the form

$$y_t = \beta_0 + \beta_1 f_1(t) + \beta_2 f_2(t) + \cdots + \beta_p f_p(t) + \epsilon_t$$

which is a multiple regression model with the functions of time $f_1(t), f_2(t), \ldots, f_p(t)$ being the independent variables and representing trend and seasonal effects, *not* time series other than y_t. An example of a time series model is

$$y_t = \beta_0 + \beta_1 t + \epsilon_t$$

Looking at the equation above, we see that here we have $f_1(t) = t$ and $f_2(t) = f_3(t) = \cdots = f_p(t) = 0$. This model says that the time series y_t can be represented by an average level that changes over time *in a linear fashion* according to the straight-line function $\beta_0 + \beta_1 t$ combined with random fluctuations (represented by ϵ_t) that cause the observations of the time series to deviate from the average level.

Since the function $f(\beta_0, \beta_1, \ldots, \beta_p; t)$ in the general time series model

$$y_t = f(\beta_0, \beta_1, \ldots, \beta_p; t) + \epsilon_t$$

does not express a relationship between the time series y_t and other time series, it makes intuitive sense that each of the estimates (for example, least squares estimates) b_0, b_1, . . . , b_p of the parameters β_0, β_1, . . . , β_p is a function of the observations gathered for n periods on y_t. Hence, the forecast

$$\hat{y}_t = f(b_0, b_1, \ldots, b_p; T)$$

of the time series in a future period T is a function of the previously observed values of the time series, y_1, y_2, . . . , y_n, and of the time T.

Since the least squares estimates b_0, b_1, . . . , b_p of the parameters β_0, β_1, . . . , β_p in the time series model

$$y_t = \beta_0 + \beta_1 f_1(t) + \beta_2 f_2(t) + \cdots + \beta_p f_p(t) + \epsilon_t$$

are those estimates that minimize $\Sigma_{t=1}^{n}(y_t - \hat{y}_t)^2$, the least squares estimates are determined by *equally* weighting each of the previously observed values of the time series, y_1, y_2, . . . , y_n. In ongoing forecasting systems, forecasts of a time series are made each period for succeeding periods. Hence, the forecasting equation and the estimates of the model parameters need to be updated at the end of each period to account for the most recent observations. This updating must take into account the fact that the model parameters β_0, β_1, . . . , β_p in the function $f(\beta_0, \beta_1, \ldots, \beta_p; t)$ *may be changing* over time; and, therefore, it is not reasonable to continually give equal weight to each of the previously observed values of the time series. If updating is done using the multiple regression approach, it is necessary to determine how many of the most remote observations should be discarded as new observations are obtained so that the estimates will be updated in an "optimal" fashion. Moreover, if there are a large number of time series to be analyzed every period, the calculations in the multiple regression approach would be extremely time consuming; and, since they require knowledge of the individual observations collected, storage of a tremendous amount of data would be required. *Exponential smoothing* is a forecasting method that weights each of the observed values of the time series *unequally,* with more recent observations being weighted more heavily than more remote observations. This procedure allows the forecaster to update the estimates of the parameters β_0, β_1, . . . , β_p of the function $f(\beta_0, \beta_1, \ldots, \beta_p; t)$ so that changes in the values of these parameters can be detected and incorporated into the forecasting system. Furthermore, as will be subsequently demonstrated, exponential smoothing is computationally easy and requires little storage of previously observed values of the time series. The unequal weighting of the previously observed values of the time series is accomplished through one or more smoothing constants, which determine how much weight is given to each observation. In addition, exponential smoothing provides ways for checking whether the forecasts are "reasonably accurate" and for correcting the forecasting equations if the need arises—often by changing the smoothing constant(s).

Part II of this book deals with the forecasting of time series described by trend and irregular components. Both the regression approach and the exponential smoothing approach are used to forecast time series with no trend, with a linear trend, and with a quadratic trend. It can be shown that the exponential smoothing approach described in Part II has a basis in statistical theory. We will not study this basis, but the interested reader is referred to Brown [3], or Johnson and Montgomery [9]. Part III deals with the forecasting of time series described by trend, seasonal, and irregular components. Both the regression approach and the exponential smoothing approach are used to forecast time series in which the "seasonal swing" is independent of the trend and in which the "seasonal swing" is proportional to the trend. With the exception of various trigonometric models at the end of Part III, the exponential smoothing approach described in Part III is intuitive and does not have a basis in statistical theory. The emphasis in Part II and Part III will be on how to intelligently use the various techniques. We will concentrate on illustrating the techniques through detailed examples, each of which begins with a data set, continues with the "fitting" of a regression model or exponential smoothing model to the data set, and concludes with forecasting future values of the time series using the fitted models.

It should be noted that, in general, all of the regression and exponential smoothing approaches we study in this book assume that the random error components in the time series model $y_t = f(\beta_0, \beta_1, \ldots, \beta_p; t) + \epsilon_t$ satisfy the assumptions discussed in Section 1–5. One of these assumptions is that the error terms $\epsilon_1, \epsilon_2, \ldots$ are not related to each other. If the error terms are only "weakly related," then these regression and exponential smoothing approaches will still probably produce fairly accurate forecasts. But, if the error terms are "strongly related," it follows that accurate forecasting of the time series can probably best be done by employing a model that expresses y_t as a function of prior random error components $\epsilon_t, \epsilon_{t-1}, \ldots$. Such a function can be written as

$$y_t = \mu + \psi_0 \epsilon_t + \psi_1 \epsilon_{t-1} + \psi_2 \epsilon_{t-2} + \cdots$$

where $\mu, \psi_0, \psi_1, \psi_2, \ldots$ are the parameters in the model. It is not immediately obvious how such models are used, but Box and Jenkins [1] have developed a systematic procedure for forecasting with these models. We will discuss this procedure in Part IV. For other discussions the interested reader is referred to Box and Jenkins [1], which is fairly mathematical and gives an excellent theoretical development with very good examples, or to Nelson [10], which is less mathematical and contains excellent computer examples.

If the successive error terms are highly related, forecasting methods based on regression and exponential smoothing may not be the best forecasting methods to use because they do not use the relationship between the errors in the most appropriate manner. In practice, regression and exponential smoothing methods are frequently applied with good results to time series with related observations, but it must be emphasized that use of the Box-Jenkins methodology is likely to produce more accurate forecasts. Some disadvantages of the Box-Jenkins methodology are that a very large amount of data and a great investment in time and other resources are required to build an accurate forecasting model. This is particularly true if a model to forecast a time series with seasonal variations is desired.

PROBLEMS

1. Table 1 gives the forecast errors produced using three different sales forecasting models.

    a. Graph these forecast errors for each model.
    b. Which error patterns appear to be random?
    c. Decide whether models A, B, and C adequately fit the data pattern characterizing sales. Explain your conclusions.

TABLE 1

Year	Quarter	Model A Forecast Error	Model B Forecast Error	Model C Forecast Error
1975	1	+25	+15	−20
	2	+12	+6	+6
	3	+7	−10	+15
	4	+5	−3	−10
1976	1	+3	+12	+8
	2	0	+4	−5
	3	−4	−7	+7
	4	−11	−1	−8
1977	1	−17	+9	+3
	2	−21	+7	+10
	3	−28	−12	−12
	4	−34	−5	+4
1978	1	−21	+17	−7
	2	−13	+3	+9
	3	−7	−9	+19
	4	−2	−3	−7
1979	1	+5	+13	+16
	2	+9	+5	−6
	3	+15	−10	−9
	4	+19	−6	+5

2. Table 2 presents predicted monthly sales and actual monthly sales for a company over the first six months of 1978.

    a. Calculate the forecast error for each month.
    b. Calculate the MAD.
    c. Calculate the MSE.

TABLE 2

	Actual Sales	Predicted Sales
January 1978	270	265
February 1978	263	268
March 1978	275	269
April 1978	262	267
May 1978	250	245
June 1978	278	275

3. Given the following sales forecasting models, determine whether these fore-casting models are time series models or causal models. Note that Sales(t) denotes sales in time t.

   a. Sales($t + 1$) = .8[Forecasted Sales(t)] + .2[Sales(t)]
   b. Sales($t + 1$) = 500 + 2.5(Advertising Expenditure in Year t)
                        + 5(Number of Customer Calls by Salesmen in Year t)
   c. Sales($t + 1$) = $\dfrac{1}{t} \sum_{i=1}^{t}$ Sales(i)

4. Table 3 contains the predicted and actual per capita income for a certain area of the United States.

TABLE 3

Year	Per Capita Income ($)	Predicted Per Capita Income ($)
1970	3074	3292
1971	3135	3250
1972	3206	3230
1973	3267	3255
1974	3310	3266
1975	3362	3283
1976	3418	3300
1977	3500	3337

   a. Calculate the forecast error for each year.
   b. Plot these forecast errors against time.
   c. Do you think the forecasting method that was used adequately fits the data pattern present? Why or why not?
   d. Calculate the MAD.
   e. Calculate the MSE.

5.   Table 4 contains actual yearly sales for a company along with the predictions of yearly sales generated using two different forecasting methods.

TABLE 4

Year	Actual Sales (in millions)	Method A Predicted Sales	Method B Predicted Sales
1974	8.0	9.0	9.5
1975	12.0	11.5	10.5
1976	14.0	14.0	12.0
1977	16.0	16.5	13.0
1978	10.0	19.0	15.0

a.   Calculate the MAD for both forecasting methods.
b.   Calculate the MSE for both forecasting methods.
b.   Explain why these measures of accuracy yield different results in this case.

Note: Problems on multiple regression analysis will be given at the end of Chapter 2, since Section 2–2 gives the reader the necessary information to work problems on multiple regression analysis.

CHAPTER 2

Building a Multiple Regression Model

2-1 INTRODUCTION

In Section 1–5 we have discussed how to fit a regression model to a set of data and how to then use the regression model to estimate the average level μ_t of a time series and to forecast a future value y_t of a time series. In this chapter we discuss how to use historical data to build a multiple regression model that is likely to provide accurate forecasts of a given time series of interest.

Section 2–2 discusses some basic tools that are useful in building an appropriate multiple regression model. These tools include graphical analysis and several basic statistical measures of the utility of the independent variables in a multiple regression model. After studying Section 2–2, which has a very applied orientation, the reader may omit the last section of this chapter, Section 2–3, and read Chapters 3 through 12 of this book without loss of understanding. Section 2–3 discusses some advanced statistical tools that can be used in building a multiple regression model. In particular, Section 2–3 relates the concept of hypothesis testing to model building. Although this concept is reviewed in Section 2–3, in order to get the most out of Section 2–3 it is probably best that the reader have some prior exposure to hypothesis testing. Section 2–2 and Chapters 3 through 12 of this book really do not require this previous exposure.

2-2 SOME BASIC PRINCIPLES OF AND TOOLS USED IN MODEL BUILDING

Constructing a multiple regression model that is likely to provide accurate forecasts of a given time series involves both specifying an appropriate set of independent variables and determining the *functional form* of the regression relationship between the dependent variable and a given set of independent variables. Determining the functional form involves making such decisions as whether the dependent variable is related to a particular independent variable in a *linear fashion* or in a *quadratic fashion*, and whether two or more independent variables should be multiplied together to form what is called an *interaction variable*. We will now discuss the meaning of these terms.

2–2.1   INTRODUCTORY CONCEPTS AND GRAPHICAL ANALYSIS

Consider the regression model

$$y_t = \mu_t + \epsilon_t$$

where μ_t is the average level of the time series at time t and ϵ_t is the random error component. The dependent variable y_t is said to be related to a single independent variable x_{t1} in a *linear fashion* if

$$\mu_t = \beta_0 + \beta_1 x_{t1}$$

This means that the time series y_t is *randomly fluctuating* around an average level μ_t that changes in a linear or straight-line fashion as x_{t1} increases. The slope of this straight line is β_1, while the intercept at $x_{t1} = 0$ is β_0. If

$$\beta_1 > 0$$

the average level μ_t of the time series increases as x_{t1} increases. If

$$\beta_1 < 0$$

the average level μ_t of the time series decreases as x_{t1} increases. Figure 2.1 illustrates how the average level changes in a linear fashion as x_{t1} increases. The dependent variable y_t is said to be related to a single independent variable x_{t1} in a *quadratic fashion* if

$$\mu_t = \beta_0 + \beta_1 x_{t1} + \beta_2 x_{t1}^2$$

This means that the time series y_t is randomly fluctuating around an average level μ_t that changes in a quadratic or curvilinear fashion as x_{t1} increases. Thus the average level μ_t of the time series either is increasing at an increasing or decreasing rate or is

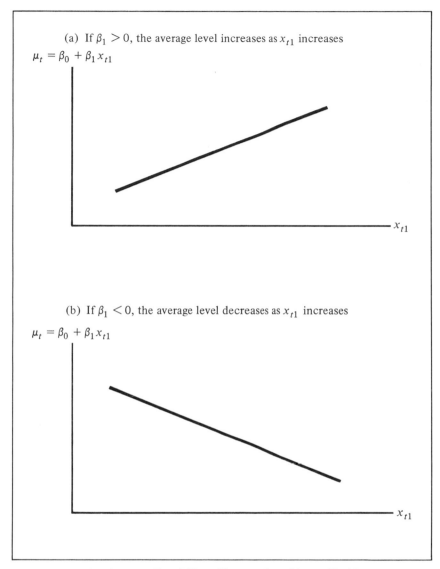

(a) If $\beta_1 > 0$, the average level increases as x_{t1} increases

$$\mu_t = \beta_0 + \beta_1 x_{t1}$$

x_{t1}

(b) If $\beta_1 < 0$, the average level decreases as x_{t1} increases

$$\mu_t = \beta_0 + \beta_1 x_{t1}$$

x_{t1}

FIGURE 2.1   *An Average Level That Changes in a Linear Fashion as x_{t1}*
Increases

decreasing at an increasing or decreasing rate as x_{t1} increases. Figure 2.2 illustrates how the average level μ_t changes in a quadratic fashion as x_{t1} increases. The particular numerical values of β_0, β_1, and β_2 determine exactly how the average level changes as x_{t1} increases. As will be illustrated in Example 2.1, a plot of the values of the dependent variable against the increasing values of each of the independent variables can be useful in determining whether the dependent variable is related to an independent variable in a linear fashion or in a quadratic fashion.

(a) The average level is
increasing at an
increasing rate as
x_{t1} increases

$$\mu_t = \beta_0 + \beta_1 x_{t1} + \beta_2 x_{t1}^2$$

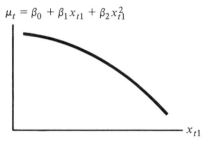

(b) The average level is
increasing at a
decreasing rate as
x_{t1} increases

$$\mu_t = \beta_0 + \beta_1 x_{t1} + \beta_2 x_{t1}^2$$

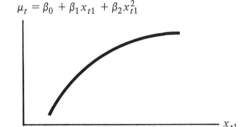

(c) The average level is
decreasing at an
increasing rate as
x_{t1} increases

$$\mu_t = \beta_0 + \beta_1 x_{t1} + \beta_2 x_{t1}^2$$

(d) The average level is
decreasing at a
decreasing rate as
x_{t1} increases

$$\mu_t = \beta_0 + \beta_1 x_{t1} + \beta_2 x_{t1}^2$$

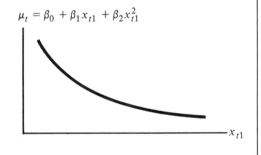

FIGURE 2.2   *An Average Level That Changes in a Quadratic Fashion as x_{t1} Increases*

*Example
2.1*   In the Fresh detergent situation of Chapter 1 we have considered the multiple
regression model

$$y_t = \mu_t + \epsilon_t$$
$$= \beta_0 + \beta_1 x_{t1} + \beta_2 x_{t2} + \beta_3 x_{t3} + \epsilon_t$$

where the variables y_t, x_{t1}, x_{t2}, and x_{t3} have been previously defined. By assuming
that

$$\mu_t = \beta_0 + \beta_1 x_{t1} + \beta_2 x_{t2} + \beta_3 x_{t3}$$

we are assuming that the dependent variable is related to each of the independent variables x_{t1}, x_{t2}, and x_{t3} in a linear fashion. If we assume that

$$\mu_t = \beta_0 + \beta_1 x_{t1} + \beta_2 x_{t2} + \beta_3 x_{t3} + \beta_4 x_{t3}^2$$

we are assuming that the dependent variable is related to each of the independent variables x_{t1} and x_{t2} in a linear fashion and to the independent variable x_{t3} in a quadratic fashion.

Before plotting the data to aid us in deciding whether the dependent variable is related to each of the independent variables in a linear fashion or in a quadratic fashion, let us consider being somewhat "inventive" and forming one independent variable from two others. That is, consider defining the independent variable

$$x_{t4} = x_{t2} - x_{t1}$$

which is the difference between x_{t2}, the average industry price, and x_{t1}, Enterprise Industries' price for Fresh. Of course, this is not the only way that x_{t1} and x_{t2} can be combined. For example,

$$x_{t5} = \frac{x_{t2}}{x_{t1}}$$

the ratio of the average industry price to Enterprise Industries' price, might be useful in forecasting sales of Fresh. In this example, we will concentrate on constructing a regression model that predicts y_t as a function of x_{t4} and x_{t3}, Enterprise Industries' advertising expenditure.

Table 2.1 lists the historical values of y_t, x_{t1}, x_{t2}, x_{t4}, and x_{t3}. First, in order to better understand the relationship between y_t and x_{t3}, let us plot the values of y_t against the increasing values of x_{t3}. This plot is shown in Figure 2.3. Note that to make this plot we have rearranged the values in Table 2.1, which appear in the order in which they were observed, so that the values of x_{t3} are in increasing order (see Table 2.2). Inspecting Figure 2.3, we see that as x_{t3} increases, y_t appears to increase in either a linear fashion with a positive slope (as illustrated in Figure 2.1a) or in a quadratic fashion (as illustrated in Figure 2.2a), which illustrates the average level increasing at an increasing rate. The plot in Figure 2.3 is probably more quadratic than linear. This would imply that if we were developing a regression model to relate y_t to the single independent variable x_{t3}, then it would be reasonable to conclude that

$$y_t = \mu_t + \epsilon_t$$
$$= \beta_0 + \beta_1 x_{t3} + \beta_2 x_{t3}^2 + \epsilon_t$$

Let us now consider forecasting y_t using this regression model. As long as we forecast a value of y_t corresponding to a value of x_{t3} between 5.25 and 7.25, the range of values of x_{t3} that are included in the historical data of Table 2.1, we would

TABLE 2.1   *The Historical Observations*

t	Demand y_t	Price x_{t1}	Average Industry Price x_{t2}	Price Difference $x_{t4} = x_{t2} - x_{t1}$	Advertising Expenditure x_{t3}
1	7.38	3.85	3.80	−.05	5.50
2	8.51	3.75	4.00	.25	6.75
3	9.52	3.70	4.30	.60	7.25
4	7.50	3.70	3.70	0	5.50
5	9.33	3.60	3.85	.25	7.00
6	8.28	3.60	3.80	.20	6.50
7	8.75	3.60	3.75	.15	6.75
8	7.87	3.80	3.85	.05	5.25
9	7.10	3.80	3.65	−.15	5.25
10	8.00	3.85	4.00	.15	6.00
11	7.89	3.90	4.10	.20	6.50
12	8.15	3.90	4.00	.10	6.25
13	9.10	3.70	4.10	.40	7.00
14	8.86	3.75	4.20	.45	6.90
15	8.90	3.75	4.10	.35	6.80
16	8.87	3.80	4.10	.30	6.80
17	9.26	3.70	4.20	.50	7.10
18	9.00	3.80	4.30	.50	7.00
19	8.75	3.70	4.10	.40	6.80
20	7.95	3.80	3.75	−.05	6.50
21	7.65	3.80	3.75	−.05	6.25
22	7.27	3.75	3.65	−.10	6.00
23	8.00	3.70	3.90	.20	6.50
24	8.50	3.55	3.65	.10	7.00
25	8.75	3.60	4.10	.50	6.80
26	9.21	3.65	4.25	.60	6.80
27	8.27	3.70	3.65	−.05	6.50
28	7.67	3.75	3.75	0	5.75
29	7.93	3.80	3.85	.05	5.80
30	9.26	3.70	4.25	.55	6.80

probably obtain a fairly accurate forecast. However, consider forecasting a demand (y_t) corresponding to an advertising expenditure of $x_{t3} = 7.75$, which is far outside the range 5.25 to 7.25. As illustrated in Figure 2.3, if we extrapolate the solid curve that we fit to the data, we will obtain a very large forecast of demand. Is this reasonable? If we assume that our advertising is extremely effective, possibly it is reasonable. But, assuming the law of diminishing returns sets in, it is just as possible that the curve describing demand will change direction, as illustrated by the dashed curve in Figure 2.3. In this case, a much lower forecast of demand is reasonable. The point is that there is no historical data to tell us which forecast is more reasonable, and that to blindly use the solid curve to forecast demand corresponding to the advertising expenditure $x_{t3} = 7.75$ is extremely dangerous. In this case, it

TABLE 2.2   *Values of x_{t3} Arranged in Increasing Order*

x_{t3}	y_t	x_{t3}	y_t
5.25	7.87	6.75	8.51
5.25	7.10	6.75	8.75
5.50	7.38	6.8	8.90
5.50	7.50	6.8	8.87
5.75	7.67	6.8	8.75
5.80	7.93	6.8	8.75
6.00	8.00	6.8	9.21
6.00	7.27	6.8	9.26
6.25	8.15	6.9	8.86
6.25	7.65	7.0	9.33
6.50	8.28	7.0	9.10
6.50	7.89	7.0	8.50
6.50	7.95	7.0	9.00
6.50	8.00	7.1	9.26
6.50	8.27	7.25	9.52

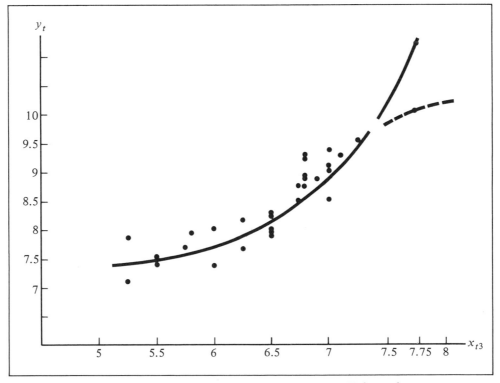

FIGURE 2.3   *The Plot of the Values of y_t against the Increasing Values of x_{t3}*

is particularly important to combine subjective judgment with our knowledge of the previously observed data pattern. This illustrates the fact that in using a regression model to predict future values of a dependent variable, it is wise, if possible, to predict values of the dependent variable corresponding to combinations of values of the independent variables within the *experimental region*. The experimental region is the range of the historical (already observed) combinations of values of the independent variables used to develop the regression model. It is, in fact, quite dangerous to extrapolate—to use a multiple regression model to predict values of the dependent variable corresponding to combinations of values of the independent variables outside of the experimental region.

It should also be mentioned that although Figure 2.3 makes it appear that

$$\mu_t = \beta_0 + \beta_1 x_{t3} + \beta_2 x_{t3}^2$$

data plots can be somewhat misleading, and it is always possible that

$$\mu_t = \beta_0 + \beta_1 x_{t3}$$

which would imply that μ_t increases in a linear fashion with increasing values of x_{t3}. Certainly, we need some more precise measures of the true data patterns. Some statistical measures to be discussed after this example will be useful in identifying true data patterns.

Let us next consider the relationship between y_t and x_{t4}. A plot of the values of y_t against the increasing values of x_{t4} is given in Figure 2.4. To make this plot we have rearranged the values of x_{t4}, which are given in Table 2.1 in the order in which they were observed, so that they are in increasing order. This rearrangement is

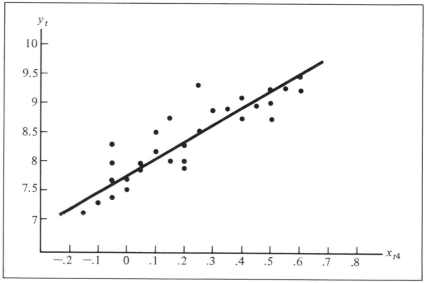

FIGURE 2.4   *The Plot of the Values of y_t against the Increasing Values of x_{t4}*

TABLE 2.3   *Values of x_{t4} Arranged in Increasing Order*

x_{t4}	y_t	x_{t4}	y_t
−.15	7.10	.20	7.89
−.10	7.27	.20	8.00
−.05	8.27	.25	8.51
−.05	7.38	.25	9.33
−.05	7.95	.30	8.87
−.05	7.65	.35	8.90
0	7.50	.40	9.10
0	7.67	.40	8.75
.05	7.87	.45	8.86
.05	7.93	.50	9.26
.10	8.15	.50	9.00
.10	8.50	.50	8.75
.15	8.75	.55	9.26
.15	8.00	.60	9.52
.20	8.28	.60	9.21

given in Table 2.3. Inspecting Figure 2.4, we see that as x_{t4} increases, y_t appears to increase in either a linear fashion with a positive slope (Figure 2.1a) or in a quadratic fashion (Figure 2.2b) at a decreasing rate. Not only does the plot look more linear than quadratic, but if the relationship were truly quadratic, one would subjectively think that as x_{t4} increases (which implies that Enterprise Industries' price is getting lower relative to the average industry price) μ_t would be increasing at an increasing, and not a decreasing, rate. Subjective judgment thus confirms that if we were developing a regression model to relate y_t to the single independent variable x_{t4}, it would be reasonable to conclude that the relationship is linear, that is, that

$$y_t = \mu_t + \epsilon_t$$
$$= \beta_0 + \beta_1 x_{t4} + \epsilon_t$$

However, it is again possible that the plot shown (Figure 2.4) is misleading, and we should probably entertain the possibility that

$$\mu_t = \beta_0 + \beta_1 x_{t4} + \beta_2 x_{t4}^2$$

Up to this point we have intuitively developed separate regression models relating y_t to x_{t3} and x_{t4}. We have concluded that probably the most reasonable regression models for the available data state that

$$y_t = \mu_t + \epsilon_t$$
$$= \beta_0 + \beta_1 x_{t3} + \beta_2 x_{t3}^2 + \epsilon_t$$

and

$$y_t = \mu_t + \epsilon_t$$
$$= \beta_0 + \beta_1 x_{t4} + \epsilon_t$$

If we wish to specify a single regression model relating y_t to both x_{t4} and x_{t3}, the model

$$y_t = \mu_t + \epsilon_t$$
$$= \beta_0 + \beta_1 x_{t4} + \beta_2 x_{t3} + \beta_3 x_{t3}^2 + \epsilon_t$$

might seem best, although other possible models could specify that

$$\mu_t = \beta_0 + \beta_1 x_{t4} + \beta_2 x_{t4}^2 + \beta_3 x_{t3} + \beta_4 x_{t3}^2$$

or

$$\mu_t = \beta_0 + \beta_1 x_{t4} + \beta_2 x_{t4}^2 + \beta_3 x_{t3}$$

or

$$\mu_t = \beta_0 + \beta_1 x_{t4} + \beta_2 x_{t3}$$

Let us further consider the most reasonable model, which is

$$y_t = \mu_t + \epsilon_t$$
$$= \beta_0 + \beta_1 x_{t4} + \beta_2 x_{t3} + \beta_3 x_{t3}^2 + \epsilon_t$$

This model does not contain what is called an *interaction variable*. To discuss the meaning of interaction, consider four time periods, which we will call time periods 1, 2, 3, and 4. Assume that the values of the independent variables and the corresponding values of μ_t given by our model for the four periods are as given in Table 2.4. Note that

$$\mu_2 - \mu_1 = \beta_2(6.8 - 6) + \beta_3[(6.8)^2 - (6)^2] = .8\beta_2 + 10.24\beta_3$$
$$\mu_4 - \mu_3 = \beta_2(6.8 - 6) + \beta_3[(6.8)^2 - (6)^2] = .8\beta_2 + 10.24\beta_3$$

The fact that

$$\mu_2 - \mu_1 = \mu_4 - \mu_3 = .8\beta_2 + 10.24\beta_3$$

says that $\mu_2 - \mu_1$, the difference in average demand levels caused by increasing advertising expenditure by .8 when the price difference x_{t4} is held constant at .1, is

TABLE 2.4   *Computation of Four Values of μ_t*

Period = t	x_{t4}	x_{t3}	x_{t3}^2	$\mu_t = \beta_0 + \beta_1 x_{t4} + \beta_2 x_{t3} + \beta_3 x_{t3}^2$
1	.10	6.0	$(6.0)^2$	$\mu_1 = \beta_0 + \beta_1(.1) + \beta_2(6) + \beta_3(6)^2$
2	.10	6.8	$(6.8)^2$	$\mu_2 = \beta_0 + \beta_1(.1) + \beta_2(6.8) + \beta_3(6.8)^2$
3	.30	6.0	$(6.0)^2$	$\mu_3 = \beta_0 + \beta_1(.3) + \beta_2(6) + \beta_3(6)^2$
4	.30	6.8	$(6.8)^2$	$\mu_4 = \beta_0 + \beta_1(.3) + \beta_2(6.8) + \beta_3(6.8)^2$

equal to $\mu_4 - \mu_3$, the difference in average demand levels caused by increasing advertising expenditure by .8 when the price difference x_{t4} is held constant at .3. In other words, our model assumes that the relationship between demand and advertising expenditure is independent of the price difference, as illustrated in Figure 2.5. In general, there is said to be *no interaction* between two independent variables if the relationship between the dependent variable and one of the independent variables is independent of the value of the other independent variable. There is said to be *interaction* between two independent variables if the relationship between the dependent variable and one of the independent variables is dependent upon the value of the other independent variable.

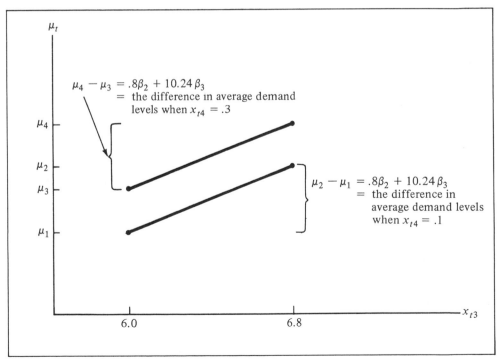

FIGURE 2.5   *No Interaction: The Difference in Average Demand Caused by Increasing x_{t3} Does Not Depend on x_{t4}*

Is it reasonable to believe there is little or no interaction between price difference and advertising expenditure in the Fresh detergent example? Probably, it is somewhat reasonable to believe this since the values of the independent variable price difference do not vary a great deal. Note from Table 2.1 that Enterprise Industries' price for Fresh is usually lower than, and is never much larger than, the average industry price. But what would happen if Enterprise Industries suddenly raised prices .50 above the average industry price? It makes sense to assume that increasing advertising expenditure would not increase demand as much if Enterprise Industries' product were at a large price disadvantage as it would if Enterprise Industries' product were not at a large price disadvantage. This is illustrated in Figure 2.6. In this case, interaction might well exist between price difference and

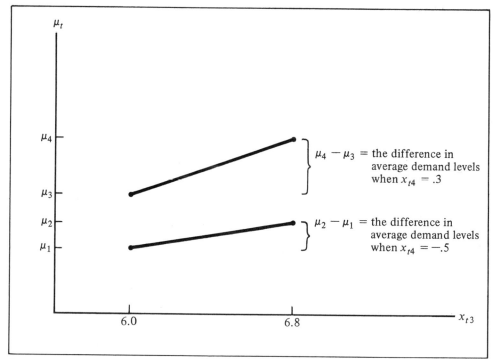

FIGURE 2.6 *Interaction: The Difference in Average Demand Caused by Increasing x_{t3} Is Less when $x_{t4} = -.5$ Than when $x_{t4} = .3$*

advertising expenditure. As long as Enterprise Industries maintains a price difference similar to previously observed price differences and maintains an advertising expenditure similar to previous advertising expenditures, it may well be appropriate to use the model that assumes there is no interaction between price difference and advertising expenditure, that is

$$y_t = \mu_t + \epsilon_t$$

$$= \beta_0 + \beta_1 x_{t4} + \beta_2 x_{t3} + \beta_3 x_{t3}^2 + \epsilon_t$$

Above, we intuitively reasoned that little or no interaction exists between the independent variables. In some situations, particularly in designed experiments, plots of the data can aid in determining whether or not interaction does exist between the independent variables. In the Fresh detergent situation, however, the data are not "structured" enough to (easily) make such plots. However, the statistical measures discussed after this example can help to determine whether interaction exists between price difference and advertising expenditure. If this interaction does exist, we can model it by multiplying the appropriate independent variables together and considering the model

$$y_t = \mu_t + \epsilon_t$$
$$= \beta_0 + \beta_1 x_{t4} + \beta_2 x_{t3} + \beta_3 x_{t3}^2 + \beta_4 x_{t4} x_{t3} + \beta_5 x_{t4} x_{t3}^2 + \epsilon_t$$

that utilizes two interaction variables—$x_{t4}x_{t3}$ and $x_{t4}x_{t3}^2$. It should be noted that in a particular regression situation it may be necessary to use anywhere from one to all of the possible interaction variables that can be formed by appropriately multiplying independent variables together. Again, the statistical measures to be discussed can aid in determining which interaction variables are important.

2–2.2   MEASURING THE CONTRIBUTION OF AN INDEPENDENT VARIABLE

As can be seen from Example 2.1, graphical analysis can be useful in determining the functional form of a regression relationship. However, usually graphical analysis is not totally conclusive, and hence it is useful to use various statistical tools to help construct an appropriate multiple regression model. We now study several useful statistical tools.

The first tool is called the t_{b_j}-*statistic* and measures the importance of a particular independent variable x_{tj} in describing the dependent variable y_t in the multiple regression model

$$y_t = \mu_t + \epsilon_t$$
$$= \beta_0 + \beta_1 x_{t1} + \cdots + \beta_{j-1} x_{t,j-1} + \beta_j x_{tj} + \beta_{j+1} x_{t,j+1} + \cdots + \beta_p x_{tp} + \epsilon_t$$

The t_{b_j}-statistic is defined by the equation

$$t_{b_j} = \frac{b_j}{s_{b_j}}$$

where b_j is the least squares estimate of β_j and s_{b_j} is a quantity called the *standard* *error of the estimate b_j.*[1] More precisely, it can be shown[2] that the t_{b_j}-statistic measures the *additional importance* of the independent variable x_{tj} over and above the combined importance of the other independent variables $x_{t1}, \ldots, x_{t,j-1}, x_{t,j+1}, \ldots, x_{tp}$ in describing the dependent variable y_t in the multiple regression model. Generally speaking, the larger t_{b_j} is, the greater is the additional importance of the independent variable x_{tj}. A rule of thumb that is sometimes used is to decide that the independent variable x_{tj} has significant additional importance if

$$|t_{b_j}| > t_{2.5}(n - (p + 1))$$

where $t_{2.5}(n - (p + 1))$ is the point on the scale of the t-distribution having $(n - (p + 1))$ degrees of freedom such that an area of .025 exists under the curve of this t-distribution between $t_{2.5}(n - (p + 1))$ and ∞. Here, n is the number of observations that have been made and p is the number of independent variables in the regression model. A table of t-distribution points is given in Appendix B. An even rougher rule of thumb that is sometimes used is to decide that the independent variable x_{tj} has significant additional importance if

$$|t_{b_j}| > 2$$

The t_{b_j}-statistic and the above rules of thumb should be used with caution. Because of sampling variation it is always possible that $|t_{b_j}|$ will be smaller than 2 even if independent variable x_{tj} really has significant additional importance in describing y_t. Conversely, even if independent variable x_{tj} really does not have significant additional importance in describing y_t, it is possible that $|t_{b_j}|$ will be nearly equal to or larger than 2. Hence, it is important to use prior belief, along with the t_{b_j}-statistic, in deciding whether to include a particular independent variable x_{tj} in a final forecasting model. For example, suppose that in a multiple regression situation, there exists an independent variable x_{tj} such that

$$|t_{b_j}| = \left|\frac{b_j}{s_{b_j}}\right| = 1.78$$

Suppose we believe on theoretical grounds that the independent variable x_{tj} is important in describing and predicting y_t. For example, it would seem that the price of an item affects the sales of the item. Then, to ignore x_{tj} and not include it in the forecasting model would contradict both our prior belief and the statistical evidence (which at least mildly supports our prior belief, since $|t_{b_j}| = 1.78$ is "nearly" equal to 2). Hence, we should probably include the variable x_{tj} in the final forecasting model. On the other hand, if there do not exist strong theoretical grounds for believing that x_{tj} is important in describing and predicting y_t and variable x_{tj} is just being "tried out" to see if it might be important, then the fact that $|t_{b_j}| = 1.78$ probably would not be enough statistical evidence to warrant including x_{tj} in a final forecasting model.

---

[1] The formula for s_{b_j} involves matrix algebra and will be given in Section *2–2. From the user's standpoint it is sufficient to know that standard multiple regression computer packages can be used to calculate s_{b_j} and the corresponding t_{b_j}-statistic.

[2] In Section *2–3 we will use the concepts of hypothesis testing to justify many of the statements, including the rules of thumb, given in this section concerning the t_{b_j} = statistic.

Indeed, if prior belief does not indicate that the variable x_{tj} is important in predicting y_t, then it might be wise to require that $|t_{b_j}|$ be substantially larger than 2 before deciding to include x_{tj} in a final forecasting model.

Another reason that the t_{b_j}-statistic should be used with caution is the problem of *multicollinearity*. When a set of independent variables in a regression analysis are related to or are dependent upon each other, multicollinearity is said to exist among the variables. For example, if we construct a model to predict weekly fuel consumption on the basis of x_{t1} = average weekly temperature and x_{t2} = the number of four-hour periods during the week in which the temperature is below freezing, one would expect that as x_{t1} decreases, x_{t2} increases, and hence that the independent variables x_{t1} and x_{t2} are related. In general, when two or more independent variables are related to each other, they contribute redundant information. That is, even though each independent variable contributes information for the prediction of the dependent variable, some of the information is overlapping. As a result, since the t_{b_j}-statistic measures the additional importance of a particular independent variable over and above the combined importance of the other independent variables in a regression model, the t_{b_j}-statistics obtained by running a regression analysis relating a dependent variable to a set of interrelated or "correlated" independent variables will probably be smaller than the t_{b_j}-statistics that would be obtained if separate regression analyses were run, where each separate regression analysis relates the dependent variable to only one of the correlated independent variables. In extreme cases, each and every t_{b_j}-statistic of the correlated independent variables can make each and every correlated independent variable look unimportant for predicting the dependent variable, when in fact the correlated independent variables taken together accurately predict the dependent variable. What we need, then, in addition to the t_{b_j}-statistic, are measures of the combined importance of all of the independent variables taken together in describing the dependent variable. Such measures will be discussed subsequently. But it should first be mentioned that one measure of the relationship or "correlation" between two independent variables, x_{ti} and x_{tj}, is called the *simple correlation coefficient*.

The simple correlation coefficient is denoted by the symbol $R_{x_{ti},x_{tj}}$, and is given by the equation

$$R_{x_{ti},x_{tj}} = \frac{\displaystyle\sum_{t=1}^{n} (x_{ti} - \bar{x}_i)(x_{tj} - \bar{x}_j)}{\left[\displaystyle\sum_{t=1}^{n} (x_{ti} - \bar{x}_i)^2 \sum_{t=1}^{n} (x_{tj} - \bar{x}_j)^2\right]^{1/2}}$$

where

$$\bar{x}_i = \frac{\displaystyle\sum_{t=1}^{n} x_{ti}}{n} \quad \text{and} \quad \bar{x}_j = \frac{\displaystyle\sum_{t=1}^{n} x_{tj}}{n}$$

It can be shown that $R_{x_{ti}, x_{tj}}$ is always between -1 and 1. A value of $R_{x_{ti}, x_{tj}}$ close to 1 indicates that the independent variables x_{ti} and x_{tj} have a strong tendency to move together in a straight-line fashion with a positive slope, and therefore that the independent variables x_{ti} and x_{tj} are highly related or *correlated*. A value of $R_{x_{ti}, x_{tj}}$ close to -1 indicates that the independent variables x_{ti} and x_{tj} have a strong tendency to move together in a straight-line fashion with a negative slope, and again that the independent variables x_{ti} and x_{tj} are highly related or correlated. A value of $R_{x_{ti}, x_{tj}}$ close to 0 indicates that x_{ti} and x_{tj} do not have much of a straight-line correlation. In this case, x_{ti} and x_{tj} may not be highly correlated, although correlations other than straight-line correlations can exist. It should be mentioned that standard multiple regression computer packages can be used to compute $R_{x_{ti}, x_{tj}}$. If strong multicollinearity exists among a set of independent variables, then one solution is to remove one or several of the highly correlated independent variables, in order to lessen the multicollinearity. However, one must be careful not to remove an independent variable that is important in predicting the dependent variable. Such a variable should not be removed unless it can be replaced by another independent variable that is equally important in predicting the dependent variable and does not cause the same strong multicollinearity. In general, strong multicollinearity does not hinder the ability of a regression model to predict the dependent variable corresponding to a particular combination of values of the independent variables, as long as the combination of values is within the experimental region. Strong multicollinearity *is* a problem when we are trying to use the t_{b_j}-statistic to assess the additional importance of each independent variable.

Before proceeding to Example 2.2, which illustrates the use of the t_{b_j}-statistic, it should be pointed out that the parameter β_0 in the general regression model

$$y_t = \mu_t + \epsilon_t$$
$$= \beta_0 + \beta_1 x_{t1} + \cdots + \beta_p x_{tp} + \epsilon_t$$

equals the average level μ_t of the time series *if all of the independent variables equal 0*. Hence, β_0 is said to measure the *intercept* of the average level. To determine the additional importance of the parameter β_0 in the regression model, we can compute the t_{b_0}-statistic, which equals

$$t_{b_0} = \frac{b_0}{s_{b_0}}$$

Here b_0 is the least squares estimate of β_0 and s_{b_0} is the standard error of the estimate b_0. The appropriate interpretation and use of this statistic is the same as the interpretation and use of the other t_{b_j} statistics.

Example 2.2   Let us use the t_{b_j}-statistic to evaluate the additional importance of the intercept and of the independent variables x_{t4}, x_{t3}, and x_{t3}^2 in the regression model

$$y_t = \beta_0 + \beta_1 x_{t4} + \beta_2 x_{t3} + \beta_3 x_{t3}^2 + \epsilon_t$$

TABLE 2.5   *The t_{b_j}-Statistics for the Model $y_t = \beta_0 + \beta_1 x_{t4} + \beta_2 x_{t3} + \beta_3 x_{t3}^2 + \epsilon_t$*

Independent Variable	j	b_j	s_{b_j}	$t_{b_j} = \dfrac{b_j}{s_{b_j}}$
Intercept	0	17.3244	5.6414	3.0709
x_{t4}	1	1.3070	.3036	4.3050
x_{t3}	2	−3.6956	1.8503	−1.9973
x_{t3}^2	3	.3486	.1512	2.3056

when this model is fit to the Fresh detergent demand data in Table 2.1. Table 2.5 summarizes the calculation of the appropriate t_{b_j}-statistics. Since there are $p = 3$ independent variables in the above model and since the historical data consists of $n = 30$ observations, we have

$$n - (p + 1) = 30 - (3 + 1) = 26$$

Referring to the t-distribution table in Appendix B, we have

$$t_{2.5}(26) = 2.056 \approx 2$$

We note that t_{b_0}, t_{b_1}, and t_{b_3} are greater in absolute value (Table 2.5) than $t_{2.5}(26)$, and so we conclude that the intercept, x_{t4}, and x_{t3}^2 have significant additional importance. Moreover, since the absolute value of t_{b_2} is nearly equal to $t_{2.5}(26)$, since we subjectively believe x_{t3} should be included in the model, and since multicollinearity might be influencing the t_{b_j}-statistics, we will also conclude that x_{t3} has significant additional importance. To summarize, we conclude that each of the independent variables in the model

$$y_t = \underset{(3.0709)}{\beta_0} + \underset{(4.3050)}{\beta_1 x_{t4}} + \underset{(-1.9973)}{\beta_2 x_{t3}} + \underset{(2.3056)}{\beta_3 x_{t3}^2} + \epsilon_t$$

has significant additional importance in describing y_t. Note that we have placed the appropriate t_{b_j}-statistics under the appropriate parameters in the model. Henceforth, we will sometimes summarize the appropriate t_{b_j}-statistics for a given regression model in this manner.

Let us next consider fitting two other multiple regression models to the Fresh detergent demand data in Table 2.1. The first model is

$$y_t = \underset{(3.3208)}{\beta_0} + \underset{(3.2950)}{\beta_1 x_{t4}} + \underset{(-1.2126)}{\beta_2 x_{t4}^2} + \underset{(-2.2818)}{\beta_3 x_{t3}} + \underset{(2.5718)}{\beta_4 x_{t3}^2} + \epsilon_t$$

Since the number of independent variables in this model is $p = 4$, from Appendix B we have

$$t_{2.5}(n - (p + 1)) = t_{2.5}(30 - (4 + 1))$$
$$= t_{2.5}(25)$$
$$= 2.060$$

Since

$$|t_{b_2}| = |-1.2126| < 2.060$$

and since our graphical analysis has indicated that y_t is probably related to x_{t4} in a linear fashion, we will conclude that x_{t4}^2 does not have significant additional importance. The second model is

$$y_t = \underset{(3.5620)}{\beta_0} + \underset{(1.1067)}{\beta_1 x_{t4}} + \underset{(-2.7713)}{\beta_2 x_{t3}} + \underset{(2.9811)}{\beta_3 x_{t3}^2} + \underset{(-.8832)}{\beta_4 x_{t4} x_{t3}} + \underset{(.7104)}{\beta_5 x_{t4} x_{t3}^2} + \epsilon_t$$

Since the number of independent variables in this model is $p = 5$, from Appendix B we have

$$t_{2.5}(n - (p + 1)) = t_{2.5}(30 - (5 + 1))$$
$$= t_{2.5}(24)$$
$$= 2.064$$

Since

$$|t_{b_4}| = |-.8832| < 2.064 \qquad \text{and} \qquad |t_{b_5}| = |.7104| < 2.064$$

and since intuitive reasoning indicates that there is little or no interaction between price difference and advertising expenditure, we will conclude that the interaction variables $x_{t4}x_{t3}$ and $x_{t4}x_{t3}^2$ do not have significant additional importance. Because the above analysis has concluded both that there is little interaction between price difference and advertising expenditure and that y_t is related to x_{t4} in a linear fashion, it seems that the model

$$y_t = \beta_0 + \beta_1 x_{t4} + \beta_2 x_{t3} + \beta_3 x_{t3}^2 + \epsilon_t$$

is a reasonable model to use to forecast future values of demand for Fresh liquid laundry detergent.

Before concluding this example, it should be stated that because multicollinearity might affect the t_{b_j}-statistics of the interaction variables $x_{t4}x_{t3}$ and $x_{t4}x_{t3}^2$ in the model

$$y_t = \beta_0 + \beta_1 x_{t4} + \beta_2 x_{t3} + \beta_3 x_{t3}^2 + \beta_4 x_{t4} x_{t3} + \beta_5 x_{t4} x_{t3}^2 + \epsilon_t$$

it would be useful to have a statistical measure of the additional combined importance of the interaction variables $x_{t4}x_{t3}$ and $x_{t4}x_{t3}^2$ *taken together* over and above the combined importance of the independent variables x_{t4}, x_{t3}, and x_{t3}^2 taken together in describing the dependent variable. We will study one such measure in Section 2-3.

2-2.3   MEASURING THE OVERALL CONTRIBUTION OF THE INDEPENDENT VARIABLES

We will now discuss several measures that describe the *overall importance* of all of the independent variables $x_{t1}, \ldots, x_{tp}$ taken together in describing the dependent variable y_t in the regression model

$$y_t = \beta_0 + \beta_1 x_{t1} + \cdots + \beta_p x_{tp} + \epsilon_t$$

Since we are describing the overall importance of all of the independent variables taken together, we say we are describing the importance or utility of the *overall model* in describing the dependent variable. Remember that the least squares estimates $b_0, b_1, \ldots, b_p$ of the parameters $\beta_0, \beta_1, \ldots, \beta_p$ in the above model give the prediction

$$\hat{y}_t = b_0 + b_1 x_{t1} + \cdots + b_p x_{tp}$$

of the time series observation y_t. Moreover, the least squares estimates are the estimates that give predictions $\hat{y}_1, \hat{y}_2, \ldots, \hat{y}_n$ of the historical time series observations $y_1, y_2, \ldots, y_n$ so that the value of the sum of the squared errors, SSE, is smaller than the value that would be given by any other estimates. The quantity

$$\mathrm{SSE} = \sum_{t=1}^{n} (y_t - \hat{y}_t)^2$$

is called the *unexplained variation* and is one measure of the combined importance of the independent variables $x_{t1}, \ldots, x_{tp}$ taken together in describing the dependent variable. The smaller SSE is, the smaller are the differences between the predicted and observed values of the historical data, and hence the greater is the utility of the overall model in describing the dependent variable. However, since the value of $\sum_{t=1}^{n}(y_t - \hat{y}_t)^2$ is dependent upon the units in which the historical observations $y_1, y_2, \ldots, y_n$ are measured, it can be somewhat difficult to tell whether a given value of $\sum_{t=1}^{n}(y_t - \hat{y}_t)^2$ is a "small" number. Hence, we need a relative measure of the utility of the overall model. That is, we need a measure that is independent of the size of the units in which the historical observations $y_1, y_2, \ldots,$ y_n are measured. Such a measure is called the *multiple coefficient of determination* and is denoted by the symbol R^2.

To introduce R^2, assume we have the historical values of the dependent variable y_t, but we do not have the historical values of the independent variables x_{t1},

..., x_{tp} with which to predict y_t. In this case, the only reasonable prediction of y_t would be

$$\bar{y} = \frac{\sum_{t=1}^{n} y_t}{n}$$

which is the average of the historical values of the dependent variable. The forecast error would then be

$$(y_t - \bar{y})$$

However, in reality we do have the historical values of the independent variables $x_{t1}, \ldots, x_{tp}$ to use to predict y_t. The prediction of y_t is

$$\hat{y}_t = b_0 + b_1 x_{t1} + \cdots + b_p x_{tp}$$

which means that the forecast error is

$$(y_t - \hat{y}_t)$$

How much has the historical information concerning the independent variables improved the forecast error? The forecast error equals $(y_t - \bar{y})$ without using the independent variables and equals $(y_t - \hat{y}_t)$ using the independent variables. So using the independent variables has decreased the forecast error from $(y_t - \bar{y})$ to $(y_t - \hat{y}_t)$, or by an amount equal to

$$(y_t - \bar{y}) - (y_t - \hat{y}_t) = (\hat{y}_t - \bar{y})$$

It can be shown that in general

$$\sum_{t=1}^{n} (y_t - \bar{y})^2 - \sum_{t=1}^{n} (y_t - \hat{y}_t)^2 = \sum_{t=1}^{n} (\hat{y}_t - \bar{y})^2$$

The quantity

$$\sum_{t=1}^{n} (y_t - \bar{y})^2$$

is called the *total variation* and equals the sum of the squared forecast errors that would be obtained if we could not use the independent variables to make forecasts. The quantity

$$\sum_{t=1}^{n} (y_t - \hat{y}_t)^2$$

is called the *unexplained variation* and equals the sum of squared forecast errors that is obtained using the independent variables. The quantity

$$\sum_{t=1}^{n} (\hat{y}_t - \bar{y})^2$$

is called the *explained variation*, which, since

$$\sum_{t=1}^{n} (\hat{y}_t - \bar{y})^2 = \sum_{t=1}^{n} (y_t - \bar{y})^2 - \sum_{t=1}^{n} (y_t - \hat{y}_t)^2$$

represents the reduction in the sum of squared forecast errors that has been accomplished by using the independent variables in making forecasts. We define the multiple coefficient of determination R^2 by the equation

$$R^2 = \frac{\displaystyle\sum_{t=1}^{n} (\hat{y}_t - \bar{y})^2}{\displaystyle\sum_{t=1}^{n} (y_t - \bar{y})^2} = \frac{\text{explained variation}}{\text{total variation}}$$

That is, R^2 is the proportion that the explained variation is of the total variation, or said another way, the proportion of the total variation that is explained by the overall regression model. Another way to express R^2 is

$$R^2 = \frac{\displaystyle\sum_{t=1}^{n} (\hat{y}_t - \bar{y})^2}{\displaystyle\sum_{t=1}^{n} (y_t - \bar{y})^2} = \frac{\displaystyle\sum_{t=1}^{n} (y_t - \bar{y})^2 - \sum_{t=1}^{n} (y_t - \hat{y}_t)^2}{\displaystyle\sum_{t=1}^{n} (y_t - \bar{y})^2}$$

$$= 1 - \frac{\displaystyle\sum_{t=1}^{n} (y_t - \hat{y}_t)^2}{\displaystyle\sum_{t=1}^{n} (y_t - \bar{y})^2} = 1 - \frac{\text{unexplained variation}}{\text{total variation}}$$

It can be shown in any regression situation that R^2 can be no smaller than 0 and no larger than 1. The nearer R^2 is to 1, the smaller is the proportion that the unexplained variation is of the total variation, and hence the smaller, relatively speaking, are the forecast errors and the greater is the utility of the overall model in describing the dependent variable.

Example
2.3

Consider again the Fresh detergent situation and the model

$$y_t = \beta_0 + \beta_1 x_{t4} + \beta_2 x_{t3} + \beta_3 x_{t3}^2 + \epsilon_t$$

The average of the 30 demand values $y_1, y_2, \ldots, y_{30}$ in Table 2.1 is

$$\bar{y} = \frac{\sum\limits_{t=1}^{30} y_t}{30} = 8.38$$

and the total variation of these 30 values is

$$\sum_{t=1}^{30} (y_t - \bar{y})^2 = 13.4586$$

Using the prediction equation

$$\begin{aligned}
\hat{y}_t &= b_0 + b_1 x_{t4} + b_2 x_{t3} + b_3 x_{t3}^2 \\
&= 17.3244 + 1.3070 x_{t4} - 3.6956 x_{t3} + .3486 x_{t3}^2
\end{aligned}$$

the unexplained variation can be computed to be

$$\sum_{t=1}^{30} (y_t - \hat{y}_t)^2 = 1.2733$$

The explained variation then equals

$$\begin{aligned}
\sum_{t=1}^{30} (\hat{y}_t - \bar{y})^2 &= \sum_{t=1}^{30} (y_t - \bar{y})^2 - \sum_{t=1}^{30} (y_t - \hat{y}_t)^2 \\
&= 13.4586 - 1.2733 = 12.1853
\end{aligned}$$

Thus, the multiple coefficient of determination for the model is

$$R^2 = \frac{\sum\limits_{t=1}^{30} (\hat{y}_t - \bar{y})^2}{\sum\limits_{t-1}^{30} (y_t - \bar{y})^2} = \frac{12.1853}{13.4586} = .9054$$

or

$$R^2 = 1 - \frac{\sum\limits_{t=1}^{30} (y_t - \hat{y}_t)^2}{\sum\limits_{t=1}^{30} (y_t - \bar{y})^2} = 1 - \frac{1.2733}{13.4586} = .9054$$

An R^2 of .9054 means that the regression model

$$y_t = \beta_0 + \beta_1 x_{t4} + \beta_2 x_{t3} + \beta_3 x_{t3}^2 + \epsilon_t$$

explains 90.54 percent of the total variation in the 30 observed demand values.

The fact that R^2 is "close" to 1 does not necessarily imply that the addition of other independent variables to the regression model would not improve the overall utility of the regression model. Basically, an *appropriate* set of independent variables must be small enough so that the costs of the forecasting process (for example, costs involved with predicting the values of the independent variables) are reasonable, and understanding and analysis of the final model are fairly easy, yet it must be large enough so that accurate description and forecasting of the time series can be achieved. We have already discussed how graphical analysis and the appropriate t_{b_j}-statistics can be used in determining an appropriate set of independent variables. Another fairly effective way to help determine an appropriate set of independent variables is to consider all reasonable regression models and compare them on the basis of various criteria. One such criterion is R^2, the multiple coefficient of determination. We have seen that the larger R^2 is for a particular regression model, the "better" is the model. However, one must balance the magnitude of R^2 or, in general, the "goodness" of any criterion, against the costs of maintaining the model and the difficulty of understanding and analyzing the model.

Remember that in choosing a forecasting technique the objective is to choose a forecasting technique which, at a reasonable cost, will produce forecasts accurate enough for the application at hand. A value of R^2 close to 1 does not necessarily signify that a regression model will provide forecasts that are "accurate enough." One way to decide whether the forecasts are accurate enough is to look at the confidence interval forecasts obtained and determine whether the confidence intervals are short enough to suit the needs of the situation. For example, it may be quite useful to be 95 percent confident that demand for Fresh liquid laundry detergent in the next sales period will be between 796,000 and 893,000 bottles. But, it may be totally worthless to be 95 percent confident that the demand will be between 596,000 and 1,093,000 bottles. A criterion that is related to the width of a confidence interval is called the *mean square error*. The mean square error is denoted by the symbol s^2 and is given by the equation

$$s^2 = \frac{\text{SSE}}{n - (p + 1)} = \frac{\sum_{t=1}^{n} (y_t - \hat{y}_t)^2}{n - (p + 1)}$$

Here n is the number of historical observations, p is the number of independent variables in the regression model

$$y_t = \beta_0 + \beta_1 x_{t1} + \cdots + \beta_p x_{tp} + \epsilon_t$$

and

$$\hat{y}_t = b_0 + b_1 x_{t1} + \cdots + b_p x_{tp}$$

is the prediction of y_t, where $b_0, b_1, \ldots, b_p$ are the least squares estimates of β_0, $\beta_1, \ldots, \beta_p$. It can be shown that the square root of s^2, which is called the *standard error* and which equals

$$s = \sqrt{\frac{\displaystyle\sum_{t=1}^{n} (y_t - \hat{y}_t)^2}{n - (p + 1)}}$$

is involved in the formula for the error $E_t(100 - \alpha)$ in a $(100 - \alpha)$ percent confidence interval for y_t. A smaller value of s usually gives a smaller value for the error $E_t(100 - \alpha)$ and hence a shorter confidence interval forecast of y_t.

It can be shown that adding any independent variable (even one unrelated to the dependent variable) to a regression model will increase the multiple coefficient of determination and decrease the unexplained variation. However, it is possible that addition of an independent variable can increase the mean square error, s^2. As we will demonstrate in Example 2.4, this can happen when the decrease in the unexplained variation caused by the addition of the extra independent variable is not enough to offset the decrease in the denominator caused by the addition of the extra independent variable. If addition of an extra independent variable increases s^2, this would probably indicate that inclusion of the variable in a final forecasting model is not warranted.

Example 2.4   Consider the Fresh detergent situation and the model

$$y_t = \beta_0 + \beta_1 x_{t4} + \beta_2 x_{t3} + \beta_3 x_{t3}^2 + \epsilon_t$$

We have found that the unexplained variation obtained when this model is fit to the data in Table 2.1 is

$$\sum_{t=1}^{30} (y_t - \hat{y}_t)^2 = 1.2733$$

Hence, the mean square error for this model is

$$s^2 = \frac{\displaystyle\sum_{t=1}^{30} (y_t - \hat{y}_t)^2}{n - (p + 1)}$$

$$= \frac{1.2733}{30 - (3 + 1)} = \frac{1.2733}{26} = .0490$$

Assume that we have also observed values of another independent variable, x_{t7}, for the 30 sales periods. Suppose that fitting the model

$$y_t = \beta_0 + \beta_1 x_{t4} + \beta_2 x_{t3} + \beta_3 x_{t3}^2 + \beta_4 x_{t7} + \epsilon_t$$

to the historical data yields an unexplained variation of

$$\sum_{t=1}^{30} (y_t - \hat{y}_t)^2 = 1.2702$$

In this case

$$s^2 = \frac{\displaystyle\sum_{t=1}^{30} (y_t - \hat{y}_t)^2}{n - (p + 1)}$$

$$= \frac{1.2702}{30 - (4 + 1)} = \frac{1.2702}{25} = .0508$$

Thus, adding x_{t7} has increased s^2 from .0490 to .0508. In this case, although addition of x_{t7} has decreased the unexplained variation, this decrease (from 1.2733 to 1.2702) has not been enough to offset the change in the denominator of s^2, which decreases from 26 to 25. Hence, inclusion of x_{t7} in a final forecasting model is probably not warranted.

Since the inclusion of an additional independent variable can actually increase the mean square error, it might be tempting to say that "we should keep adding independent variables until the mean square error is not decreased." However, since it can be shown that the mean square error can be made equal to 0, just as the multiple coefficient of determination can be made equal to 1, by including enough independent variables in the regression model, one must proceed with caution when using the "mean square error" as a criterion. In general, before deciding to include an additional independent variable in a final prediction model, one must decide whether the decrease in the mean square error caused by including the additional independent variable is enough to justify the extra "cost" involved in obtaining a future value of the independent variable for use in predicting a corresponding future value of the dependent variable.

Example 2.5   Consider once again the Fresh detergent situation. In Table 2.6 we list some different regression models relating the dependent variable y_t to the independent variables x_{t1}, x_{t2}, x_{t3}, and x_{t4}. For each model we also list the corresponding values of the multiple coefficient of determination, R^2, and the mean square error, s^2, that result if the model is fitted to the data in Table 2.1. We have not listed every model that could be considered, but those we have listed suffice to show how R^2 and s^2 can change as we add various independent variables.

TABLE 2.6 *Various Regression Models That Can Be Used to Predict Demand*

Model	R^2	s^2
$y_t = \beta_0 + \beta_1 x_{t1} + \epsilon_t$	.2202	.3784
$y_t = \beta_0 + \beta_1 x_{t1} + \beta_2 x_{t1}^2 + \epsilon_t$	.2286	.3845
$y_t = \beta_0 + \beta_1 x_{t2} + \epsilon_t$	.5490	.2168
$y_t = \beta_0 + \beta_1 x_{t2} + \beta_2 x_{t2}^2 + \epsilon_t$	.5590	.2198
$y_t = \beta_0 + \beta_1 x_{t3} + \epsilon_t$	.7673	.1119
$y_t = \beta_0 + \beta_1 x_{t3} + \beta_2 x_{t3}^2 + \epsilon_t$	.8380	.0808
$y_t = \beta_0 + \beta_1 x_{t4} + \epsilon_t$	.7915	.1002
$y_t = \beta_0 + \beta_1 x_{t4} + \beta_2 x_{t4}^2 + \epsilon_t$	.8043	.0975
$y_t = \beta_0 + \beta_1 x_{t1} + \beta_2 x_{t2} + \epsilon_t$	.8288	.0854
$y_t = \beta_0 + \beta_1 x_{t1} + \beta_2 x_{t3} + \epsilon_t$	.7717	.1138
$y_t = \beta_0 + \beta_1 x_{t2} + \beta_2 x_{t3} + \epsilon_t$	.8377	.0809
$y_t = \beta_0 + \beta_1 x_{t1} + \beta_2 x_{t2} + \beta_3 x_{t3} + \epsilon_t$	.8936	.0551
$y_t = \beta_0 + \beta_1 x_{t1} + \beta_2 x_{t2} + \beta_3 x_{t3} + \beta_4 x_{t3}^2 + \epsilon_t$	.9084	.0493
$y_t = \beta_0 + \beta_1 x_{t1} + \beta_2 x_{t1}^2 + \beta_3 x_{t2} + \beta_4 x_{t2}^2 + \beta_5 x_{t3} + \beta_6 x_{t3}^2 + \epsilon_t$	.9161	.0491
$y_t = \beta_0 + \beta_1 x_{t4} + \beta_2 x_{t3} + \epsilon_t$	.8860	.0568
$y_t = \beta_0 + \beta_1 x_{t4} + \beta_2 x_{t3} + \beta_3 x_{t3}^2 + \epsilon_t$	.9054	.0490
$y_t = \beta_0 + \beta_1 x_{t4} + \beta_2 x_{t4}^2 + \beta_3 x_{t3} + \beta_4 x_{t3}^2 + \epsilon_t$	.9106	.0481
$y_t = \beta_0 + \beta_1 x_{t4} + \beta_2 x_{t3} + \beta_3 x_{t3}^2 + \beta_4 x_{t4} x_{t3} + \beta_5 x_{t4} x_{t3}^2 + \epsilon_t$	.9225	.0434

We have concluded by previous discussions that the model

$$y_t = \beta_0 + \beta_1 x_{t4} + \beta_2 x_{t3} + \beta_3 x_{t3}^2 + \epsilon_t$$

is a reasonable model to use to forecast future values of demand for Fresh detergent. The results given in Table 2.6 further verify that this model is reasonable. Computed least squares estimates (b_j) of the parameters of this model are given in Table 2.5 and provide the forecasting equation

$$\hat{y}_t = 17.3244 + 1.3070 x_{t4} - 3.6956 x_{t3} + .3486 x_{t3}^2$$

for the time series observation y_t. For example, if in sales period $t = 31$ it is known that the price of Fresh will be $x_{t1} = 3.80$, the average industry price will be $x_{t2} = 3.90$, and Enterprise Industries' advertising expenditure for Fresh will be $x_{t3} = 6.80$, then the price difference x_{t4} equals

$$x_{t4} = x_{t2} - x_{t1} = 3.90 - 3.80 = .10$$

Hence, a point forecast made by the above forecasting equation of demand for Fresh in period 31 is

$$\hat{y}_t = 17.3244 + 1.3070(.10) - 3.6956(6.80) + .3486(6.80)^2$$
$$= 8.445$$

TABLE 2.7 Comparison of the Point Forecasts and Confidence Interval Forecasts Made of y_{31}, y_{32}, y_{33}, and ($y_{31} + y_{32} + y_{33}$) by Two Regression Models

Model: $y_t = \beta_0 + \beta_1 x_{t1} + \beta_2 x_{t2} + \beta_3 x_{t3} + \epsilon_t$

Period t	Independent Variables			Point Forecast $\hat{y}_t = 7.5891 - 2.3577 x_{t1}$ $+ 1.6122 x_{t2} + .5012 x_{t3}$	Error $E_t(95)$	Confidence Interval $[\hat{y}_t - E_t(95), \hat{y}_t + E_t(95)]$
	x_{t1}	x_{t2}	x_{t3}			
31	3.80	3.90	6.80	8.325	.523	[7.80, 8.85]
32	3.75	3.85	6.85	8.388	.517	[7.87, 8.90]
33	3.85	3.90	6.90	8.257	.557	[7.70, 8.81]

The point forecast and confidence interval forecast of $\sum_{t=31}^{33} y_t$ are, respectively, $\sum_{t=31}^{33} \hat{y}_t = 24.97$ and

$$\left[\sum_{t=31}^{33} \hat{y}_t - E_{(31,33)}(95), \sum_{t=31}^{33} \hat{y}_t + E_{(31,33)}(95) \right] = [24.97 - 1.062, 24.97 + 1.062] = [23.91, 26.03]$$

Model: $y_t = \beta_0 + \beta_1 x_{t4} + \beta_2 x_{t3} + \beta_3 x_{t3}^2 + \epsilon_t$

Period t	Independent Variables				Point Forecast $\hat{y}_t = 17.3244 + 1.3070 x_{t4}$ $- 3.5956 x_{t3} + .3486 x_{t3}^2$	Error $E_t(95)$	Confidence Interval $[\hat{y}_t - E_t(95), \hat{y}_t - E_t(95)]$
	x_{t1}	x_{t2}	$x_{t4} = x_{t2} - x_{t1}$	x_{t3}			
31	3.80	3.90	.10	6.80	8.445	.483	[7.96, 8.93]
32	3.75	3.85	.10	6.85	8.498	.489	[8.01, 8.99]
33	3.85	3.90	.05	6.90	8.488	.507	[7.98, 9.00]

The point forecast and confidence interval forecast of $\sum_{t=31}^{33} y_t$ are, respectively, $\sum_{t=31}^{33} \hat{y}_t = 25.43$ and

$$\left[\sum_{t=31}^{33} \hat{y}_t - E_{(31,33)}(95), \sum_{t=31}^{33} \hat{y}_t + E_{(31,33)}(95) \right] = [25.43 - .970, 25.43 + .970] = [24.46, 26.40]$$

In Table 2.7 on page 75 we compare the point forecasts and confidence interval forecasts of y_{31}, y_{32}, y_{33}, and $(y_{31} + y_{32} + y_{33})$ made by the model

$$y_t = \beta_0 + \beta_1 x_{t4} + \beta_2 x_{t3} + \beta_3 x_{t3}^2 + \epsilon_t$$

which we have used the techniques of this section to build, with the point forecasts and confidence interval forecasts of y_{31}, y_{32}, y_{33}, and $(y_{31} + y_{32} + y_{33})$ made by the model

$$y_t = \beta_0 + \beta_1 x_{t1} + \beta_2 x_{t2} + \beta_3 x_{t3} + \epsilon_t$$

which is the simple model used to illustrate the forecasting process in Section 1–5. Note that the model

$$y_t = \beta_0 + \beta_1 x_{t4} + \beta_2 x_{t3} + \beta_3 x_{t3}^2 + \epsilon_t$$

provides somewhat larger point forecasts and somewhat shorter confidence interval forecasts than does the model

$$y_t = \beta_0 + \beta_1 x_{t1} + \beta_2 x_{t2} + \beta_3 x_{t3} + \epsilon_t$$

In the examples concerning Fresh detergent we have not shown a plot for the values of the dependent variable y_t against increasing values of time, t. If such a plot is made, it would reveal that the values of y_t do not appear to possess a trend (to increase or decrease) or to vary in a seasonal fashion as time, t, increases. A plot should always be made of the values of a time series against increasing values of time. If the values of the time series appear to possess trend and/or seasonal variation, it is necessary to take into account the trend and/or seasonal variation when forecasting future values of the time series. How to model trend and/or seasonal variation will be discussed in Chapters 3–12.

*2–2 THE REGRESSION CALCULATIONS

Consider the regression model

$$y_t = \beta_0 + \beta_1 x_{t1} + \cdots + \beta_j x_{tj} + \cdots + \beta_p x_{tp} + \epsilon_t$$

The t_{b_j}-statistic is calculated by the formula

$$t_{b_j} = \frac{b_j}{s_{b_j}}$$

Here b_j is the least squares estimate of β_j, and s_{b_j} is calculated by the formula

$$s_{b_j} = s\sqrt{c_{jj}}$$

where

$$s = \left[\frac{\displaystyle\sum_{t=1}^{n} (y_t - \hat{y}_t)^2}{n - (p + 1)} \right]^{1/2}$$

and c_{jj} is the (j,j)th element of the matrix $(\mathbf{X'X})^{-1}$ used in the calculation of the least squares estimates. We can "picture" c_{jj} and the other elements of $(\mathbf{X'X})^{-1}$ as shown below.

$$(\mathbf{X'X})^{-1} = \begin{bmatrix} c_{00} & c_{01} & \cdot & \cdot & \cdot & c_{0j} & \cdot & \cdot & \cdot & c_{0p} \\ c_{10} & c_{11} & \cdot & \cdot & \cdot & c_{1j} & \cdot & \cdot & \cdot & c_{1p} \\ \cdot & \cdot & \cdot & \cdot & \cdot & \cdot & \cdot & \cdot & & \cdot \\ \cdot & \cdot & \cdot & \cdot & \cdot & \cdot & \cdot & \cdot & & \cdot \\ \cdot & \cdot & \cdot & \cdot & \cdot & \cdot & \cdot & \cdot & & \cdot \\ c_{j0} & c_{j1} & \cdot & \cdot & \cdot & c_{jj} & \cdot & \cdot & \cdot & c_{jp} \\ \cdot & \cdot & \cdot & \cdot & \cdot & \cdot & \cdot & \cdot & & \cdot \\ \cdot & \cdot & \cdot & \cdot & \cdot & \cdot & \cdot & \cdot & & \cdot \\ \cdot & \cdot & \cdot & \cdot & \cdot & \cdot & \cdot & \cdot & & \cdot \\ c_{p0} & c_{p1} & \cdot & \cdot & \cdot & c_{pj} & \cdot & \cdot & \cdot & c_{pp} \end{bmatrix}$$

Remember (Section *1–5) that a computational formula for the unexplained variation is

$$\sum_{t=1}^{n} (y_t - \hat{y}_t)^2 = \sum_{t=1}^{n} y_t^2 - \mathbf{b'X'y}$$

It can also be shown that a computational formula for the explained variation is

$$\sum_{t=1}^{n} (\hat{y}_t - \bar{y})^2 = \mathbf{b'X'y} - n\bar{y}^2$$

Example
2.6   Again consider the Fresh detergent situation. Using the data of Table 2.1 we see that if we wish to find the least squares estimates of β_0, β_1, β_2, and β_3 in the regression model

$$y_t = \beta_0 + \beta_1 x_{t4} + \beta_2 x_{t3} + \beta_3 x_{t3}^2 + \epsilon_t$$

we use

$$
y = \begin{bmatrix} 7.38 \\ 8.51 \\ \cdot \\ \cdot \\ \cdot \\ 9.26 \end{bmatrix}
\qquad
X = \begin{bmatrix}
1 & -.05 & 5.50 & (5.50)^2 \\
1 & .25 & 6.75 & (6.75)^2 \\
\cdot & \cdot & \cdot & \cdot \\
\cdot & \cdot & \cdot & \cdot \\
\cdot & \cdot & \cdot & \cdot \\
1 & .55 & 6.80 & (6.80)^2
\end{bmatrix}
$$

It can be shown that

$$
(X'X)^{-1} = \begin{bmatrix}
c_{00} & c_{01} & c_{02} & c_{03} \\
c_{10} & c_{11} & c_{12} & c_{13} \\
c_{20} & c_{21} & c_{22} & c_{23} \\
c_{30} & c_{31} & c_{32} & c_{33}
\end{bmatrix}
$$

$$
= \begin{bmatrix}
649.884 & -11.2114 & -212.766 & 17.2917 \\
-11.2114 & 1.8823 & 4.1237 & -.3766 \\
-212.766 & 4.1237 & 69.9072 & -5.7016 \\
17.2917 & -.3766 & -5.7016 & .4667
\end{bmatrix}
$$

so the least squares estimates of β_0, β_1, β_2, and β_3 are

$$
b = \begin{bmatrix} b_0 \\ b_1 \\ b_2 \\ b_3 \end{bmatrix} = (X'X)^{-1}X'y = \begin{bmatrix} 17.3244 \\ 1.3070 \\ -3.6956 \\ .3486 \end{bmatrix}
$$

Moreover, the explained variation is

$$
\sum_{t=1}^{30} (\hat{y}_t - \bar{y})^2 = b'X'y - n\bar{y}^2 = 2118.9173 - 30(8.38)^2
$$
$$
= 2118.9173 - 2106.732
$$
$$
= 12.1853
$$

and

$$
s = \left[\frac{\displaystyle\sum_{t=1}^{30} (y_t - \hat{y}_t)^2}{n - (p + 1)} \right]^{1/2} = \left[\frac{1.2733}{30 - (3 + 1)} \right]^{1/2} = .2213
$$

Hence, the appropriate t_{b_i}-statistics are calculated as shown below.

Since $s_{b_0} = s\sqrt{c_{00}} = .2213\sqrt{649.884} = 5.6414,\ t_{b_0} = \dfrac{b_0}{s_{b_0}} = \dfrac{17.3244}{5.6414} = 3.0709$

Since $s_{b_1} = s\sqrt{c_{11}} = .2213\sqrt{1.8823}\ = .3036,\ t_{b_1} = \dfrac{b_1}{s_{b_1}} = \dfrac{1.3070}{.3036} = 4.3050$

Since $s_{b_2} = s\sqrt{c_{22}} = .2213\sqrt{69.9072}\ = 1.8503,\ t_{b_2} = \dfrac{b_2}{s_{b_2}} = \dfrac{-3.6956}{1.8503} = -1.9973$

Since $s_{b_3} = s\sqrt{c_{33}} = .2213\sqrt{.4667}\ = .1512,\ t_{b_3} = \dfrac{b_3}{s_{b_3}} = \dfrac{.3486}{.1512} = 2.3056$

Since $x'_{31} = [1\ .1\ (6.8)\ (6.8)^2]$,

$$x'_{31}(X'X)^{-1}x_{31} = .1294$$

Hence, a 95 percent confidence interval for y_{31} is

$$[\hat{y}_{31} - E_{31}(95),\ \hat{y}_{31} + E_{31}(95)]$$

where

$$\hat{y}_{31} = 17.3244 + 1.3070(.1) - 3.6956(6.8) + .3486(6.8)^2$$
$$= 8.445$$

and

$$E_{31}(95) = t_{2.5}(30 - (3 + 1))s\sqrt{1 + x'_{31}(X'X)^{-1}x_{31}}$$
$$= 2.056(.2213)\sqrt{1 + .1294}$$
$$= .483$$

2–3 HYPOTHESES IN REGRESSION ANALYSIS

We have said that the t_{b_j}-statistic measures the additional importance of the independent variable x_{tj} over and above the combined importance of the other independent variables $x_{t1}, \ldots, x_{t,j-1}, x_{t,j+1}, \ldots, x_{tp}$ in describing the dependent variable y_t in the multiple regression model

$$y_t = \beta_0 + \beta_1 x_{t1} + \cdots + \beta_{j-1} x_{t,j-1} + \beta_j x_{tj}$$
$$+ \beta_{j+1} x_{t,j+1} + \cdots + \beta_p x_{tp} + \epsilon_t$$

The t_{b_j}-statistic can be related to the concept of an hypothesis. In this section we will discuss the use of hypotheses in regression analysis. However, we wish to give this discussion in a more general context than is provided by the t_{b_j}-statistic. Hence, we will first discuss a measure that describes the additional combined importance of a given set of independent variables $x_{t,g+1}, \ldots, x_{tp}$ taken together over and above the combined importance of another set of independent variables $x_{t1}, \ldots, x_{tg}$ taken together in describing the dependent variable y_t. In Example 2.2 we have seen that such a measure would be useful in assessing the additional combined importance of the interaction variables $x_{t4}x_{t3}$ and $x_{t4}x_{t3}^2$ taken together over and above the combined importance of the independent variables x_{t4}, x_{t3}, and x_{t3}^2 taken together in describing the dependent variable y_t in the regression model

$$y_t = \beta_0 + \beta_1 x_{t4} + \beta_2 x_{t3} + \beta_3 x_{t3}^2 \\ + \beta_4 x_{t4} x_{t3} + \beta_5 x_{t4} x_{t3}^2 + \epsilon_t$$

2–3.1 MEASURING THE CONTRIBUTION OF A SET OF INDEPENDENT VARIABLES

To begin the discussion, we will consider the two models given below.

$$\text{Model 1:} \quad y_t = \beta_0 + \beta_1 x_{t1} + \cdots + \beta_g x_{tg} + \epsilon_t$$

$$\text{Model 2:} \quad y_t = \beta_0 + \beta_1 x_{t1} + \cdots + \beta_g x_{tg} + \beta_{g+1} x_{t,g+1} \\ + \cdots + \beta_p x_{tp} + \epsilon_t$$

Let us first consider Model 1. Remember that the least squares estimates b_0, b_1, $\ldots$, b_g of the parameters $\beta_0, \beta_1, \ldots, \beta_g$ in the above model give the prediction

$$\hat{y}_t = b_0 + b_1 x_{t1} + \cdots + b_g x_{tg}$$

of the time series observation y_t. Remember also that the least squares estimates are the estimates that give predictions $\hat{y}_1, \hat{y}_2, \ldots, \hat{y}_n$ of the historical time series observations $y_1, y_2, \ldots, y_n$ so that the value of the *unexplained variation*

$$\text{SSE}_1 = \sum_{t=1}^{n} (y_t - \hat{y}_t)^2$$

is smaller than the value that would be given by any other estimates. This is one measure of the combined importance of the independent variables $x_{t1}, \ldots, x_{tg}$ in describing the dependent variable. The smaller SSE_1 is, the smaller are the differences between the predicted and observed values of the historical data, and hence the greater is the combined importance of the independent variables $x_{t1}, \ldots, x_{tg}$ in describing the dependent variable.

To assess the additional combined importance of the independent variables $x_{t,g+1}, \ldots, x_{tp}$ over and above the combined importance of the independent variables $x_{t1}, \ldots, x_{tg}$, consider Model 2:

$$y_t = \beta_0 + \beta_1 x_{t1} + \cdots + \beta_g x_{tg} + \beta_{g+1} x_{t,g+1} + \cdots + \beta_p x_{tp} + \epsilon_t$$

Letting $b_0, b_1, \ldots, b_g, b_{g+1}, \ldots, b_p$ be the least squares estimates of the parameters $\beta_0, \beta_1, \ldots, \beta_g, \beta_{g+1}, \ldots, \beta_p$ and letting

$$\hat{y}_t = b_0 + b_1 x_{t1} + \cdots + b_g x_{tg} + b_{g+1} x_{t,g+1} + \cdots + b_p x_{tp}$$

be the prediction of y_t, the smaller that

$$\text{SSE}_2 = \sum_{t=1}^{n} (y_t - \hat{y}_t)^2$$

is, the greater is the combined importance of $x_{t1}, \ldots, x_{tg}, x_{t,g+1}, \ldots, x_{tp}$ taken together in describing the dependent variable. If

$$\{\text{SSE}_1 - \text{SSE}_2\}$$

is "much larger" than 0, SSE_2 is much smaller than SSE_1. Therefore, the additional combined importance of the independent variables $x_{t,g+1}, \ldots, x_{tp}$ over and above the combined importance of the independent variables $x_{t1}, \ldots, x_{tg}$ in describing the dependent variable is substantial. Now, the "extra" independent variables $x_{t,g+1}, \ldots, x_{tp}$ will always make $\{\text{SSE}_1 - \text{SSE}_2\}$ somewhat larger than 0. The question is whether this difference is large enough to conclude that the independent variables $x_{t,g+1}, \ldots, x_{tp}$ have significant additional combined importance in describing y_t. Since the value of $\{\text{SSE}_1 - \text{SSE}_2\}$ is dependent upon the units in which the observed time series values $y_1, y_2, \ldots, y_n$ are measured, it can be difficult to tell whether $\{\text{SSE}_1 - \text{SSE}_2\}$ is significantly larger than 0. We need a "relative measure" of the additional combined importance of $x_{t,g+1}, \ldots, x_{tp}$. That is, we need a measure that is independent of the magnitude of the units in which y_1, $y_2, \ldots, y_n$ are measured. Such a measure is called the $F(x_{t,g+1}, \ldots, x_{tp} \mid x_{t1}, \ldots, x_{tg})$-statistic and is defined by the equation

$$F(x_{t,g+1}, \ldots, x_{tp} \mid x_{t1}, \ldots, x_{tg}) = \frac{\{\text{SSE}_1 - \text{SSE}_2\}/(p - g)}{\text{SSE}_2/(n - (p + 1))}$$

Here $(p - g)$ equals the number of additional independent variables $x_{t,g+1}, \ldots, x_{tp}$ that are included in Model 2 but not in Model 1. Generally speaking, the larger this statistic is, the greater is the additional combined importance of $x_{t,g+1}, \ldots, x_{tp}$. A rule of thumb[1] that is sometimes used is to decide that the independent variables $x_{t,g+1}, \ldots, x_{tp}$ have *significant* additional combined importance if

$$F(x_{t,g+1}, \ldots, x_{tp} \mid x_{t1}, \ldots, x_{tg}) > f_5(p - g, n - (p + 1))$$

where $f_5(p - g, n - (p + 1))$ is the point on the scale of the F-distribution having $(p - g)$ and $(n - (p + 1))$ degrees of freedom such that an area of .05 exists under the curve of this F-distribution between $f_5(p - g, n - (p + 1))$ and ∞. It should be noted that a table of F-distribution points is given in Appendix B. Before proceeding to Example 2.7, it should be noted that we will show in our discussion of hypotheses that a value of $F(x_{t,g+1}, \ldots, x_{tp} \mid x_{t1}, \ldots, x_{tg})$ that indicates that the variables $x_{t,g+1}, \ldots, x_{tp}$ have significant additional combined importance does not necessarily imply that each of the variables $x_{t,g+1}, \ldots, x_{tp}$ has significant additional importance on an individual basis, but, as will be shown, it does imply that at least one of the variables $x_{t,g+1}, \ldots, x_{tp}$ has significant additional importance.

Example 2.7

Consider the Fresh detergent situation again and the following two models:

$$\text{Model 3:} \quad y_t = \beta_0 + \beta_1 x_{t4} + \beta_2 x_{t3} + \beta_3 x_{t3}^2 + \epsilon_t$$

for which it can be shown that the unexplained variation is

$$\text{SSE}_3 = \sum_{t=1}^{30} (y_t - \hat{y}_t)^2 = 1.2733$$

and

$$\text{Model 4:} \quad y_t = \beta_0 + \beta_1 x_{t4} + \beta_2 x_{t3} + \beta_3 x_{t3}^2$$
$$+ \beta_4 x_{t4} x_{t3} + \beta_5 x_{t4} x_{t3}^2 + \epsilon_t$$

for which it can be shown that

$$\text{SSE}_4 = \sum_{t=1}^{30} (y_t - \hat{y}_t)^2 = 1.0425$$

In Model 4 $n - (p + 1) = 30 - (5 + 1) = 24$, and so

$$\frac{\text{SSE}_4}{n - (p + 1)} = \frac{1.0425}{30 - (5 + 1)} = \frac{1.0425}{24} = .0434$$

A statistical measure of the additional combined importance of the interaction variables $x_{t4} x_{t3}$ and $x_{t4} x_{t3}^2$ taken together over and above the combined importance of the independent variables x_{t4}, x_{t3}, and x_{t3}^2 taken together in describing the dependent variable is

---

[1] In Section 2–3.2 we will use the concepts of hypothesis testing to justify the rule of thumb given in this section.

$$F(x_{t4}x_{t3}, x_{t4}x_{t3}^2 \mid x_{t4}, x_{t3}, x_{t3}^9) = \frac{(SSE_3 - SSE_4)/(p - g)}{SSE_4/(n - (p + 1))}$$

$$= \frac{(1.2733 - 1.0425)/2}{1.0425/24} = \frac{.1154}{.0434}$$

$$= 2.659$$

Since this value is somewhat less than

$$f_5(2,24) = 3.40$$

which is obtained from Appendix B, we conclude that the interaction variables $x_{t4}x_{t3}$ and $x_{t4}x_{t3}^2$ probably do not have significant additional combined importance over and above the variables x_{t4}, x_{t3}, and x_{t3}^2 in describing y_t.

Consider Model 3 again:

$$y_t = \beta_0 + \beta_1 x_{t4} + \beta_2 x_{t3} + \beta_3 x_{t3}^2 + \epsilon_t$$

For this model it can be shown that

$$SSE_3 = 1.2733$$

Here $n - (p + 1) = 30 - (3 + 1) = 26$, and so

$$\frac{SSE_3}{n - (p + 1)} = \frac{1.2733}{26} = .0490$$

Consider also

$$\text{Model 5:} \quad y_t = \beta_0 + \beta_1 x_{t4} + \epsilon_t$$

For this model it can be shown that

$$SSE_5 = 2.8059$$

A statistical measure of the additional combined importance of the independent variables x_{t3} and x_{t3}^2 over and above the importance of the independent variable x_{t4} is

$$F(x_{t3}, x_{t3}^2 \mid x_{t4}) = \frac{(SSE_5 - SSE_3)/(p - g)}{SSE_3/(n - (p + 1))}$$

$$= \frac{(2.8059 - 1.2733)/2}{1.2733/26} = \frac{.7663}{.0490}$$

$$= 15.6388$$

Since this value is substantially larger than

$$f_5(2,26) = 3.37$$

which is obtained from Appendix B, we conclude that the independent variables x_{t3} and x_{t3}^2 have significant additional combined importance over and above x_{t4} in describing y_t.

Last, consider

$$\text{Model 6:} \quad y_t = \beta_0 + \beta_1 x_{t4} + \beta_2 x_{t3} + \epsilon_t$$

For this model it can be shown that

$$\text{SSE}_6 = 1.5337$$

A statistical measure of the additional importance of the independent variable x_{t3}^2 over and above the combined importance of the independent variables x_{t4} and x_{t3} in describing the dependent variable is

$$F(x_{t3}^2 \mid x_{t4}, x_{t3}) = \frac{(\text{SSE}_6 - \text{SSE}_3)/(p - g)}{\text{SSE}_3/(n - (p + 1))}$$

$$= \frac{(1.5337 - 1.2733)/1}{1.2733/26} = \frac{.2604}{.0490}$$

$$= 5.3143$$

Since this value is substantially larger than

$$f_5(1,26) = 4.23$$

which is obtained from Appendix B, we conclude that the independent variable x_{t3}^2 has significant additional importance over and above the combined importance of x_{t4} and x_{t3} in describing y_t.

Recall that another measure of the additional importance of the independent variable x_{t3}^2 over and above the combined importance of the independent variables x_{t4} and x_{t3} in describing the dependent variable in Model 3,

$$y_t = \beta_0 + \beta_1 x_{t4} + \beta_2 x_{t3} + \beta_3 x_{t3}^2 + \epsilon_t$$

is the t_{b_3}-statistic, which has been computed (Table 2.5) to be

$$t_{b_3} = \frac{b_3}{s_{b_3}} = 2.3056$$

Since we saw in Example 2.2 that $t_{b_3} = 2.3056$ is larger than

$$t_{2.5}(n - (p + 1)) = t_{2.5}(30 - (3 + 1))$$
$$= t_{2.5}(26)$$
$$= 2.056$$

we concluded from the t_{b_3}-statistic that the independent variable x_{t3}^2 has significant additional importance over and above the combined importance of the variables x_{t4} and x_{t3} in describing y_t.

Hence, to summarize, we have made the same conclusions concerning the additional importance of x_{t3}^2 using both the $F(x_{t3}^2 \mid x_{t4}, x_{t3})$-statistic and the t_{b_3}-statistic. This is because

$$F(x_{t3}^2 \mid x_{t4}, x_{t3}) = 5.3143 > 4.23 = f_5(1,26)$$

and

$$t_{b_3} = 2.3056 > 2.056 = t_{2.5}(26)$$

Note that, within rounding error,

$$(t_{b_3})^2 = (2.3056)^2 = 5.3158$$

is equal to

$$F(x_{t3}^2 \mid x_{t4}, x_{t3}) = 5.3143$$

and that, again within rounding error,

$$[t_{2.5}(26)]^2 = (2.056)^2 = 4.2271$$

is equal to

$$f_5(1,26) = 4.23$$

In fact, it can be shown theoretically that

$$(t_{b_3})^2 = F(x_{t3}^2 \mid x_{t4}, x_{t3})$$

and

$$[t_{2.5}(26)]^2 = f_5(1,26)$$

Hence, the conclusion that

$$|t_{b_3}| > t_{2.5}(26)$$

will be made if and only if

$$t_{b_3}^2 = F(x_{t3}^2 \mid x_{t4}, x_{t3}) > f_5(1,26)$$

Thus, the rules of thumb using the t_{b_3}- statistic and the $F(x_{t3}^2 \mid x_{t4}, x_{t3})$-statistic are equivalent.

2–3.2   USING TYPE I AND TYPE II ERRORS IN MODEL BUILDING

We now consider how to decide by hypothesis testing whether the single independent variable x_{tj} has significant additional importance over and above the combined importance of the independent variables $x_{t1}, \ldots, x_{t,j-1}, x_{t,j+1}, \ldots, x_{tp}$ in describing the dependent variable y_t in the regression model

$$y_t = \beta_0 + \beta_1 x_{t1} + \cdots + \beta_{j-1} x_{t,j-1} + \beta_j x_{tj} + \beta_{j+1} x_{t,j+1} + \cdots + \beta_p x_{tp} + \epsilon_t$$

Intuitively, it would seem reasonable to decide that the independent variable x_{tj} has significant additional importance if we reject the *null hypothesis*

$$H_0 : \quad \beta_j = 0$$

which says that changing x_{tj} does not change y_t and hence x_{tj} has no effect upon y_t, in favor of the *alternative hypothesis*

$$H_1 : \quad \beta_j \neq 0$$

which says that changing x_{tj} does change y_t and hence x_{tj} has an effect upon y_t. In the classical theory of hypothesis testing, the decision rule used to decide when to reject H_0 is frequently based on specifying a fairly small probability of a *Type I error*. A Type I error is said to be committed if a true null hypothesis is rejected. We denote the probability of a Type I error by the symbol α^*. Although we wish α^* to be small, making α^* "too small" may result in committing a *Type II error*. A Type II error is said to be committed if a false null hypothesis is not rejected. A frequently used convention is to set α^* at .05. That is, we specify a decision rule used to decide when to reject H_0 based on specifying a .05 probability of a Type I error. In general, it can be shown that if the decision rule is based on setting the probability of a Type I error equal to α^*, the decision rule that should be used is to reject the null hypothesis

$$H_0 : \quad \beta_j = 0$$

in favor of the alternative hypothesis

$$H_1 : \quad \beta_j \neq 0$$

and hence decide that the independent variable x_{tj} has significant additional importance, if

$$|t_{b_j}| > t_{100(\alpha^*/2)}(n - (p + 1))$$

where $t_{100(\alpha^*/2)}(n - (p + 1))$ is the point on the scale of the t-distribution having
$(n - (p + 1))$ degrees of freedom such that an area of $\alpha^*/2$ exists under the curve
of the t-distribution between $t_{100(\alpha^*/2)}(n - (p + 1))$ and ∞. For example, if we set α^*
at .05, we have

$$t_{100(\alpha^*/2)}(n - (p + 1)) = t_{100(.05/2)}(n - (p + 1))$$
$$= t_{2.5}(n - (p + 1))$$

Hence, if the decision rule is based on setting α^* at .05, the rule to use is to reject the
null hypothesis

$$H_0 : \quad \beta_j = 0$$

in favor of the alternative hypothesis

$$H_1 : \quad \beta_j \neq 0$$

if

$$|t_{b_j}| > t_{2.5}(n - (p + 1))$$

Note that this is precisely the previously discussed rule of thumb used to decide
whether the independent variable x_{tj} has significant additional importance. Thus,
our previously discussed rule of thumb is based on setting α^* at .05. Very impor-
tantly, it should be noted that if, in a particular situation,

$$|t_{b_j}| \leq t_{2.5}(n - (p + 1))$$

this does not mean that we believe that the null hypothesis

$$H_0 : \quad \beta_j = 0$$

is literally true. One reason for this is that setting α^* at .05 means that the statistical
evidence must be very strong before we reject H_0. We set α^* equal to .05 so that if

$$|t_{b_j}| > t_{2.5}(n - (p + 1))$$

then we can be quite sure that

$$H_1: \quad \beta_j \neq 0$$

is true, and hence that the independent variable x_{tj} does have significant additional importance. Hence, if $|t_{b_j}| \leq t_{2.5}(n - (p + 1))$, x_{tj} may not be enough additionally important to warrant including it in a final forecasting model. Remember, however, that because of sampling variation and multicollinearity, prior belief and practical considerations should be used, along with the t_{b_j}-statistic, in deciding whether to include the independent variable x_{tj} in a final forecasting model.

We saw in Example 2.7 that the rules of thumb based on the t_{b_3}-statistic and on the $F(x_{t3}^2 \mid x_{t4}, x_{t3})$-statistic were equivalent. In general, it can be shown that

$$t_{b_j}^2 = F(x_{tj} \mid x_{t1}, \ldots, x_{t,j-1}, x_{t,j+1}, \ldots, x_{tp})$$

and

$$t_{100(\alpha^*/2)}^2(n - (p + 1)) = f_{100\alpha^*}(1, n - (p + 1))$$

where $f_{100\alpha^*}(1, n - (p + 1))$ is the point on the scale of the F-distribution having 1 and $(n - (p + 1))$ degrees of freedom such that an area of α^* exists under the curve of the F-distribution between $f_{100\alpha^*}(1, n - (p + 1))$ and ∞. Hence, the decision rule

$$|t_{b_j}| > t_{100(\alpha^*/2)}(n - (p + 1))$$

is equivalent to the decision rule

$$t_{b_j}^2 \left(= F(x_{tj}^2 \mid x_{t1}, \ldots, x_{t,j-1}, x_{t,j+1}, \ldots, x_{tp}) \right) > f_{100\alpha^*}(1, n - (p + 1))$$

We state this fact because in the subsequent parts of this section we wish to use the F-distribution to give a unified treatment of hypothesis testing. To summarize what we will need: the decision rule, based on setting the probability of a Type I error equal to α^*, is to reject the null hypothesis

$$H_0: \quad \beta_j = 0$$

in favor of the alternative hypothesis

$$H_1: \quad \beta_j \neq 0$$

and hence decide that the independent variable x_{tj} has significant additional importance, if

$$t_{b_j}^2 \left(= F(x_{tj} \mid x_{t1}, \ldots, x_{t,j-1}, x_{t,j+1}, \ldots, x_{tp}) \right) > f_{100\alpha^*}(1, n - (p + 1))$$

We next consider how to decide by hypothesis testing whether the set of independent variables $x_{t,g+1}, \ldots, x_{tp}$ taken together has significant additional combined importance over and above the combined importance of the set of independent variables $x_{t1}, x_{t2}, \ldots, x_{tg}$ taken together in describing the dependent variable y_t in the model

$$y_t = \beta_0 + \beta_1 x_{t1} + \cdots + \beta_g x_{tg} + \beta_{g+1} x_{t,g+1} + \cdots + \beta_p x_{tp} + \epsilon_t$$

It seems reasonable to decide that the set of independent variables $x_{t,g+1}, \ldots, x_{tp}$ has significant additional combined importance if we reject the null hypothesis

$$H_0: \quad \beta_{g+1} = \beta_{g+2} = \cdots = \beta_p = 0$$

which says that none of the variables $x_{t,g+1}, \ldots, x_{tp}$ has an effect upon y_t, in favor of the alternative hypothesis

$$H_1: \quad \text{At least one of } \beta_{g+1}, \beta_{g+2}, \ldots, \beta_p \text{ does not equal 0}$$

which says that at least one of the independent variables $x_{t,g+1}, \ldots, x_{tp}$ has an effect upon y_t. In general, the decision rule based on setting the probability of a Type I error equal to α^* is to reject the null hypothesis H_0 in favor of the alternative hypothesis H_1 if

$$F(x_{t,g+1}, \ldots, x_{tp} \mid x_{t1}, \ldots, x_{tg}) > f_{100\alpha^*}(p - g, n - (p + 1))$$

where $f_{100\alpha^*}(p - g, n - (p + 1))$ is the point on the scale of the F-distribution having $p - g$ and $(n - (p + 1))$ degrees of freedom such that an area of α^* exists under the curve of this F-distribution between $f_{100\alpha}(p - g, n - (p + 1))$ and ∞. The point $f_{100\alpha^*}(p - g, n - (p + 1))$ is called the *rejection point of the hypothesis test*. Note that if we set α^* at .05, then

$$f_{100\alpha^*}(p - g, n - (p + 1)) = f_{100(.05)}(p - g, n - (p + 1))$$
$$= f_5(p - g, n - (p + 1))$$

and thus the decision rule is to reject H_0 in favor of H_1 if

$$F(x_{t,g+1}, \ldots, x_{tp} \mid x_{t1}, \ldots, x_{tg}) > f_5(p - g, n - (p + 1))$$

This is precisely the previously discussed rule of thumb used to decide whether the set of independent variables $x_{t,g+1}, \ldots, x_{tp}$ has significant additional combined importance. The use of the "general" rejection point

$$f_{100\alpha^*}(p - g, n - (p + 1))$$

which is based on setting the probability of a Type I error equal to α^*, is illustrated in Figure 2.7a.

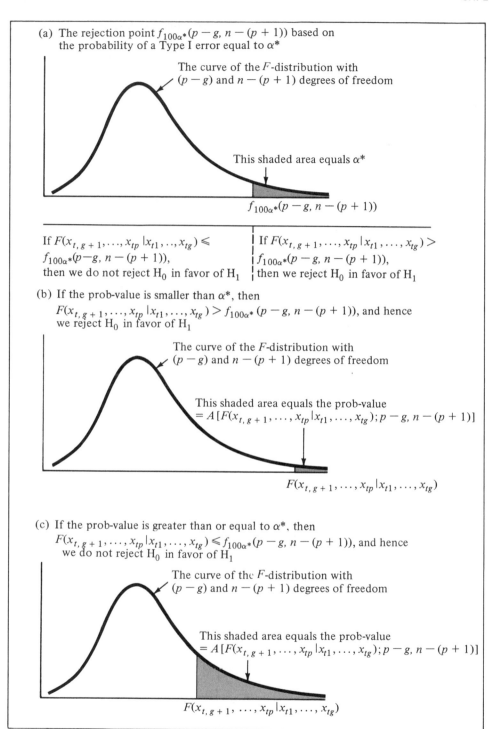

(a) The rejection point $f_{100\alpha*}(p-g, n-(p+1))$ based on the probability of a Type I error equal to $\alpha*$

The curve of the F-distribution with $(p-g)$ and $n-(p+1)$ degrees of freedom

This shaded area equals $\alpha*$

$f_{100\alpha*}(p-g, n-(p+1))$

If $F(x_{t,g+1}, \ldots, x_{tp} \mid x_{t1}, \ldots, x_{tg}) \leqslant$ $f_{100\alpha*}(p-g, n-(p+1))$, then we do not reject H_0 in favor of H_1

If $F(x_{t,g+1}, \ldots, x_{tp} \mid x_{t1}, \ldots, x_{tg}) >$ $f_{100\alpha*}(p-g, n-(p+1))$, then we reject H_0 in favor of H_1

(b) If the prob-value is smaller than $\alpha*$, then
$$F(x_{t,g+1}, \ldots, x_{tp} \mid x_{t1}, \ldots, x_{tg}) > f_{100\alpha*}(p-g, n-(p+1)), \text{ and hence}$$
we reject H_0 in favor of H_1

The curve of the F-distribution with $(p-g)$ and $n-(p+1)$ degrees of freedom

This shaded area equals the prob-value $= A[F(x_{t,g+1}, \ldots, x_{tp} \mid x_{t1}, \ldots, x_{tg}); p-g, n-(p+1)]$

$F(x_{t,g+1}, \ldots, x_{tp} \mid x_{t1}, \ldots, x_{tg})$

(c) If the prob-value is greater than or equal to $\alpha*$, then
$$F(x_{t,g+1}, \ldots, x_{tp} \mid x_{t1}, \ldots, x_{tg}) \leqslant f_{100\alpha*}(p-g, n-(p+1)), \text{ and hence}$$
we do not reject H_0 in favor of H_1

The curve of the F-distribution with $(p-g)$ and $n-(p+1)$ degrees of freedom

This shaded area equals the prob-value $= A[F(x_{t,g+1}, \ldots, x_{tp} \mid x_{t1}, \ldots, x_{tg}); p-g, n-(p+1)]$

$F(x_{t,g+1}, \ldots, x_{tp} \mid x_{t1}, \ldots, x_{tg})$

FIGURE 2.7   *Hypothesis Testing and Prob-Values*

2–3.3   USING PROB-VALUES IN MODEL BUILDING

The decision rule to reject H_0 in favor of H_1 if

$$F(x_{t,g+1}, \ldots, x_{tp} \mid x_{t1}, \ldots, x_{tg}) > f_{100\alpha^*}(p - g, n - (p + 1))$$

can be expressed in terms of what is called a *prob-value*. The prob-value is defined to be the probability of observing a sample result more contradictory to H_0 than the observed sample result, if H_0 is true. Since our decision rule says to reject H_0 in favor of H_1 if

$$F(x_{t,g+1}, \ldots, x_{tp} \mid x_{t1}, \ldots, x_{tg}) > f_{100\alpha^*}(p - g, n - (p + 1))$$

it follows that large values of $F(x_{t,g+1}, \ldots, x_{tp} \mid x_{t1}, \ldots, x_{tg})$ are contradictory to H_0. Hence, the prob-value is calculated as follows.

The prob-value equals the probability of observing a value of

$$F(x_{t,g+1}, \ldots, x_{tp} \mid x_{t1}, \ldots, x_{tg})$$

greater than or equal to the actual value observed, if the null hypothesis H_0 is true. This probability equals the area under the curve of the F-distribution with $(p - g)$ and $(n - (p + 1))$ degrees of freedom between the observed value of

$$F(x_{t,g+1}, \ldots, x_{tp} \mid x_{t1}, \ldots, x_{tg})$$

and ∞.

We will henceforth denote the area under the curve of the F-distribution with ν_1 and ν_2 degrees of freedom between an arbitrary point F^* on the scale of the F-distribution and ∞ in the following manner:

$$A[F^*; \nu_1, \nu_2]$$

This area is illustrated in Figure 2.8. The calculation of the prob-value can be summarized as

$$\begin{aligned} \text{prob-value} = A[&F(x_{t,g+1}, \ldots, x_{tp} \mid x_{t1}, \ldots, x_{tg}); \\ &p - g, n - (p + 1)]. \end{aligned}$$

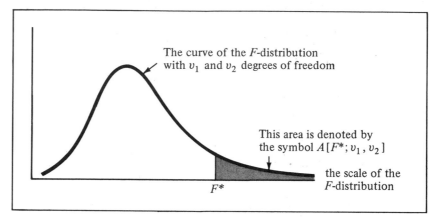

FIGURE 2.8 *The Area under the Curve of the F-Distribution with ν_1 and ν_2 Degrees of Freedom Is Denoted by the Symbol $A[F^*; \nu_1, \nu_2]$*

Two prob-values are illustrated in Figures 2.7b and 2.7c. It is important to know that most modern, regression computer packages present the results of hypothesis tests by computing the related prob-value. From Figures 2.7b and 2.7c it is clear that the prob-value can be used as a decision rule, based on setting the probability of a Type I error equal to α^*, for deciding whether to reject H_0. As can be seen from inspecting Figure 2.7b, if the prob-value is smaller than α^*, then

$$F(x_{t,g+1}, \ldots, x_{tp} \mid x_{t1}, \ldots, x_{tg}) > f_{100\alpha^*}(p - g, n - (p + 1))$$

and we reject H_0 in favor of H_1. As can be seen from inspecting Figure 2.7c, if the prob-value is greater than or equal to α^*, then

$$F(x_{t,g+1}, \ldots, x_{tp} \mid x_{t1}, \ldots, x_{tg}) \le f_{100\alpha^*}(p - g, n - (p + 1))$$

so we do not reject H_0 in favor of H_1. To summarize: we may use the prob-value as a decision rule based on specifying the probability of a Type I error equal to α^* for deciding whether to reject H_0 in favor of H_1—

> If the prob-value is less than α^*, then we reject H_0. Otherwise, we do not reject H_0.

As a decision rule, the prob-value is in a sense easier to use than the classical method, illustrated in Figure 2.7a, of looking up the rejection point $f_{100\alpha^*}(p - g, n - (p + 1))$ and rejecting H_0 if

$$F(x_{t,g+1}, \ldots, x_{tp} \mid x_{t1}, \ldots, x_{tg}) > f_{100\alpha^*}(p - g, n - (p + 1))$$

For example, if there were several hypothesis testers, all of whom wished to use different values of α^*, then, if the classical rejection point method were used, each hypothesis tester would have to look up a different rejection point $f_{100\alpha^*}(p - g,\ n - (p + 1))$ to decide whether to reject H_0 at the hypothesis tester's particular chosen value of α^*. However, if prob-values are used, only the prob-value needs to be calculated, and each hypothesis tester knows that if the prob-value is less than the particular chosen value of α^*, then H_0 should be rejected. Otherwise, H_0 should not be rejected. For example, assume the prob-value equals .026. If one hypothesis tester sets α^* at .10, then, since

$$\text{prob-value} = .026 < .10 = \alpha^*$$

the hypothesis tester should reject H_0. If another hypothesis tester sets α^* at .05, then, since

$$\text{prob-value} = .026 < .05 = \alpha^*$$

that hypothesis tester should also reject H_0. If another hypothesis tester sets α^* at .01, then, since

$$\text{prob-value} = .026 > .01 = \alpha^*$$

that hypothesis tester should not reject H_0.

Consider again the regression model

$$y_t = \beta_0 + \beta_1 x_{t1} + \cdots + \beta_{j-1} x_{t,j-1} + \beta_j x_{tj} + \beta_{j+1} x_{t,j+1} + \cdots + \beta_p x_{tp} + \epsilon_t$$

Since the test of the null hypothesis

$$H_0: \quad \beta_j = 0$$

against the alternative hypothesis

$$H_1: \quad \beta_j \neq 0$$

utilizes the t_{b_j}-statistic and since

$$t_{b_j}^2 = F(x_{tj} \mid x_{t1}, \ldots, x_{t,j-1}, x_{t,j+1}, \ldots, x_{tp})$$

it follows that the related prob-value is calculated as shown below.

$$\text{prob-value} = A(t_{b_j}^2;\ 1,\ n - (p + 1))$$

$$= \text{Area under the curve of the } F\text{-distribution with 1}$$
$$\text{and } n - (p + 1) \text{ degrees of freedom between } t_{b_j}^2$$
$$\text{and } \infty$$

Based on setting the probability of a Type I error equal to α^*, we would reject H_0 in favor of H_1 if the above prob-value is less than α^*.

Up to this point we have considered the prob-value as a decision rule for deciding when to reject the null hypothesis H_0 in favor of the alternative hypothesis H_1. However, the prob-value has a more important use than its use as a decision rule. Although it is a frequently used convention to set α^* at .05, this choice or any choice of α^* is really quite arbitrary. This has led many statisticians to use the prob-value to determine the amount of doubt cast upon H_0 by

$$F(x_{t,g+1}, \ldots, x_{tp} \mid x_{t1}, \ldots, x_{tg})$$

and hence the weight of evidence it supplies for H_1, instead of first arbitrarily choosing a value of α^*. This will be illustrated in Example 2.8, where it will be demonstrated that

> the smaller the prob-value is, the more doubt is cast upon the validity of the null hypothesis H_0 and hence the more evidence is supplied for the validity of the alternative hypothesis H_1.

Since the area under the entire curve of the F-distribution is 1, the largest a prob-value can be is 1. In general the nearer the prob-value is to 1, the less doubt is cast upon the null hypothesis; and the nearer the prob-value is to 0, the more doubt is cast upon the null hypothesis and hence the more the statistical evidence supports the validity of the alternative hypothesis. Before proceeding to Example 2.8, however, it should be stated that since the prob-value is superior to the rejection-point method both as a decision rule and as a supplier of statistical evidence, we will henceforth report the results of hypothesis tests by reporting the appropriate prob-values.

Example 2.8   Let us again consider the Fresh detergent situation and

$$\text{Model 3:} \quad y_t = \beta_0 + \beta_1 x_{t4} + \beta_2 x_{t3} + \beta_3 x_{t3}^2 + \epsilon_t$$

Consider testing the null hypothesis

$$H_0 : \quad \beta_2 = \beta_3 = 0$$

against the alternative hypothesis

$$H_1 : \quad \text{At least one of } \beta_2 \text{ or } \beta_3 \text{ does not equal } 0$$

The $F(x_{t3}, x_{t3}^2 \mid x_{t4})$-statistic and related prob-value are given in Table 2.8 and are calculated on page 96.

TABLE 2.8 *Hypotheses and Prob-Values*

Model	Unexplained Variation	Null Hypothesis	$F(x_{t,g+1}, \ldots, x_{tp} \mid x_{t1}, \ldots, x_{tg})$	Prob-Value = $A[F(x_{t,g+1}, \ldots, x_{tp} \mid x_{t1}, \ldots, x_{tg}); p - g, n - (p + 1)]$
Model 4: $y_t = \beta_0 + \beta_1 x_{t4} + \beta_2 x_{t3} + \beta_3 x_{t3}^2 + \beta_4 x_{t4} x_{t3} + \beta_5 x_{t4} x_{t3}^2 + \epsilon_t$	$SSE_4 = 1.0425$			
Model 3: $y_t = \beta_0 + \beta_1 x_{t4} + \beta_2 x_{t3} + \beta_3 x_{t3}^2 + \epsilon_t$	$SSE_3 = 1.2733$	$H_0:\ \beta_4 = \beta_5 = 0$ in Model 4	$F(x_{t4} x_{t3}, x_{t4} x_{t3}^2 \mid x_{t4}, x_{t3}, x_{t3}^2)$ $= \dfrac{(SSE_3 - SSE_4)/(p - g)}{SSE_4/(n - (p + 1))}$ $= \dfrac{(1.2733 - 1.0425)/2}{1.0425/24}$ $= \dfrac{.1154}{.0434} = 2.659$	$A[2.659; 2,24] = .09073$
Model 5: $y_t = \beta_0 + \beta_1 x_{t4} + \epsilon_t$	$SSE_5 = 2.8059$	$H_0:\ \beta_2 = \beta_3 = 0$ in Model 3	$F(x_{t3}, x_{t3}^2 \mid x_{t4})$ $= \dfrac{(SSE_5 - SSE_3)/(p - g)}{SSE_3/(n - (p + 1))}$ $= \dfrac{(2.8059 - 1.2733)/2}{1.2733/26}$ $= \dfrac{.7663}{.0490} = 15.6388$	$A[15.6388; 2,26] = .00003$
Model 6: $y_t = \beta_0 + \beta_1 x_{t4} + \beta_2 x_{t3} + \epsilon_t$	$SSE_6 = 1.5337$	$H_0:\ \beta_3 = 0$ in Model 3	$F(x_{t3}^2 \mid x_{t4}, x_{t3})$ $= \dfrac{(SSE_6 - SSE_3)/(p - g)}{SSE_3/(n - (p + 1))}$ $= \dfrac{(1.5337 - 1.2733)/1}{1.2733/26}$ $= \dfrac{.2604}{.0490} = 5.3143 = (t_{b_3})^2$	$A[5.3143; 1,26] = .02934$

$$F(x_{t3}, x_{t3}^2 \mid x_{t4}) = \frac{(\text{SSE}_5 - \text{SSE}_3)/(p - g)}{\text{SSE}_3/(n - (p + 1))}$$

$$= \frac{(2.8059 - 1.2733)/2}{1.2733/26}$$

$$= 15.6388$$

$$\text{prob-value} = A[15.6388; 2,26]$$

$$= .00003$$

Since the prob-value equals .00003, if H_0 is true, the probability of observing a value of

$$F(x_{t3}, x_{t3}^2 \mid x_{t4})$$

greater than or equal to the value observed (15.6388) is .00003. Hence, two conclusions are possible. The first conclusion is that H_0 is true, and a sample result has occurred that is so rare that there is only a .00003 chance of observing a sample result more contradictory to H_0. The second conclusion is that H_0 is not true and that H_1 is true. A reasonable person would probably believe the second conclusion. It should be noted that if we set $\alpha^* = .05$, then since

$$\text{prob-value} = .00003 < .05 = \alpha^*$$

we would reject H_0. The prob-value itself, however, makes us very sure that the null hypothesis is false, because no reasonable person would believe that, if H_0 were true, a sample result would occur that is so rare that there is only a .00003 chance of observing a sample result more contradictory to H_0.

Consider again

$$\text{Model 3:} \quad y_t = \beta_0 + \beta_1 x_{t4} + \beta_2 x_{t3} + \beta_3 x_{t3}^2 + \epsilon_t$$

and testing the null hypothesis

$$H_0: \quad \beta_3 = 0$$

against the alternative hypothesis

$$H_1: \quad \beta_3 \neq 0$$

The statistic

$$(t_{b_3})^2 = F(x_{t3}^2 \mid x_{t4}, x_{t3})$$

and related prob-value are given in Table 2.8 and are calculated on page 97.

$$F(x_{t3}^L \mid x_{t4}, x_{t3}) = \frac{(SSE_6 - SSE_3)/(p - g)}{SSE_3/(n - (p + 1))}$$

$$= \frac{(1.5337 - 1.2733)/1}{1.2733/26}$$

$$= 5.3143 = (t_{b_3})^2$$

$$\text{prob-value} = A[5.3143; 1,26] = .02934$$

If we set $\alpha^* = .05$, then since

$$\text{prob-value} = .02934 < .05 = \alpha^*$$

we would reject H_0. Furthermore, the fact that the prob-value is quite small by itself makes us fairly certain that H_0 is false.

Consider next

Model 4 : $y_t = \beta_0 + \beta_1 x_{t4} + \beta_2 x_{t3} + \beta_3 x_{t3}^2 + \beta_4 x_{t4} x_{t3} + \beta_5 x_{t4} x_{t3}^2 + \epsilon_t$

and testing the null hypothesis

$$H_0: \quad \beta_4 = \beta_5 = 0$$

against the alternative hypothesis

$$H_1: \quad \text{At least one of } \beta_4 \text{ or } \beta_5 \text{ does not equal } 0$$

The $F(x_{t4}x_{t3}, x_{t4} x_{t3}^2 \mid x_{t4}, x_{t3}, x_{t3}^2)$-statistic and the related prob-value are given in Table 2.8 and are calculated below.

$$F(x_{t4}x_{t3}, x_{t4}x_{t3}^2 \mid x_{t4}, x_{t3}, x_{t3}^2) = \frac{(SSE_3 - SSE_4)/(p - g)}{SSE_4/(n - (p + 1))}$$

$$= \frac{(1.2733 - 1.0425)/2}{1.0425/24}$$

$$= 2.659$$

$$\text{prob-value} = A[2.659; 2,24] = .09073$$

If we set $\alpha^* = .05$, then since

$$\text{prob-value} = .09073 > .05 = \alpha^*$$

we would not reject H_0. By itself the prob-value casts some doubt on H_0, since it says that, if H_0 is true, there is only a .09073 chance of observing a sample result more contradictory to H_0 than the actual observed sample result. However, since we have intuitively reasoned that there may well be little or no interaction between x_{t4} and x_{t3}, perhaps the prob-value is not small enough to warrant including the interaction terms $x_{t4}x_{t3}$ and $x_{t4}x_{t3}^2$ in a final forecasting model.

2–3.4   AN OVERALL F-STATISTIC

Remember that the multiple coefficient of determination, R^2, is one measure of the overall utility of a regression model. We will now develop another such measure, a measure that is related to hypothesis testing. To begin, consider

$$\text{Model 7:} \quad y_t = \beta_0 + \epsilon_t$$

It can be shown that the least squares estimate of β_0 is $b_0 = \bar{y}$, the average of the n historical observations $y_1, y_2, \ldots, y_n$. The prediction of y_t is therefore

$$\hat{y}_t = b_0 = \bar{y}$$

and the unexplained variation is

$$\text{SSE}_7 = \sum_{t=1}^{n} (y_t - \hat{y}_t)^2$$

$$= \sum_{t=1}^{n} (y_t - \bar{y})^2$$

Consider next

$$\text{Model 8:} \quad y_t = \beta_0 + \beta_1 x_{t1} + \cdots + \beta_p x_{tp} + \epsilon_t$$

for which the unexplained variation is

$$\text{SSE}_8 = \sum_{t=1}^{n} (y_t - \hat{y}_t)^2$$

where $\hat{y}_t = b_0 + b_1 x_{t1} + \cdots + b_p x_{tp}$, and $b_0, b_1, \ldots, b_p$ are the least squares estimates of $\beta_0, \beta_1, \ldots, \beta_p$. In order to test the null hypothesis

$$\text{H}_0: \quad \beta_1 = \beta_2 = \cdots = \beta_p = 0$$

against the alternative hypothesis

$$\text{H}_1: \quad \text{At least one of } \beta_1, \beta_2, \ldots, \beta_p \text{ does not equal 0}$$

the $F(x_{t,g+1}, \ldots, x_{tp} \mid x_{t1}, \ldots, x_{tg})$-statistic that would be used is

$$F(r_{t1}, r_{t2}, \ldots, r_{tp}) = \frac{(\text{SSE}_7 - \text{SSE}_8)/p}{\text{SSE}_8/(n - (p + 1))}$$

Since

$$\text{SSE}_7 - \text{SSE}_8 = \sum_{t=1}^{n} (y_t - \bar{y})^2 - \sum_{t=1}^{n} (y_t - \hat{y}_t)^2$$

and since we saw in Section 2–2 that

$$\sum_{t=1}^{n} (\hat{y}_t - \bar{y})^2 = \sum_{t=1}^{n} (y_t - \bar{y})^2 - \sum_{t=1}^{n} (y_t - \hat{y}_t)^2$$

it follows that

$$F(x_{t1}, x_{t2}, \ldots, x_{tp}) = \frac{\displaystyle\sum_{t=1}^{n} (\hat{y}_t - \bar{y})^2/p}{\displaystyle\sum_{t=1}^{n} (y_t - \hat{y}_t)^2/(n - (p + 1))}$$

$$= \frac{[\text{explained variation}]/p}{[\text{unexplained variation}]/(n - (p + 1))}$$

Hence, the smaller is

$$\text{prob-value} = A[F(x_{t1}, \ldots, x_{tp}); p, n - (p + 1)]$$

the more evidence there is for the support of the alternative hypothesis

$$H_1: \quad \text{At least one of } \beta_1, \ldots, \beta_p \text{ does not equal } 0$$

that is, the more evidence there is that the overall model significantly describes the dependent variable. It should of course be noted that just because we decide the overall model significantly describes the dependent variable, this does not mean that "better" models do not exist.

Example 2.9   Consider the Fresh detergent situation and the model

$$y_t = \beta_0 + \beta_1 x_{t4} + \beta_2 x_{t3} + \beta_3 x_{t3}^2 + \epsilon_t$$

For this model it can be shown that

$$F(x_{t4}, x_{t3}, x_{t3}^2) = \frac{\sum\limits_{t=1}^{n} (\hat{y}_t - \bar{y})^2/p}{\sum\limits_{t=1}^{n} (y_t - \hat{y}_t)^2/(n - (p + 1))}$$

$$= \frac{12.1853/3}{1.2733/(30 - (3 + 1))}$$

$$= \frac{12.1853/3}{1.2733/26} = \frac{4.0618}{.0490}$$

$$= 82.8939$$

Hence,

$$\text{prob-value} = A[82.8939; 3{,}26]$$

which can be shown to be less than .000005. This extremely small prob-value convinces us that we should reject H_0 and conclude that the overall model significantly describes the dependent variable.

PROBLEMS

Section 2–2   1.   Consider the Fresh detergent situation and the model

$$y_t = \beta_0 + \beta_1 x_{t4} + \beta_2 x_{t3} + \beta_3 x_{t3}^2 + \beta_4 x_{t4}x_{t3} + \beta_5 x_{t4}x_{t3}^2 + \epsilon_t$$

Assume that when this model is fit to the data in Table 2.1 the least squares estimates of the parameters of this model and the standard errors of the estimates are as given in the table below. Assume it is also true that

$$\sum_{t=1}^{30} (y_t - \hat{y}_t)^2 = 1.0425$$

Independent Variable	j	b_j	s_{b_j}
Intercept	0	27.7622	7.7940
x_{t4}	1	30.3586	27.4312
x_{t3}	2	−7.1443	2.5779
x_{t3}^2	3	.6321	.2120
$x_{t4}x_{t3}$	4	−7.4555	8.4416
$x_{t4}x_{t3}^2$	5	.4628	.6514

a. Calculate the appropriate t_{b_i}-statistics.
b. Calculate the multiple coefficient of determination, R^2.
c. Calculate the mean square error, s^2.

*2. Consider Problem 1.

a. Specify the **X**-matrix that would be used to calculate the least squares
estimates of the parameters of the model

$$y_t = \beta_0 + \beta_1 x_{t4} + \beta_2 x_{t3} + \beta_3 x_{t3}^2 + \beta_4 x_{t4} x_{t3} + \beta_5 x_{t4} x_{t3}^2 + \epsilon_t$$

if this model is fit to the data of Table 2.1.
b. If the above model is to be used to find a 95 percent confidence interval
for y_{31}, and if $x_{31,4} = .1$ and $x_{31,3} = 6.8$, find the row vector $\mathbf{x}'_{31}$ that would
be used to calculate the error $E_{31}(95)$ in the confidence interval.

3. Consider the Fresh detergent situation and the model

$$y_t = \beta_0 + \beta_1 x_{t4} + \beta_2 x_{t3} + \beta_3 x_{t3}^2 + \epsilon_t$$

Assuming that the values of the independent variables in periods 32 and 33 are
as given in Table 2.7, use the prediction equation

$$\hat{y}_{31} = 17.3244 + 1.3070 x_{t4} - 3.6956 x_{t3} + .3486 x_{t3}^2$$

to verify that the point forecasts of y_{32} and y_{33} given in Table 2.7 are correct.

*4. Consider Problem 3.

a. Specify the row vectors $\mathbf{x}'_{32}$, $\mathbf{x}'_{33}$, and $\sum_{t=31}^{33} \mathbf{x}'_t$ that would be used to calculate the errors $E_{32}(95)$, $E_{33}(95)$, and $E_{(31,33)}(95)$ given in Table 2.7.
b. If $\mathbf{x}'_{32}(\mathbf{X}'\mathbf{X})^{-1}\mathbf{x}_{32} = .1556$, $\mathbf{x}'_{33}(\mathbf{X}'\mathbf{X})^{-1}\mathbf{x}_{33} = .2435$, and

$$\left(\sum_{t=31}^{33} \mathbf{x}'_t \right) (\mathbf{X}'\mathbf{X})^{-1} \left(\sum_{t=31}^{33} \mathbf{x}_t \right) = 1.5420$$

verify that the errors $E_{32}(95)$, $E_{33}(95)$, and $E_{(31,33)}(95)$ given in Table 2.7
are correct.

Section
2–3

5. Consider Problem 1 and the model

$$y_t = \beta_0 + \beta_1 x_{t4} + \beta_2 x_{t3} + \beta_3 x_{t3}^2 + \beta_4 x_{t4} x_{t3} + \beta_5 x_{t4} x_{t3}^2 + \epsilon_t$$

* Starred problems are only for students who have read Section *2–2.

a. The prob-value relating to the null hypothesis

$$H_0 : \quad \beta_3 = \beta_4 = \beta_5 = 0$$

equals .02388. Demonstrate how this prob-value is calculated. In your answer be sure to calculate the statistic

$$F(x_{t3}^2, \ x_{t4}x_{t3}, \ x_{t4}x_{t3}^2 \mid x_{t4}, \ x_{t3})$$

which measures the additional combined importance of the independent variables x_{t3}^2, $x_{t4}x_{t3}$, and $x_{t4}x_{t3}^2$ over and above the combined importance of the independent variables x_{t3} and x_{t4} in describing the dependent variable. Would you reject H_0 based on setting $\alpha^* = .05$? Would you reject H_0 based on setting $\alpha^* = .01$? Give your reasoning.

Hint: To do this problem you need to know that if the model

$$y_t = \beta_0 + \beta_1 x_{t4} + \beta_2 x_{t3} + \epsilon_t$$

is fit to the data of Table 2.1, then the unexplained variation that results is

$$\sum_{t=1}^{30} (y_t - \hat{y}_t)^2 = 1.5337$$

b. The prob-value relating to the null hypothesis

$$H_0 : \quad \beta_1 = \beta_2 = \beta_3 = \beta_4 = \beta_5 = 0$$

is less than .000005. Demonstrate how this prob-value is calculated. In your answer be sure to calculate the statistic

$$F(x_{t4}, \ x_{t3}, \ x_{t3}^2, \ x_{t4}x_{t3}, \ x_{t4}x_{t3}^2)$$

measuring the utility of the overall model. Would you reject H_0 based on setting $\alpha^* = .001$? Give your reasoning.

PART I SUMMARY

Many of the decisions made by organizations are based on forecasts of future events. Businesses require forecasts in marketing, finance, personnel management, production scheduling, process control, and long-range planning.

In Chapter 1 you learned that these forecasts are often based on historical data that is in the form of a time series. This historical data is used to identify a pattern, which is then extrapolated into the future on the assumption that the observed pattern will continue in the future. In order to identify data patterns, you have seen that it is convenient to think of a time series as consisting of four components— trend, cycle, seasonal variation, and irregular fluctuations. Because these components can occur in any combination, no single best forecasting method exists. Indeed, it is very important to match the forecasting technique to the pattern of data that characterizes a time series.

You have also learned that errors are made in forecasting. Such errors are caused by the irregular component and our inability to perfectly predict the trend, seasonal, and cyclical components of time series. Because of these errors in forecasting, we often require some estimate of how wrong our forecasts might be. You have seen that confidence interval forecasts provide such estimates. You have also considered the problem of measuring forecast errors. You learned that graphically plotting forecast errors against time can indicate whether the forecasting method being used is appropriate, and that the magnitude of forecasting errors can be measured by either the mean absolute deviation (MAD) or the mean squared error (MSE). You learned how to calculate both the MAD and the MSE. Measures such as the MAD and MSE are used to aid in selecting a forecasting technique and to monitor forecasting techniques in an attempt to determine when something has "gone wrong."

A multitude of forecasting techniques exist. Qualitative forecasting methods use the opinions of experts to subjectively predict future events, and we briefly discussed three of these methods—subjective curve fitting, the Delphi Method, and the method of time independent technological comparisons. Quantitative forecasting techniques involve the analysis of historical data in an attempt to predict future values of a variable of interest. Quantitative forecasting techniques can be classified into two basic types—time series models and causal models. You have learned that time series models predict future values of the variable of interest solely on the basis of the historical pattern of that variable, assuming that the historical pattern will continue, and that, therefore, these predictions will not be influenced by proposed changes in management policies. Causal models predict future values of the variable of interest based on the relationship between that variable and other variables. You learned that causal models are helpful in evaluating the impact of alternative management policies being considered.

We have also discussed the choice of a forecasting method. In making this choice the form of the forecast, time frame, pattern of data, cost of forecasting, desired accuracy, availability of data, and ease of operation must all be considered. You have seen that the best forecasting method is not necessarily the most accurate method. Rather, the forecasting method that should be used is the one that meets the needs of the situation in terms of accuracy at the least cost and inconvenience.

We concluded Chapter 1 by previewing the topics to be discussed in the remainder of this book. We will discuss three basic types of quantitative techniques—regression approaches, exponential smoothing approaches, and the Box-Jenkins approach.

You studied one type of forecasting model in detail in Part I. This type of model is called a multiple regression model. You saw that the least squares estimates of the parameters in a multiple regression model can be used to produce point forecasts and confidence interval forecasts of future values of a time series. In Chapter 2 you learned several techniques that can be used to help in building a multiple regression model that is likely to produce accurate forecasts. You saw that constructing a multiple regression model involves specifying an appropriate set of independent variables and determining the functional form of the regression relationship between the dependent variable and a given set of independent variables. You learned that graphical analysis is often useful in determining the functional form of a regression relationship, but, you also saw that this type of analysis is seldom totally conclusive. Thus, several statistical tools are also helpful in building a multiple regression model. You learned how to use several of these tools—the t_{b_j}-statistic, the simple correlation coefficient, the multiple coefficient of determination (R^2), the mean square error (s^2) and standard error (s), the $F(x_{t,g+1}, \ldots, x_{tp} \mid x_{t1}, \ldots, x_{tg})$-statistic, and the $F(x_{t1}, x_{t2}, \ldots, x_{tp})$-statistic. Last, you have seen that the concept of hypothesis testing is often used in building an appropriate multiple regression model.

EXERCISES

1. When basing a forecast on historical data, what basic assumption is made?

2. Name and explain the four components of a time series.

3. Why are many different forecasting methods used to predict future events?

4. Explain two sources of errors in forecasting.

5. Suppose that we have large forecast errors. Give two possible reasons for the existence of these large errors.

6. Explain the difference between a point forecast and a confidence interval forecast.

7. Why do forecasters examine actual forecast errors in an attempt to identify patterns?

8. In measuring forecast accuracy, why isn't the sum of the forecast errors used?

9. What is the basic difference between the MSE and the MAD?

10. Identify two ways in which measures of forecasting errors are used.

11. Explain the basic difference between qualitative and quantitative forecasting methods.

12. Explain the differences between time series and causal models.

13. Name the factors that should be considered when choosing a forecasting technique.

14. Is the most accurate forecasting method always the best forecasting method?

15. Intuitively explain why the "error of the sum" is less than the "sum of the errors."

16. Explain what is meant by the term "least squares estimates."

17. Does a large value of R^2 (relative to 1) mean that the predictions produced by a regression model will be accurate?

18. What is the basic difference between regression models and exponential smoothing models?

19. Explain the difference between immediate, short-term, medium, and long-term time frames.

20. Give several decisions in your everyday life that required the use of a formal or intuitive forecast.

21. Discuss the difference between a confidence interval for the average level μ_t of a time series and a confidence interval for the actual value y_t of a time series.

22. Discuss the inference assumptions.

23. Discuss how graphical analysis can be used to help determine the functional form of a regression relationship.

24. What does the t_{b_j}-statistic measure?

25. What does R^2 measure?

26. Why is s^2 a fairly good measure of the overall utility of a regression model?

27. Discuss the effects of multicollinearity.

28. Discuss the meaning of interaction.

29. What does the $F(x_{t,g+1}, \ldots, x_{tp} \mid x_{t1}, \ldots, x_{tg})$-statistic measure?

30. Discuss the use of prob-values in evaluating hypotheses.

VOCABULARY LIST

Forecasting	Quantitative forecasting method
Time series	Time series model
Trend	Causal model
Cycle	Time frame
Seasonal variations	Multiple regression model
Irregular fluctuations	Least squares estimates
Point forecast	Exponential smoothing models
Confidence interval forecast	Box-Jenkins models
Forecast error	Linear fashion
Absolute deviation	Quadratic fashion
Mean absolute deviation	Interaction variable
Squared error	t_{b_j}-statistic
Mean squared error	Multiple coefficient of determination, R^2
Qualitative forecasting method	Total variation
Subjective curve fitting	Explained variation
Delphi Method	Unexplained variation
Technological comparisons	Mean square error, s^2

QUIZ

Answer TRUE if the statement is always true. If the statement is not true, replace the underlined word(s) with a word(s) that makes the statement true.

1. Forecasting is important because predictions of future events are needed in the decision-making process.

2. A chronological sequence of observations on a variable of interest is called a <u>forecasting</u> <u>method</u>.

3. The four components of a time series are <u>trend</u>, <u>cycle</u>, <u>seasonal</u> <u>variations</u>, and <u>irregular</u> <u>fluctuations</u>.

4. Increases in total population would most likely cause <u>cyclical</u> movements in industry sales.

5. A <u>seasonal</u> pattern completes itself within a calendar year.

6. The four components of a time series <u>cannot</u> occur at the same time.

7. There <u>is</u> a single best forecasting technique.

8. If we require an estimate of how wrong our forecast might be, a <u>point</u> <u>forecast</u> is appropriate.

9. The pattern of forecast errors is examined in an attempt to determine whether <u>the</u> <u>forecasting</u> <u>methodology</u> <u>is</u> <u>appropriate</u>.

10. The <u>MAD</u> penalizes a forecasting method relatively more for large errors than does the <u>MSE</u>.

11. <u>Qualitative</u> forecasting methods predict future events subjectively.

12. When we wish to analyze the impact of changes in management policies, <u>time</u> <u>series</u> <u>models</u> are most appropriate.

13. <u>Causal</u> <u>models</u> exploit the relationship that exists between other variables and the variable to be forecast.

14. The most accurate forecasting method is <u>always</u> the best forecasting method.

15. When forecasts are generated for points in time from one to three months in the future, the time frame is called <u>immediate</u>.

16. <u>Measurement</u> <u>error</u> takes into account the effect on the dependent variable of all independent variables other than those explicitly included in the regression model.

17. A large value of R^2 by itself <u>does</u> <u>not</u> indicate that the regression model will yield accurate predictions.

18. Multiple regression models <u>do</u> <u>not</u> yield confidence interval predictions for future events.

19. The "sum of the errors" <u>is</u> <u>less</u> <u>than</u> the "error of the sum."

20. Exponential smoothing models weight recent observations <u>more</u> <u>heavily</u> than distantly past observations.

21. A confidence interval for μ_t is usually <u>wider</u> than a confidence interval for $y_t = \mu_t + \epsilon_t$.

22. If a dependent variable y_t increases in a decreasing fashion as an independent variable x_t increases, then y_t is related to x_t in a <u>linear</u> fashion.

23. The inference assumptions relate to the validity of various <u>confidence</u> <u>interval</u> formulas.

24. The t_{b_j}-statistics in a regression analysis can be affected by <u>multicollinearity</u>.

25. Generally speaking, the smaller that s^2 is, the <u>shorter</u> are the confidence intervals computed in a regression analysis.

26. The t_{b_j}-statistic <u>cannot</u> be used to help measure the presence of interaction.

27. The $F(x_{t,g+1}, \ldots, x_{tp} \mid x_{t1}, \ldots, x_{tg})$-statistic is <u>always</u> <u>equivalent</u> to the t_{b_j}-statistic.

28. The prob-value and the probability of a Type I error are <u>equivalent</u> concepts.

29. The smaller the prob-value, the <u>more</u> we believe the null hypothesis.

30. The $F(x_{t1}, \ldots, x_{tp})$-statistic measures <u>the</u> <u>overall</u> <u>utility</u> of a regression model.

REFERENCES

1. Box, G. E. P., and G. M. Jenkins, *Time Series Analysis, Forecasting and Control,* Holden-Day, Inc., San Francisco, 1970.

2. Brown, Bernice B., *Delphi Process: A Methodology Used for the Elicitation of Opinion of Experts,* P-3925, RAND Corporation, Santa Monica, California, September 1968.

3. Brown, R. G., *Statistical Forecasting for Inventory Control,* McGraw-Hill Book Company, New York, 1959.

4. Dalkey, Norman C., *Delphi,* P-3704, RAND Corporation, Santa Monica, California, October 1967.

5. Dalkey, Norman C., *The Delphi Method: An Experimental Study of Group Opinion,* RM-5888-PR, RAND Corporation, Santa Monica, California, June 1969.

6. Draper, N. R., and H. Smith, *Applied Regression Analysis,* John Wiley & Sons, Inc., New York, 1966.

7. Gerstenfeld, Arthur, "Technological Forecasting," *Journal of Business,* vol. 44, no. 1, January 1971.

8. Gordon, T. J., and H. Hayward, "Initial Experiments with the Cross-Impact Method of Forecasting," *Futures,* vol. 1, no. 2, December 1968.

9. Johnson, L. A., and D. C. Montgomery, *Forecasting and Time Series Analysis,* McGraw-Hill Book Company, New York, 1976.

10. Nelson, C. R., *Applied Time Series Analysis for Managerial Forecasting,* Holden-Day, Inc., San Francisco, 1973.

11. Neter, John, and William Wasserman, *Applied Linear Statistical Models,* R. D. Irwin, Homewood, Illinois, 1974.

12. Sigford, J. V., and R. H. Parvin, "Project PATTERN: A Methodology for Determining Relevance in Complex Decision Making," *IEEE Transactions on Engineering Management,* vol. 12, no. 1, March 1965.

13. Zwicky, Fritz, "Morphology of Propulsive Power," *Monographs on Morphological Research No. 1,* Society of Morphological Research, Pasadena, California, 1962.

Forecasting Time Series Described by Trend and Irregular Components

CONTENTS

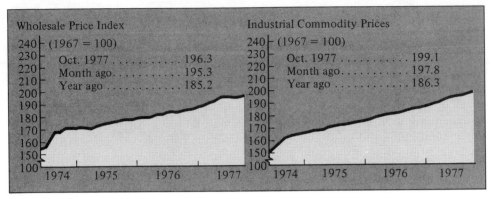

Price Trends

Source: *Industry Week,* November 21, 1977. Data from U.S. Bureau of Labor Statistics. Reprinted by permission of *Industry Week.*

OBJECTIVES FOR PART II

Many time series exhibit very little or no seasonal variation. Such time series can often be adequately described and forecasted using trend and irregular components. The objective of Part II is to study the analysis and forecasting of such time series. In particular, regression and single, double, and triple exponential smoothing techniques will be presented. As examples of using these techniques, we will forecast the monthly catch of cod by the Bay City Seafood Company, the demand for the Bismark X-12 electronic calculator by customers shopping in Smith's Department Stores, and monthly loan requests made to the State University Credit Union.

INTRODUCTION TO PART II

Part II discusses the forecasting of time series that can be adequately described by trend and irregular components. The time series models studied in this chapter are special cases of the model

$$y_t = TR_t + \epsilon_t$$

where TR_t is the trend factor of the time series in time period t; and ϵ_t is the irregular factor of the time series in time period t. Specifically, we will study forecasting time series for which TR_t can be assumed to be given by one of the following three equations.

1.
$$TR_t = \beta_0$$

so that

$$y_t = TR_t + \epsilon_t$$
$$= \beta_0 + \epsilon_t$$

in which case we are assuming there is *no trend*. This means that the time series is "randomly fluctuating" around an average level, β_0, that does not change, or, changes very slowly over time. Figure II.1 illustrates that the average level of a time series with no trend does not change as time advances.

2.
$$TR_t = \beta_0 + \beta_1 t$$

so that

$$y_t = TR_t + \epsilon_t$$
$$= \beta_0 + \beta_1 t + \epsilon_t$$

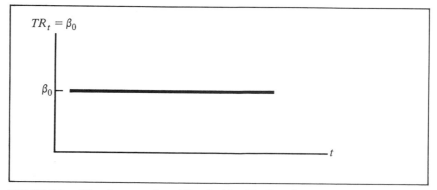

FIGURE II.1 *The Average Level Implied by No Trend*

in which case we are assuming that there is a *linear trend*. This means that the time series is "randomly fluctuating" around an average level that changes in a linear or straight-line fashion over time. The slope of this straight-line relationship is β_1, while the intercept at time 0 is β_0. If the slope β_1 of the trend line is greater than 0, the average level of the time series increases as time advances, whereas if the slope β_1 is less than 0, the average level of the time series decreases as time advances. Figure II.2 illustrates how the average level changes according to a linear trend.

3.
$$TR_t = \beta_0 + \beta_1 t + \beta_2 t^2$$

so that

$$y_t = TR_t + \epsilon_t$$
$$= \beta_0 + \beta_1 t + \beta_2 t^2 + \epsilon_t$$

which implies that there is a *quadratic trend*. This means that the time series is "randomly fluctuating" around an average level that changes in a quadratic or curvilinear fashion over time. Thus the average level of the time series either is increas-

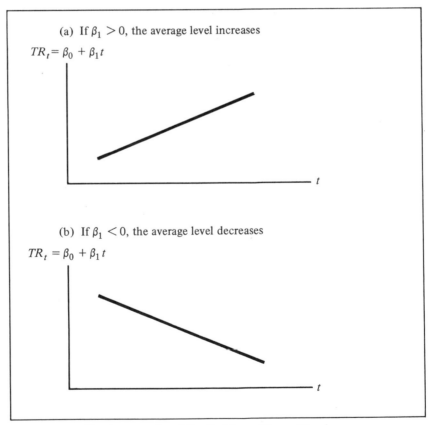

FIGURE II.2   *The Average Level Implied by a Linear Trend*

ing at an increasing or decreasing rate or is decreasing at an increasing or decreasing rate. Figure II.3 illustrates how the average level changes according to a quadratic trend, the exact nature of which is determined by the particular numerical values of β_0, β_1, and β_2.

   Both the regression approach and the exponential smoothing approach will be discussed and used to forecast time series with no trend (in Chapter 3), with a linear trend (in Chapter 4), and with a quadratic trend (in Chapter 5). As explained in Chapter 2, exponential smoothing achieves unequal weighting of the previously observed values of the time series by using one or more smoothing constants.

   Before continuing, it should be recalled from Chapter 1 that the models and methods discussed in Part II assume that the error terms ϵ_1, ϵ_2, . . . in the model

$$y_t = TR_t + \epsilon_t$$

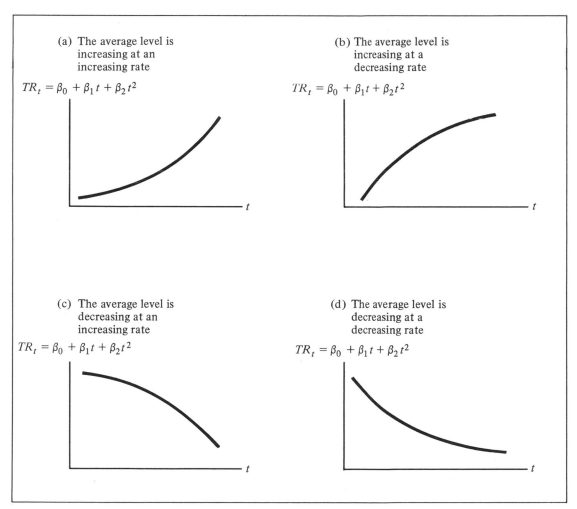

FIGURE II.3   *The Average Level Implied by a Quadratic Trend*

adhere to the assumptions discussed in Section 1–5. One of these assumptions basically states that the observations are not related to each other. If the observations are only "weakly related," then the methods discussed in Part II will still probably produce fairly accurate forecasts. But, if the observations are "strongly related," the Box-Jenkins methodology discussed in Part IV uses the dependency to produce forecasts that are likely to be more accurate than the forecasts produced by the methods in Part II.

CHAPTER 3

An Introduction to Exponential Smoothing

3–1 INTRODUCTION

In this chapter we discuss the forecasting of time series with *no trend*. Both regression and simple exponential smoothing can be used to forecast such time series. Both approaches are presented in Section 3–2.

The implementation of an exponential smoothing procedure consists of several stages. One of these stages involves the choice of an "appropriate" smoothing constant. This choice is discussed in Section 3–3. The overall implementation of an exponential smoothing procedure is discussed in Section 3–4.

If an exponential smoothing forecasting procedure is not providing accurate forecasts, perhaps because of a change in the underlying pattern of the time series, this condition must be detected, and corrective action must be taken. Such corrective action often involves changing the smoothing constant. Section 3–5 discusses the analysis of forecast errors in order to determine when corrective action is necessary and presents adaptive control processes that appropriately change the smoothing constant when the need arises.

While the material in Sections 3–3, 3–4, and 3–5 is discussed in terms of "simple exponential smoothing," it also pertains more generally to all of the exponential smoothing models discussed in this book.

3–2 FORECASTING TIME SERIES WITH NO TREND

In this chapter we will discuss a situation in which we assume that the average level of the time series is not changing over time, or a situation in which we assume that the average level is changing over time very slowly. Thus an appropriate model for the time series might be

$$y_t = \beta_0 + \epsilon_t$$

In this case, the time series can be described by an average value or level, β_0, which does not change, or, changes very slowly over time, combined with random fluctuations, which cause the observations in the time series to deviate from the average level. These random fluctuations are reflected through the random error component ϵ_t.

Example 3.1   The Bay City Seafood Company, which owns a fleet of fishing trawlers and operates a fish processing plant, wishes to make monthly predictions of its catch of cod (measured in tons). In order to forecast its minimum and maximum possible revenues from cod sales and in order to plan the operations of its fish processing plant, the Bay City Seafood Company desires to make both point forecasts and confidence interval forecasts of its monthly cod catch. The point forecast is the best guess at the monthly cod catch, while the confidence interval forecast states the smallest and largest tonnage that the monthly cod catch can reasonably be expected to be. The Bay City Seafood Company has recorded monthly cod catch in tons for the previous two years, which we will call year 1 and year 2. The cod catch history is given in Table 3.1. When this data is plotted, it appears to randomly fluctuate around a constant average level, as indicated in Figure 3.1. Since the company subjectively believes that this data pattern will continue in the future, it seems reasonable to use the model

$$y_t = \beta_0 + \epsilon_t$$

to forecast cod catch in future months.

TABLE 3.1   *Cod Catch (in tons)*

	Year 1	Year 2
Jan.	362	276
Feb.	381	334
Mar.	317	394
Apr.	297	334
May	399	384
June	402	314
July	375	344
Aug.	349	337
Sept.	386	345
Oct.	328	362
Nov.	389	314
Dec.	343	365

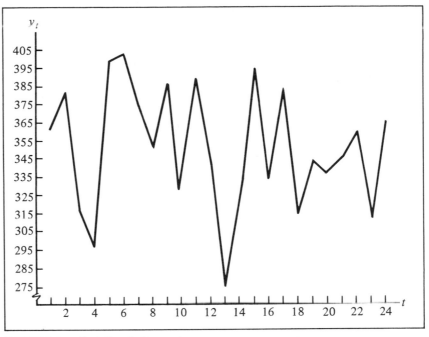

FIGURE 3.1   *Cod Catch (in tons)*

3–2.1   THE REGRESSION APPROACH

The least squares estimate of β_0 is $b_0 = \bar{y}$, the average of the n observations y_1, y_2, . . . , y_n that have been previously collected. Hence, a forecast of any future value y_t of the time series is

$$\hat{y}_t = b_0 = \bar{y}$$

Example 3.2

Using the cod catch data in Table 3.1, the least squares estimate of β_0 is

$$b_0 = \bar{y} = \frac{\displaystyle\sum_{t=1}^{24} y_t}{24}$$

$$= \frac{362 + 381 + \cdots + 365}{24}$$

$$= 351.29$$

Thus, the forecast of cod catch for any future month is 351.29 or 351 if the forecast is rounded to a whole number. It can be shown that a 95 percent confidence interval for the cod catch in any future month is

$$[351.29 - 71.37, 351.29 + 71.37] \quad \text{or} \quad [279.29, 422.66]$$

*3–2.1   THE REGRESSION CALCULATIONS

Using the data in Table 3.1, we see that if we wish to find the least squares estimate of β_0 in the regression model

$$y_t = \beta_0 + \epsilon_t$$

we use

$$\mathbf{y} = \begin{bmatrix} 362 \\ 381 \\ \cdot \\ \cdot \\ \cdot \\ 365 \end{bmatrix} \quad \text{and} \quad \mathbf{X} = \begin{bmatrix} 1 \\ 1 \\ \cdot \\ \cdot \\ \cdot \\ 1 \end{bmatrix}$$

which implies that

$$(\mathbf{X'X})^{-1} = \frac{1}{24} \left(= \frac{1}{n}\right); \quad \mathbf{X'y} = 362 + 381 + \cdots + 365 = \sum_{t=1}^{24} y_t$$

Thus, the least squares estimate of β_0 is

$$b_0 = (\mathbf{X'X})^{-1} \mathbf{X'y} = \frac{\sum_{t=1}^{n} y_t}{n} = \bar{y} = \frac{362 + 381 + \cdots + 365}{24}$$
$$= 351.29$$

Moreover, a 95 percent confidence interval for y_t is

$$[\hat{y}_t - E_t(95), \hat{y}_t + E_t(95)] \quad \text{or} \quad [279.29, 422.66]$$

where

$$\hat{y}_t = b_0 = 351.29$$

$$s = \left[\frac{\sum_{t=1}^{24} (y_t - \hat{y}_t)^2}{24 - 1}\right]^{1/2} = 33.82$$

$$t_{5/2}(24 - 1) = t_{5/2}(23) = 2.069$$

$$\mathbf{x}'_t = [1]$$

$$f_t = \sqrt{1 + \mathbf{x}'_t(\mathbf{X}'\mathbf{X})^{-1}\mathbf{x}_t} = \sqrt{1 + 1/n} = \sqrt{1 + 1/24} = 1.04$$

$$E_t(95) = t_{5/2}(23) \cdot s \cdot f_t = 71.37$$

3–2.2   THE EXPONENTIAL SMOOTHING APPROACH

To begin the exponential smoothing approach, let us now suppose that at the end of a particular time period, which we shall call time period $T - 1$, we have obtained a set of observations for the time series, which we shall denote y_1, y_2, ..., y_{T-1}. Given these observations we wish to estimate β_0, the average level of the time series. As just stated, the least squares estimate of β_0, which we now denote as $b_0(T - 1)$ to emphasize the fact that the most recent observation in the time series corresponds to time period $T - 1$, is

$$b_0(T \quad 1) - \bar{y} = \sum_{t=1}^{T-1} y_t/(T - 1)$$

So the estimate of β_0 is simply the average of the observations in the time series through period $T - 1$. Given this estimate, the forecast for any future time period, say period $T - 1 + \tau$, where τ is a positive integer, is $b_0(T - 1)$. Thus the forecast for any future period is the average of the observations in the time series through period $T - 1$.

Now suppose that we obtain a new observation y_T at the end of the next time period, period T. We would like to incorporate this new observation into the estimate of β_0. That is, we wish to obtain an updated estimate of β_0 that is based on the new observation y_T as well as the old observations y_1, y_2, ..., y_{T-1}. We shall call this new estimate $b_0(T)$ to indicate that the most recent observation in the time series corresponds to time period T. Again, regression may be used to obtain such an estimate. This estimate is given by

$$b_0(T) = \bar{y} = \sum_{t=1}^{T} y_t/T$$

So this new estimate can be found by simply recalculating $\bar{y}$, which is now the average of the observations in the time series through period T. This new estimate $b_0(T)$ is then the new forecast for the average level of the time series in any future time period.

Another way to incorporate this new observation into the estimate of β_0 is known as *simple exponential smoothing*.[1] This approach generates the new estimate $b_0(T)$ in a different way. It seems intuitive to change the old estimate $b_0(T - 1)$ by some fraction of the forecast error which resulted when the old estimate was

---

[1] The discussion in this paragraph is adapted from a similar discussion in Johnson and Montgomery [8].

used to forecast the value of the time series for the present period. This forecast error is given by

$$e_T = y_T - b_0(T - 1)$$

and is the difference between the observed value in period T and the forecast made for period T in period $T - 1$. If the fraction we use is α, then the updated estimate is given by

$$b_0(T) = b_0(T - 1) + \alpha[y_T - b_0(T - 1)]$$

Thus the new estimate is partly based on the old estimate $b_0(T - 1)$. If the old estimate yielded a forecast for period T that was too low, then the new estimate is higher. If the old estimate yielded a forecast for period T that was too high, then the new estimate is lower. The magnitude of the adjustment up or down is determined by the magnitude of the forecast error. A large error leads to a large adjustment while a small error leads to a small adjustment.

In order to simplify notation at this point we shall define $S_T = b_0(T)$. So the equation we use to update the estimate can be rewritten as

$$S_T = S_{T-1} + \alpha(y_T - S_{T-1}) = S_{T-1} + \alpha y_T - \alpha S_{T-1}$$

$$\boxed{S_T = \alpha y_T + (1 - \alpha)S_{T-1}}$$

This equation defines the updating procedure called *simple exponential smoothing*. We call S_T the *smoothed estimate* or *smoothed statistic*. The fraction α is called the *smoothing constant*.

Examining the smoothing equation

$$S_T = \alpha y_T + (1 - \alpha)S_{T-1}$$

we see that the smoothed estimate is simply an estimate based on the observations $y_1, y_2, \ldots, y_T$. This is true since S_{T-1} or $b_0(T - 1)$ is the average of the observations $y_1, y_2, \ldots, y_{T-1}$. Now, let us change the time origin so that the initial estimate of β_0 is assumed to be generated in time period zero. We shall call this initial estimate S_0. In practice, the estimate S_0 would be obtained by calculating the average of an initial set of observations in the time series. If such an initial set of observations is not available, S_0 is commonly set equal to the first observed value of the time series. Let us also assume that the smoothing equation

$$S_t = \alpha y_t + (1 - \alpha)S_{t-1}$$

has been used to update the estimates for each time period t from period 1 to period T, the present time period. In this situation, the smoothed estimate for period T, that is, S_T, can be shown to be a linear combination of all the past observations. To see this we consider the smoothed estimate

$$S_T = \alpha y_T + (1 - \alpha)S_{T-1}$$

from which we can see that

$$S_{T-1} = \alpha y_{T-1} + (1 - \alpha)S_{T-2}$$

Substitution therefore gives us

$$S_T = \alpha y_T + (1 - \alpha)[\alpha y_{T-1} + (1 - \alpha)S_{T-2}]$$
$$= \alpha y_T + \alpha(1 - \alpha)y_{T-1} + (1 - \alpha)^2 S_{T-2}$$

Again, we can see that

$$S_{T-2} = \alpha y_{T-2} + (1 - \alpha)S_{T-3}$$

and substituting again we have

$$S_T = \alpha y_T + \alpha(1 - \alpha)y_{T-1} + (1 - \alpha)^2[\alpha y_{T-2} + (1 - \alpha)S_{T-3}]$$
$$= \alpha y_T + \alpha(1 - \alpha)y_{T-1} + \alpha(1 - \alpha)^2 y_{T-2} + (1 - \alpha)^3 S_{T-3}$$

Substituting recursively for S_{T-3}, S_{T-4}, . . . , S_2, and S_1 we obtain

$$S_T = \alpha y_T + \alpha(1 - \alpha)y_{T-1} + \alpha(1 - \alpha)^2 y_{T-2} + \cdots + \alpha(1 - \alpha)^{T-1}y_1 + (1 - \alpha)^T S_0$$

Thus we see that S_T, the estimate of β_0 in time period T, can be expressed in terms of the observations $y_1, y_2, \ldots, y_T$, and the initial estimate S_0. The coefficients of the observations, $\alpha, \alpha(1 - \alpha), \alpha(1 - \alpha)^2, \ldots, \alpha(1 - \alpha)^{T-1}$, measure the contributions that the observations $y_T, y_{T-1}, y_{T-2}, \ldots, y_1$ make to the most recent estimate S_T. It can be seen that these coefficients decrease geometrically with the age of the observations. For example, if the smoothing constant α is 0.1, then these coefficients are 0.1, 0.09, 0.081, 0.0081, and so on. The updating procedure we have described is called *simple exponential smoothing* because these coefficients decrease exponentially.

Since these coefficients are decreasing, the most recent observation y_T makes the largest contribution to the current estimate of β_0. Older observations make smaller and smaller contributions to the current estimate of β_0 at each successive time point. Thus, remote observations are "dampened out" of the current estimate of β_0 as time advances. The rate at which remote observations are dampened out depends on the smoothing constant α. For values of α near 1, remote observations are dampened out quickly, while for values of α near 0, remote observations are dampened out more slowly. For example, for $\alpha = .9$ we obtain coefficients

$$.9, .09, .009, .0009, \ldots$$

while for $\alpha = .1$ we obtain coefficients

$$.1, .09, .081, .0081, \ldots$$

So the choice of the smoothing constant has a great bearing on the estimate S_T. In general, when the time series is quite volatile, that is, when the random component ϵ_t has a large variance, we would select a small smoothing constant so that the smoothed estimate S_T will weight S_{T-1}, the smoothed estimate for the previous period, to a greater degree than it weights the observation y_T. For a more stable time series, in which the random component ϵ_t has a smaller variance, we could select a larger smoothing constant. We shall return to the problem of the choice of a smoothing constant in Section 3–3.

We now turn to the problem of making a forecast. Suppose we are in period T and the current estimate of β_0 is $S_T = b_0(T)$. We wish to forecast the time series for a future period $T + \tau$. Since the model is $y_t = \beta_0 + \epsilon_t$, the forecast is simply $\hat{y}_{T+\tau}(T) = S_T$, the current estimate of β_0. Here the use of the hat (ˆ) indicates that $\hat{y}_{T+\tau}(T)$ is a forecast value rather than an observation of the time series. The notation $\hat{y}_{T+\tau}(T)$ is used to emphasize that this forecast is being made for period $T + \tau$ and is being made in period T.

Example 3.3   We now consider the use of simple exponential smoothing as a forecasting tool to be used in forecasting the cod catch time series in Example 3.1. In this example we shall select a smoothing constant of $\alpha = .02$. The reason for this selection will be explained in the next section. The initial estimate of β_0 is obtained by averaging the monthly cod catch over years 1 and 2. Thus we obtain $S_0 = 351.29$. At this point we might generate a forecast for February of year 3, based on both the initial estimate $S_0 = 351.29$ and a new observation of cod catch for January of year 3, by using the smoothing equation

$$S_1 = \alpha y_1 + (1 - \alpha)S_0$$

However, if this procedure is followed, the initial estimate S_0 weights each observation from years 1 and 2 equally. It seems more reasonable to generate a forecast for February of year 3 which weights recent observations more heavily than distant observations. For example, it would be logical to weight observations from the last months of year 2 more heavily than observations from the first several months of year 1. This can be accomplished by starting with the initial estimate S_0 and using the smoothing equation to sequentially update this estimate by smoothing the observations from years 1 and 2. The updated estimate for December of year 2 will then be used along with the observation for January of year 3 to obtain a forecast for February of year 3. So, we begin the smoothing procedure with $S_0 = 351.29$ and the observation for January of year 1 which is denoted $y_1 = 362$. Thus at the end of January of year 1 the updated estimate of β_0 is given by

$$S_1 = \alpha y_1 + (1 - \alpha)S_0$$
$$= .02(362) + (1 - .02)(351.29)$$
$$= 351.50$$

We may now use $y_2 = 381$, the observation for February of year 1, to again update the estimate of β_0. We obtain

$$S_2 = \alpha y_2 + (1 - \alpha)S_1$$

$$= .02(381) + .98(351.50)$$

$$= 352.09$$

Continuing in this manner, we repeatedly update this estimate using the balance of the observations in years 1 and 2. The results of these calculations are summarized in Table 3.2. We find that the updated estimate, often called the smoothed estimate, for December of year 2 is $S_{24} = 350.97$. This means that at the end of year 2 the forecast for any future month is $S_{24} = 350.97$ (351 when rounded). Now, as time

TABLE 3.2   *Updating the Smoothed Statistic Using Simple Exponential Smoothing with $\alpha = .02$ on the Historical Cod Catch Time Series*

Year	Month	Period Number T	Actual Cod Catch y_T	Smoothed or Updated Estimate S_T
				($S_0 = 351.29$)
1	Jan.	1	362	351.50
	Feb.	2	381	352.09
	Mar.	3	317	351.39
	Apr.	4	297	350.30
	May	5	399	351.28
	June	6	402	352.29
	July	7	375	352.75
	Aug.	8	349	352.67
	Sept.	9	386	353.34
	Oct.	10	328	352.83
	Nov.	11	389	353.55
	Dec.	12	343	353.34
2	Jan.	13	276	351.79
	Feb.	14	334	351.44
	Mar.	15	394	352.29
	Apr.	16	334	351.92
	May	17	384	352.56
	June	18	314	351.79
	July	19	344	351.64
	Aug.	20	337	351.34
	Sept.	21	345	351.22
	Oct.	22	362	351.43
	Nov.	23	314	350.68
	Dec.	24	365	350.97

goes on, suppose that the cod catch in January of year 3 is observed to be $y_{25} = 392$. We again use the smoothing equation to obtain an updated estimate of β_0:

$$S_{25} = \alpha y_{25} + (1 - \alpha)S_{24}$$
$$= .02(392) + .98(350.97)$$
$$= 351.79$$

The forecast of cod catch for any future month is now $S_{25} = 351.79$ (352 when rounded). Table 3.3 presents actual cod catch for each month in the next two years along with the smoothed estimate for each period and the rounded forecast made one month before the current period. A plot of the actual cod catch and the forecasted cod catch is given in Figure 3.2.

TABLE 3.3   *One-Period-Ahead Forecasts of Future Values of Cod Catch Using Simple Exponential Smoothing with $\alpha = .02$*

Year	Month	Period Number T	Actual Cod Catch y_T	Smoothed Estimate S_T	Forecast Made Last Period S_{T-1} (rounded)
				($S_{24} = 350.97$)	
3	Jan.	25	392	351.79	351
	Feb.	26	294	350.63	352
	Mar.	27	337	350.36	351
	Apr.	28	313	349.61	350
	May	29	361	349.84	350
	June	30	322	349.28	350
	July	31	410	350.50	349
	Aug.	32	356	350.61	350
	Sept.	33	365	350.90	351
	Oct.	34	343	350.74	351
	Nov.	35	365	351.02	351
	Dec.	36	348	350.96	351
4	Jan.	37	400	351.94	351
	Feb.	38	330	351.50	352
	Mar.	39	357	351.61	352
	Apr.	40	343	351.44	352
	May	41	418	352.77	351
	June	42	337	352.46	353
	July	43	341	352.23	352
	Aug.	44	355	352.28	352
	Sept.	45	336	351.96	352
	Oct.	46	378	352.48	352
	Nov.	47	350	352.43	352
	Dec.	48	320	351.78	352

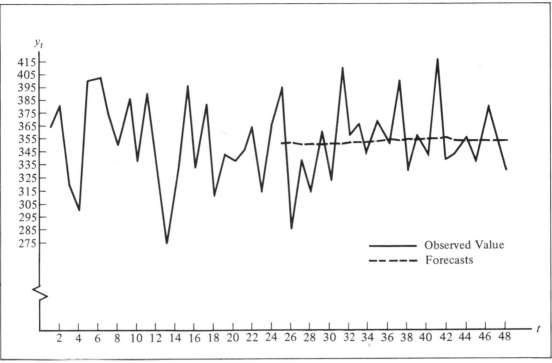

FIGURE 3.2   *One-Period-Ahead Forecasts of Future Values of Cod Catch Using Simple Exponential Smoothing with $\alpha = .02$*

3–3   DETERMINATION OF AN APPROPRIATE SMOOTHING CONSTANT

When an exponential smoothing procedure is used as a forecasting tool, the forecaster must specify a value for the smoothing constant α to be used. As noted previously, the smoothing constant determines the extent to which past observations influence the forecast. Since small values of the smoothing constant α dampen out remote observations in the time series slowly, a small α results in a slow response to changes in the parameters describing the average level of the time series. On the other hand, large values of α quickly dampen out remote observations in the time series. Thus a large α gives a larger weight to the more recent observations in the time series and results in a more rapid response to changes in the time series. Unfortunately, this rapid response can cause the forecasting procedure to respond to irregular movements in the time series that do not reflect changes in the parameters describing the time series. Obviously, such a situation is no better than a situation in which the forecasting procedure reacts very slowly to changes in the parameters of the time series. In practice, it has been found that values of α from .01 to .30 work quite well. One way of choosing the smoothing constant involves the use of simulation. This procedure consists of simulating a set of historical data using different values of the smoothing constant α. That is, for each value of α a set of forecasts is generated using the appropriate exponential smoothing procedure. These forecasts are then compared with the actual observations in the time series. The value of α that yields the "best" forecasts is the value of the smoothing constant to be used in generating forecasts for future values of the time series. Which set of forecasts is "best" is determined by considering the forecast errors obtained in the simulations. The forecast error in a particular time period is simply the difference between the actual observation in the time series and the forecast made for that period. The set of forecasts that has the smallest sum of squared forecast errors is the set deemed to be "best."

Example 3.4   Let us consider Example 3.3 in which we wished to forecast monthly cod catch using simple exponential smoothing. In this example we selected a smoothing constant of $\alpha = .02$. Suppose we now wish to determine whether a smoothing constant of $\alpha = .20$ would be a better choice. In order to make this determination we shall simulate the two years of historical data (years 1 and 2) using simple exponential smoothing. First, we shall perform this simulation using the exponential smoothing procedure with a smoothing constant of $\alpha = .02$. We initialize the smoothing procedure by using the first few data points in the historical time series to estimate the parameter β_0. It is customary to use about 6 data points in obtaining this estimate. The first 6 observations in this time series are: 362, 381, 317, 297, 399, 402. Since our model is $y_t = \beta_0 + \epsilon_t$, our estimate of β_0 is simply the average of these six observations, which is 359.67. Thus we begin the exponential smoothing procedure with $S_0 = 359.67$. We next generate forecasts for all months in years 1 and 2. Given these forecasts, we can calculate the forecast error for each period, the squared forecast error for each period, and the sum of squared forecast errors for all periods combined. These results are shown in Table 3.4. We see that using the smoothing constant $\alpha = .02$ yields a sum of squared forecast errors of 27,744.

Next we repeat this procedure using a smoothing constant of $\alpha = .20$, again beginning the smoothing procedure with the initial estimate of β_0, 359.67. The results of this simulation are shown in Table 3.5. We see that using the smoothing constant $\alpha = .20$ yields a sum of squared forecast errors of 30,774.

When we compare the sum of squared forecast errors obtained in these two simulations we find that a smoothing constant of $\alpha = .02$ performs better than a smoothing constant of $\alpha = .20$. If we assume that the time series will behave in the future as it has in the past, a smoothing constant of $\alpha = .02$ will likely yield more accurate forecasts than a smoothing constant of $\alpha = .20$.

Clearly, however, there is no reason to restrict our attention to the values $\alpha = .02$ and $\alpha = .20$. As stated previously, values of α from .01 to .30 are most commonly used. Thus we might perform this simulation for any number of values in this range. For example, if, in the smoothing problem above, simulation had been performed for values of α from .02 to .30 in increments of .02, the resulting sums of squared forecast errors would be as shown in Table 3.6. It can be seen from the table that a smoothing constant of $\alpha = .02$ yields the minimum sum of squared forecast errors. If the time series continues to behave as it has in the past, a smoothing constant of $\alpha = .02$ will likely produce more accurate forecasts than the other values of α listed in Table 3.6.

TABLE 3.4   *Simulating One-Period-Ahead Forecasting of the Historical Cod Catch Time Series Using Simple Exponential Smoothing with* $\alpha = .02$

Year	Month	Actual Cod Catch y_T	Smoothed Estimate S_T	Forecast Made Last Period	Forecast Error	Squared Forecast Error
			$(S_0 = 359.67)$			
1	Jan.	362	359.72	360	2	4
	Feb.	381	360.14	360	21	441
	Mar.	317	359.28	360	−43	1849
	Apr.	297	358.03	359	−62	3844
	May	399	358.85	358	41	1681
	June	402	359.71	359	43	1849
	July	375	360.02	360	15	225
	Aug.	349	359.80	360	−11	121
	Sept.	386	360.32	360	26	676
	Oct.	328	359.68	360	−32	1024
	Nov.	389	360.26	360	29	841
	Dec.	343	359.92	360	−17	289
2	Jan.	276	358.24	360	−84	7056
	Feb.	334	357.75	358	−24	576
	Mar.	394	358.48	358	36	1296
	Apr.	334	357.99	358	−24	576
	May	384	358.51	358	26	676
	June	314	357.62	359	−45	2025
	July	344	357.35	358	−14	196
	Aug.	337	356.94	357	−20	400
	Sept.	345	356.70	357	−12	144
	Oct.	362	356.81	357	5	25
	Nov.	314	355.95	357	−43	1849
	Dec.	365	356.13	356	9	81

Sum of squared errors = 27,744

In Example 3.4 and in the rest of the examples in this book the simulations to determine the optimal smoothing constant in an exponential smoothing procedure are performed by ranging the smoothing constant from .02 to .30 in increments of .02. In practice, the smoothing constant is determined by setting values of α from .01 to .30 in increments smaller than .02 to determine the α value that minimizes the sum of squared forecast errors. However, using an increment of .02 will suffice for our purpose, which is to demonstrate how the sum of squared forecast errors changes as the smoothing constant changes.

Note that the simulation of historical data in an effort to determine an appropriate smoothing constant might indicate that the exponential smoothing procedure being used is not the appropriate model for the time series being

TABLE 3.5   *Simulating One-Period-Ahead Forecasting of the Historical Cod Catch Time Series Using Simple Exponential Smoothing with $\alpha = .20$*

Year	Month	Actual Cod Catch y_T	Smoothed Estimate S_T	Forecast Made Last Period	Forecast Error	Squared Forecast Error
			($S_0 = 359.67$)			
1	Jan.	362	360.14	360	2	4
	Feb.	381	364.31	360	21	441
	Mar.	317	354.85	364	−47	2209
	Apr.	297	343.28	355	−58	3364
	May	399	354.42	343	56	3136
	June	402	363.94	354	48	2304
	July	375	366.15	364	11	121
	Aug.	349	362.72	366	−17	289
	Sept.	386	367.37	363	23	529
	Oct.	328	359.50	367	−39	1521
	Nov.	389	365.40	359	30	900
	Dec.	343	360.92	365	−22	484
2	Jan.	276	343.94	361	−85	7225
	Feb.	334	341.95	344	−10	100
	Mar.	394	352.36	342	52	2704
	Apr.	334	348.69	352	−18	324
	May	384	355.75	349	35	1225
	June	314	347.40	356	−42	1764
	July	344	346.72	347	−3	9
	Aug.	337	344.77	347	−10	100
	Sept.	345	344.82	345	0	0
	Oct.	362	348.26	345	17	289
	Nov.	314	341.40	348	−34	1156
	Dec.	365	346.12	341	24	576

Sum of squared errors = 30,774

analyzed. This is probably the case if simulation indicates that the "best" smoothing constant is a value greater than .3. In such a case, it may be that values in the time series are dependent on one another, that is, the observations may be *autocorrelated*. Although exponential smoothing is often used successfully in such situations, other methods of analysis, such as the Box-Jenkins methodology, may be more appropriate. A "best" smoothing constant greater than .3 may also indicate cyclical or seasonal behavior in the time series, which can be best handled using models that explicitly include such factors in the analysis of the time series. Several methods of this type will be discussed in Part III.

TABLE 3.6   *The Sums of Squared Forecast Errors for Different Values of α*

Smoothing Constant α	Sum of Squared Errors
.02	27744
.04	28021
.06	28036
.08	28459
.10	28792
.12	28987
.14	29496
.16	29756
.18	30274
.20	30774
.22	31269
.24	31725
.26	32203
.28	32596
.30	33316

3–4 THE OVERALL IMPLEMENTATION OF AN EXPONENTIAL SMOOTHING PROCEDURE

At this point it may be helpful to outline the sequence of events involved in the implementation of a simple exponential smoothing procedure. First, an initial estimate of β_0 should be calculated using the first several, say six, observations in the time series. In our example we obtained an estimate of $S_0 = 359.67$ from the first six observations. Next, starting with this initial estimate, the smoothing equation is used to simulate all the historical data available using a particular value of the smoothing constant α. The sum of squared forecast errors is then calculated for the forecasts generated using this smoothing constant. This simulation process is repeated for other values of the smoothing constant α and the sum of squared forecast errors is calculated for each α. In our example, the results of these simulations are summarized in Table 3.6. The value α that minimizes the sum of squared forecast errors is the value that should be chosen for actual forecasting. Now that the "best" smoothing constant has been determined, the smoothing process may be used to forecast future values of the time series. In order to generate these forecasts, a new initial estimate of β_0 is calculated using all of the historical data. In our example, this initial estimate was the average of the entire two-year cod catch history and was found to be $S_0 = 351.29$. This new initial value and the optimal smoothing constant are now used to perform the smoothing operation upon the historical data. This procedure yields an updated estimate of β_0, which is a forecast for any future value of the time series. In our example, the initial estimate of $S_0 = 351.29$ and the optimal smoothing constant of $\alpha = .02$ were used to obtain the updated estimates found in Table 3.2. The updated estimate of β_0 at the end of December of year 2 was found to be 350.97. This value is the forecast for January of year 3 or for any other future value of the time series. The smoothing equation may now be used with the optimal smoothing constant to generate forecasts for future values of the time series as new data are obtained. In our example, these forecasts are summarized in Table 3.3.

The general procedure summarized in the previous paragraph will be followed in the implementation of the more complicated double and triple smoothing calculations described in later chapters. It is important to note that two aspects of this general procedure are arbitrary. To begin our discussion of these aspects, let us define n to be the number of historical observations of the time series that are available for use in developing a forecasting model. In our simple exponential smoothing example we have $n = 24$. The first arbitrary aspect of the general procedure is the choice of n_1, which we define to be the number of observations chosen from the n total observations to be used to calculate the values of the "smoothed statistic(s)" initializing the simulations which determine an appropriate smoothing constant. In the foregoing simple exponential smoothing example we use $n_1 = 6$ observations to determine the smoothed statistic $S_0 = 359.07$ that initialize the simulations. The second arbitrary aspect is the choice of n_2, which we define to be the number of observations used to calculate the initial values of the smoothed statistics used to begin actual forecasting after an appropriate smoothing constant has been determined. In our simple exponential smoothing example we use $n_2 = 24$ observations to calculate the initial smoothed statistic $S_0 = 351.29$ used to begin actual forecasting. In the double exponential smoothing example of Section 4–2.2 where $n = 24$, we will choose $n_1 = 6$ and $n_2 = 24$. In the triple exponential smoothing example of Section 5–2.2 where $n = 24$, we will choose $n_1 = 12$ and $n_2 = 24$. Note that in all examples we choose n_2 equal to n, and we choose n_1 substantially smaller than n. The basic reason n_1 is chosen to be substantially less than n is to insure that there are a substantial number of observations, other than the n_1 observations used to obtain the smoothed statistics initializing the simulations, remaining to be simulated. In general, the question of determining n_1 and n_2 is fairly complex. The reader should accept for the moment that our choices of n_1 and n_2 are reasonable in the simple, double, and triple exponential smoothing examples of Chapters 3, 4, and 5. Section 5–3 will provide a more complete discussion of the problem of choosing n_1 and n_2.

3–5 THE ANALYSIS OF FORECAST ERRORS AND ADAPTIVE CONTROL PROCESSES

A forecasting system will never produce perfect forecasts of a time series. In this section we discuss some methods that can be used to determine when something is wrong with the forecasting system. That is, we wish to determine if the forecast errors are larger than an "accurate" forecasting system can reasonably be expected to produce. To do this, assume we have a history of T single-period-ahead forecast errors, $e_1(\alpha), e_2(\alpha), \ldots, e_T(\alpha)$. Here, (α) denotes the fact that the single-period-ahead forecast errors are obtained from an exponential smoothing forecasting system employing a smoothing constant of α. We next define the following sum (Y) of the single-period-ahead forecast errors

$$Y(\alpha,T) = \sum_{t=1}^{T} e_t(\alpha)$$

It is obvious that

$$Y(\alpha,T) = Y(\alpha, T-1) + e_T(\alpha)$$

and we define the following mean absolute deviation (D)

$$D(\alpha,T) = \frac{\sum_{t=1}^{T} |e_t(\alpha)|}{T}$$

Then, the *tracking signal* $TS(\alpha,T)$ is defined as

$$TS(\alpha,T) = \left| \frac{Y(\alpha,T)}{D(\alpha,T)} \right|$$

If $TS(\alpha,T)$ is "large," this means that $Y(\alpha,T)$ is large relative to the mean absolute deviation $D(\alpha,T)$, which in turn says that, since

$$Y(\alpha,T) = \sum_{t=1}^{T} e_T(\alpha)$$

the forecasting system is producing errors that are either consistently positive or consistently negative. That is, a large value of $TS(\alpha,T)$ implies that the forecasting system is producing forecasts that are either consistently smaller or consistently larger than the time series values that are being forecasted. Since an "accurate" forecasting system should be producing roughly one half positive errors and one half negative errors, a large value of $TS(\alpha,T)$ indicates that the forecasting system is not performing accurately. In practice, if $TS(\alpha,T)$ exceeds a control limit, denoted by K_1, for two or more consecutive periods, this is taken as a strong indication that the forecast errors have been larger than an accurate forecasting system can reasonably be expected to produce. The control limit K_1 is generally taken to be between 4 and 6 for most exponential smoothing models.

More precise methods for determining the control limits for particular exponential smoothing models can be found by the interested reader in Johnson and Montgomery [8], which also discusses tracking signals slightly different from the one discussed above. If the tracking signal exceeds a large control limit (say near 6), it is a very strong indication that the forecasting system is not performing accurately. If the tracking signal indicates that corrective action is needed, several possibilities exist. One possibility is that the model needs to be changed. To do this, variables may be added or deleted to obtain a better representation of the time series. Another possibility is that the model being used does not need to be changed, but the estimates of the parameter(s) of the model do need to be changed. This can be accomplished by changing the smoothing constant, a subject which is further discussed in the next paragraphs. Before continuing to these paragraphs, we discuss the initial values to be used for $Y(\alpha,T)$ and $D(\alpha,T)$ when starting the forecasting procedure. Since it is reasonable to assume that the original model is correct, $Y(\alpha,0) = 0$ is the starting value for $Y(\alpha,T)$. Since there is some random variation in the process, however, it is not reasonable to use $D(\alpha,0) = 0$ as the starting value for $D(\alpha,T)$. One possibility is to calculate an initial value $D(\alpha,0)$ by employing the one-step-ahead forecast errors used in the simulation of historical data to determine an appropriate smoothing constant.

It is common procedure to use different values of the smoothing constant at different times in the analysis of a time series. For example, a large value of α may be appropriate at the start of an exponential smoothing procedure when the initial values used to initialize the procedure are based on only a few observations. This allows new observations to weigh heavily in early forecasts. Later, a smaller value of α may be used since more information concerning the behavior of the time series has built up. It is also common to use a small value of α when a time series is stable and to use a larger value of α when the parameters in the time series are suspected to be changing. Forecasters often use procedures that automatically change the value of the smoothing constant. These procedures are known as *adaptive-control procedures* since they allow the smoothing constant to adapt itself to changes in the parameters of the time series.

Several techniques have been developed to automatically control the values of smoothing constants. We discuss a method developed by W. M. Chow [5]. This method involves the use of three values for the smoothing constant. Let us suppose that a value for the smoothing constant in our smoothing procedure has been determined using simulation. We shall call this value α. Chow's method employs two other values of the smoothing constant: an upper value α_U and a lower value α_L such that

$$\alpha_U = \alpha + d$$

$$\alpha_L = \alpha - d$$

where d is a positive constant. Chow suggests that a value of $d = .05$ be used. The smoothing procedure is now performed using each of these values for the smoothing constant. That is, three different forecasts are generated in each period, one for each of the smoothing constants α_L, α, and α_U. The forecast generated using the smoothing constant α is the forecast to be used in practice. The other two forecasts are used to monitor the time series in an attempt to detect substantial changes in it. This is done in the following manner. In each time period three streams of forecasts exist. These three streams are generated by the smoothing constants α_L, α, and α_U. For each of these streams, the mean absolute deviation of forecast errors can be calculated. We define $D(\alpha_L,T)$, $D(\alpha,T)$, and $D(\alpha_U,T)$ to be the mean absolute deviations of forecast errors calculated in period T using the streams of forecasts generated using, respectively, the smoothing constants α_L, α, and α_U. Whenever

$$D(\alpha,T) < \begin{cases} D(\alpha_L,T) \\ D(\alpha_U,T) \end{cases}$$

the forecast errors generated using α have been smaller than the forecast errors generated using α_L and α_U, and it is, therefore, reasonable to continue using the smoothing constant α to generate forecasts. If, however,

$$D(\alpha_L,T) < \begin{cases} D(\alpha,T) \\ D(\alpha_U,T) \end{cases}$$

the smoothing constant to be used in forecasting should be reduced by setting α at α_L and adding and subtracting d to fix new limits, or, if

$$D(\alpha_U,T) < \begin{cases} D(\alpha,T) \\ D(\alpha_L,T) \end{cases}$$

α should be increased to α_U and new limits fixed by adding and subtracting d. Each time the smoothing constant is changed, the mean absolute deviations $D(\alpha_L,T)$, $D(\alpha,T)$, and $D(\alpha_U,T)$ are set equal to zero and the monitoring procedure is started over.

Chow reported good results with his method. However, a disadvantage of this method is that three forecasts must be computed each period, and hence a large amount of data must be stored in the information system maintained by the forecaster.

Chow's procedure can be extended to deal with exponential smoothing procedures that employ several smoothing constants. The interested reader is again referred to Johnson and Montgomery [8], which also discusses other adaptive control procedures.

PROBLEMS

1.   Consider the Bay City Seafood Company cod catch data analyzed in Example 3.3 and presented in Table 3.1. Verify that the smoothed statistic for March of

year 1 (period 3) is 351.39 as presented in Table 3.2. Use a smoothing constant of $\alpha = .02$.

2.   Consider the Bay City Seafood Company cod catch data analyzed in Example 3.3 and presented in Table 3.1. Verify that the smoothed statistic for April of year 1 (period 4) is 350.30 as presented in Table 3.2. Use a smoothing constant of $\alpha = .02$.

3.   Consider the data presented in Table 3.3. Verify that the cod catch forecast made in February of year 3 (period 26) for any future period is 351 (rounded).

4.   Consider the data presented in Table 3.3. Verify that the cod catch forecast made in March of year 3 (period 27) for any future period is 350 (rounded).

5.   Simulate one-period-ahead forecasts for the historical cod catch data presented in Table 3.1 using a smoothing constant of $\alpha = .10$. Verify that the sum of squared forecast errors obtained using this smoothing constant is 28,792.

CHAPTER 4

Double Exponential Smoothing

4–1 INTRODUCTION

In this chapter we discuss the forecasting of time series with a *linear trend*. Both the regression and exponential smoothing approaches can be used to forecast such time series. In Section 4–2.1 we present the regression approach. In Section 4–2.2 we present the exponential smoothing approach that is called double exponential smoothing.

Point forecasts and confidence interval forecasts can be obtained using either the regression approach or the exponential smoothing approach. In Section 4–3 we will discuss how confidence interval forecasts can be found when single and double exponential smoothing are used as forecasting techniques. The confidence interval forecasts obtained through exponential smoothing are particularly advantageous because they can easily be updated in light of new observations of the time series.

4–2 FORECASTING TIME SERIES WITH A LINEAR TREND

We next consider a situation in which the average level of the time series changes over time. Specifically, we assume that the average level changes over time in a linear fashion. Thus an appropriate model for the time series might be

$$y_t = \beta_0 + \beta_1 t + \epsilon_t$$

In this model the average level of the time series changes in a linear fashion over time since the expression $\beta_0 + \beta_1 t$ indicates a straight line relationship between the average level of the time series and time. The slope of this relationship is β_1 while the intercept at time 0 is β_0. The time series can then be described by the trend implied by this straight line combined with random fluctuations which cause observations of the time series to deviate from the trend line. If the slope β_1 of the trend line is greater than 0, this implies that the average level of the time series increases as time advances, whereas if the slope β_1 is less than 0, this implies that the average level of the time series decreases as time advances.

Example 4.1   Smith's Department Stores, Inc. owns and operates three stores in Central City. This small chain of stores, often referred to as ''Smith's,'' has sold personal electronic calculators since the early 1970s. For the last two years Smith's has carried a relatively new type of electronic calculator called the Bismark X-12. Sales of the Bismark X-12 have been generally increasing over these two years. Periodically, Smith's orders a supply of the Bismark X-12 from Bismark Electronics Company and then distributes the supply to its three stores. Smith's uses an inventory policy that attempts to insure that Smith's stores will have enough Bismark X-12 calculators to meet practically all of the demand for the Bismark X-12, while at the same time insuring that Smith's does not needlessly tie up its money by ordering many more calculators than can be reasonably expected to be sold. In order to implement its inventory policy, Smith's must have both a point forecast and a confidence interval forecast of total monthly demand for the Bismark X-12. The point forecast is the best guess at monthly demand for the Bismark X-12, while the confidence interval forecast gives the smallest and largest values that monthly demand for the Bismark X-12 can reasonably be expected to be. Smith's has recorded monthly sales of the Bismark X-12 for the previous two years, which we will call year 1 and year 2. The calculator sales history is given in Table 4.1.

It is of course true that ''sales'' and ''demand'' are not the same. If Smith's has a Bismark X-12 electronic calculator demanded but does not have any such calculators in stock, and if the customer who desires to buy the calculator will not wait for Smith's to order the calculator and instead goes elsewhere to buy the calculator, then the demand for the calculator does not result in a sale. However, since Smith's has a policy of having enough calculators in stock to insure that it very seldom loses a potential sale, we will in this example consider sales and demand to be the same thing.

TABLE 4.1   *Calculator Sales*

	Year 1	Year 2
Jan.	197	296
Feb.	211	276
Mar.	203	305
Apr.	247	308
May	239	356
June	269	393
July	308	363
Aug.	262	386
Sept.	258	443
Oct.	256	308
Nov.	261	358
Dec.	288	384

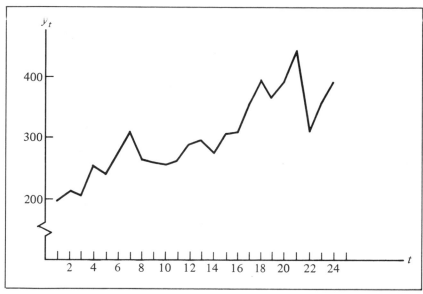

FIGURE 4.1   *Calculator Sales*

If the data in Table 4.1 is plotted it appears to randomly fluctuate around an average level that increases over time in a linear or straight-line fashion, as indicated in Figure 4.1. Since Smith's subjectively believes that this straight-line trend indicating generally increasing sales will continue for at least the next two years, it seems reasonable to use the model

$$y_t = \beta_0 + \beta_1 t + \epsilon_t$$

to forecast monthly calculator sales in the next two years.

4–2.1   THE REGRESSION APPROACH

The least squares estimates of β_1 and β_0 are, respectively,

$$b_1 = \frac{n \sum_{t=1}^{n} t y_t - \left(\sum_{t=1}^{n} t\right)\left(\sum_{t=1}^{n} y_t\right)}{n \sum_{t=1}^{n} t^2 - \left(\sum_{t=1}^{n} t\right)^2}$$

and

$$b_0 = \frac{\sum_{t=1}^{n} y_t}{n} - b_1 \left(\frac{\sum_{t=1}^{n} t}{n}\right)$$

where $y_1, y_2, \ldots, y_n$ are the n observations that have been previously collected. Hence, the forecast of a future value y_t of the time series is

$$\hat{y}_t = b_0 + b_1 t$$

Example Using the calculator sales data in Table 4.1 the least squares estimates of β_1
4.2 and β_0 are, respectively,

$$b_1 = \frac{24 \sum\limits_{t=1}^{24} t y_t - \left(\sum\limits_{t=1}^{24} t \right) \left(\sum\limits_{t=1}^{24} y_t \right)}{24 \sum\limits_{t=1}^{24} t^2 - \left(\sum\limits_{t=1}^{24} t \right)^2} = 8.07$$

and

$$b_0 = \frac{\sum\limits_{t=1}^{24} y_t}{24} - 8.07 \left(\frac{\sum\limits_{t=1}^{24} t}{24} \right) = 198.03$$

It can be shown that $R^2 = .77$. Hence, the forecast of the future value y_t of the time series is

$$\hat{y}_t = b_0 + b_1 t = 198.03 + 8.07t$$

For example, a prediction of calculator sales in January of year 3 is

$$\hat{y}_{25} = 198.03 + 8.07(25) = 399.89$$

It can also be shown that a 95 percent confidence interval for calculator sales in January of year 3 is

$$[399.89 - 71.28, 399.89 + 71.28] \quad \text{or} \quad [328.61, 471.17]$$

As another example, a prediction of calculator sales in February of year 3 is

$$\hat{y}_{26} = 198.03 + 8.07(26) = 407.96$$

It can be shown that a 95 percent confidence interval for calculator sales in February of year 3 is

$$[407.96 - 71.96, 407.96 + 71.96] \quad \text{or} \quad [336.00, 479.92]$$

*4–2.1   THE REGRESSION CALCULATIONS

Using the data in Table 4.1 we see that if we wish to find the least squares estimates of β_0 and β_1 in the regression model

$$y_t = \beta_0 + \beta_1 t + \epsilon_t$$

then we use

$$\mathbf{y} = \begin{bmatrix} 197 \\ 211 \\ 203 \\ . \\ . \\ . \\ 384 \end{bmatrix} \qquad \mathbf{X} = \begin{bmatrix} 1 & 1 \\ 1 & 2 \\ 1 & 3 \\ . & . \\ . & . \\ . & . \\ 1 & 24 \end{bmatrix}$$

Thus, the least squares estimates of β_0 and β_1 are

$$\begin{bmatrix} b_0 \\ b_1 \end{bmatrix} = (\mathbf{X'X})^{-1}\mathbf{X'y} = \begin{bmatrix} 198.03 \\ 8.07 \end{bmatrix}$$

Moreover, a 95 percent confidence interval for y_t is

$$[\hat{y}_t - E_t(95), \ \hat{y}_t + E_t(95)]$$

where

$$\hat{y}_t = b_0 + b_1 t = 198.03 + 8.07t = \begin{cases} 399.89 & \text{if } t = 25 \\ 407.96 & \text{if } t = 26 \end{cases}$$

$$s = \left(\frac{\sum\limits_{t=1}^{24} (y_t - \hat{y}_t)^2}{24 - 2} \right)^{1/2} = 31.67$$

$$t_{5/2}(24 - 2) = t_{5/2}(22) = 2.074$$

$$\mathbf{x}_t' = [1 \ t] = \begin{cases} [1, 25] & \text{if } t = 25 \\ [1, 26] & \text{if } t = 26 \end{cases}$$

$$f_t = (1 + \mathbf{x}_t' (\mathbf{X'X})^{-1}\mathbf{x}_t)^{1/2} = \begin{cases} 1.085 & \text{if } t = 25 \\ 1.096 & \text{if } t = 26 \end{cases}$$

$$E_t(95) = t_{5/2}(22) \cdot s \cdot f_t = \begin{cases} 71.28 & \text{if } t = 25 \\ 71.96 & \text{if } t = 26 \end{cases}$$

Before concluding this starred subsection it should be remarked that an alternative formula for f_t that does not make use of matrix algebra is

$$f_t = \left(1 + \frac{1}{n} + \frac{(t - \bar{x})^2}{\sum\limits_{i=1}^{n} (i - \bar{x})^2} \right)^{1/2}, \qquad \text{where } \bar{x} = \frac{\left(\sum\limits_{i=1}^{n} i \right)}{n}$$

Here it should be noted that n is the number of observations that have been collected for the time series, while t is the time index corresponding to the period for which the forecast is being made.

4–2.2   THE EXPONENTIAL SMOOTHING APPROACH

To begin the exponential smoothing approach, let us now suppose that at the end of time period $T - 1$ we have available a set of observations of the time series, which we shall denote $y_1, y_2, \ldots, y_{T-1}$. Given these observations we wish to estimate the parameters β_0 and β_1 and use these estimates to generate forecasts for future values of the time series. The least squares estimates of β_1 and β_0, which we now denote $b_1 (T - 1)$ and $b_0 (T - 1)$ to emphasize the fact that the most recent observation in the time series corresponds to time period $T - 1$, are

$$b_1(T - 1) = \frac{(T - 1) \sum\limits_{t=1}^{T-1} ty_t - \left(\sum\limits_{t=1}^{T-1} t \right) \left(\sum\limits_{t=1}^{T-1} y_t \right)}{(T - 1) \sum\limits_{t=1}^{T-1} t^2 - \left(\sum\limits_{t=1}^{T-1} t \right)^2}$$

and

$$b_0(T - 1) = \frac{\sum\limits_{t=1}^{T-1} y_t}{T - 1} - b_1(T - 1) \left(\frac{\sum\limits_{t=1}^{T-1} t}{T - 1} \right)$$

Given these estimates, the forecast made in period $T - 1$ for any future time period, say period $(T - 1) + \tau$ where τ is a positive integer, is

$$\hat{y}_{T-1+\tau}(T - 1) = b_0(T - 1) + b_1(T - 1) (T - 1 + \tau)$$

Suppose next that at the end of period T we obtain a new observation y_T. We would like to incorporate this new observation into the estimates of β_0 and β_1. That is, we wish to obtain updated estimates of β_0 and β_1 that are based on the new observation y_T as well as the old observations y_1, y_2, . . . , y_{T-1}. These updated estimates, which we denote $b_0(T)$ and $b_1(T)$, can be found by computing least squares regression estimates using the observations y_1, y_2, . . . , y_T. This would be done by using the formulas previously given with $T - 1$ replaced by T.

Another approach often used to determine updated estimates of β_0 and β_1 is known as *double exponential smoothing*. This approach gives an updated estimate for β_1 at the end of period T, which we will denote $b_1(T)$ and which is given by the equation

$$b_1(T) = \frac{\alpha}{1 - \alpha}(S_T - S_T^{[2]})$$

Here S_T is the *single smoothed estimate* or *single smoothed statistic*, which is found using the familiar smoothing equation $S_T = \alpha y_T + (1 - \alpha)S_{T-1}$. The expression $S_T^{[2]}$ is known as the *double smoothed statistic*. It is found by applying the smoothing operation to the output of the single smoothing equation. That is, the series of $S_T^{[2]}$ values is obtained by smoothing the series of S_T values using the double smoothing equation

$$S_T^{[2]} = \alpha S_T + (1 - \alpha)S_{T-1}^{[2]}$$

Note again that $S_T^{[2]}$ is not the square of the single smoothed statistic, but is rather the double smoothed statistic. The constant α, which is between 0 and 1, is again known as the *smoothing constant*. The double smoothing procedure gives an updated estimate for β_0 at the end of period T, which we will denote $b_0(T)$ and which is given by the equation

$$b_0(T) = 2S_T - S_T^{[2]} - Tb_1(T)$$

$$= 2S_T - S_T^{[2]} - T\left[\frac{\alpha}{1 - \alpha}(S_T - S_T^{[2]})\right]$$

The derivations of the equations for the updated estimates $b_0(T)$ and $b_1(T)$ are not at all intuitive and will be omitted here. The use of these equations will be illustrated in the example at the end of this section.

Let us now suppose that we have available data for all periods up to and including period T. We wish to make a forecast for time period $T + \tau$. The forecast is given by the equation

$$\hat{y}_{T+\tau}(T) = b_0(T) + b_1(T)(T + \tau)$$

$$= [b_0(T) + b_1(T)T] + b_1(T)\tau$$

$$= a_0(T) + b_1(T)\tau$$

Here

$$a_0(T) = b_0(T) + b_1(T)T$$

The expression $b_0(T) + b_1(T)T$ is simply the estimate of the y-intercept of the trend line with origin considered at time zero plus the estimate of the slope of the trend line multiplied by T, the estimates having been calculated at time T. Thus $a_0(T)$ is the y-intercept of the updated trend line when the time origin is considered to be at time T. This shift of the origin allows us to express the trend line on a "current origin" basis. That is, the shift of the origin allows us to express the trend line with the origin considered to be at time T, the current time period. This shift of the origin is illustrated in Figure 4.2. Such shifts of the origin of time are common practice in forecasting.

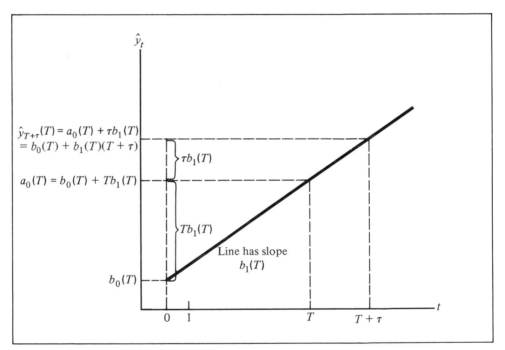

FIGURE 4.2   *The Shift of Origin*

Using the equation for $b_0(T)$ we have

$$
\begin{aligned}
a_0(T) &= b_0(T) + b_1(T)T \\
&= [2S_T - S_T^{[2]} - Tb_1(T)] + b_1(T)T \\
&= 2S_T - S_T^{[2]}
\end{aligned}
$$

then, using the equation for $b_1(T)$, the equation that yields the forecast for period $T + \tau$ becomes

$$\hat{y}_{T+\tau}(T) = a_0(T) + b_1(T)\tau$$

$$= 2S_T - S_T^{[2]} + \frac{\alpha}{1 - \alpha}(S_T - S_T^{[2]})\tau$$

$$= \left(2 + \frac{\alpha\tau}{(1 - \alpha)}\right) S_T - \left(1 + \frac{\alpha\tau}{(1 - \alpha)}\right) S_T^{[2]}$$

In order to begin the double exponential smoothing procedure we must have initial values for S_0 and $S_0^{[2]}$. Because these smoothed statistics are not at all intuitive, the direct assignment of values to these quantities is usually not possible. Initial values of S_0 and $S_0^{[2]}$ can be found by applying regression analysis to historical data and obtaining estimates of the coefficients β_0 and β_1. If such data does not exist, estimates of β_0 and β_1 must be obtained subjectively. Let us denote these initial estimates of β_0 and β_1 by $b_0(0)$ and $b_1(0)$. We know that for any time period T,

$$b_1(T) = \frac{\alpha}{1 - \alpha}(S_T - S_T^{[2]})$$

Thus for time period $T = 0$ we have

$$b_1(0) = \frac{\alpha}{1 - \alpha}(S_0 - S_0^{[2]})$$

We also know that for any time period T,

$$b_0(T) = 2S_T - S_T^{[2]} - T\frac{\alpha}{1 - \alpha}(S_T - S_T^{[2]})$$

Thus for time period $T = 0$ we have

$$b_0(0) = 2S_0 - S_0^{[2]}$$

So at time period $T = 0$ the equations

$$b_1(0) = \frac{\alpha}{1 - \alpha}(S_0 - S_0^{[2]}) \qquad \text{and} \qquad b_0(0) = 2S_0 - S_0^{[2]}$$

may be solved for S_0 and $S_0^{[2]}$ in terms of the initial estimates $b_0(0)$ and $b_1(0)$:

$$S_0 = b_0(0) - \left(\frac{1 - \alpha}{\alpha}\right) b_1(0)$$

$$S_0^{[2]} = b_0(0) - 2\left(\frac{1 - \alpha}{\alpha}\right) b_1(0)$$

The solution of these equations produces the initial values of the smoothed statistics, S_0 and $S_0^{[2]}$, needed to begin the smoothing procedure. If initial estimates of β_0 and β_1 are not available, these equations obviously cannot be used to determine S_0 and $S_0^{[2]}$. In such cases, it is common practice to assign a value equal to the first observation in the time series to both S_0 and $S_0^{[2]}$.

Given the initial estimates of S_0 and $S_0^{[2]}$ we can then perform the smoothing operation in each period using the smoothing equations

$$S_T = \alpha y_T + (1 - \alpha)S_{T-1} \quad \text{and} \quad S_T^{[2]} = \alpha S_T + (1 - \alpha)S_{T-1}^{[2]}$$

We can then make forecasts for any period $T + \tau$ using the forecasting equation

$$\hat{y}_{T+\tau}(T) = \left(2 + \frac{\alpha\tau}{(1 - \alpha)}\right) S_T - \left(1 + \frac{\alpha\tau}{(1 - \alpha)}\right) S_T^{[2]}$$

Simulation of an historical time series may be used to determine the "best" smoothing constant to be used in a double smoothing procedure. As in the case of single smoothing, a set of forecasts is generated for each of a set of values of α. These forecasts are then compared to the actual observations in the historical time series. The value of the smoothing constant minimizing the sum of squared forecast errors is the value of α to be used in the smoothing procedure. These simulations are initialized with starting values for the single and double smoothed statistics. Again, the first few observations in the time series, say the first $n_1 = 6$ observations, are used to obtain initial estimates of β_0 and β_1 via regression analysis. These initial estimates $b_0(0)$ and $b_1(0)$ can then be used to find initial values of the smoothed statistics S_0 and $S_0^{[2]}$ through use of the equations

$$S_0 = b_0(0) - \left(\frac{1 - \alpha}{\alpha}\right) b_1(0) \quad \text{and} \quad S_0^{[2]} = b_0(0) - 2\left(\frac{1 - \alpha}{\alpha}\right) b_1(0)$$

These equations will yield initial values of S_0 and $S_0^{[2]}$ for any value of α that might be considered, given the least squares estimates $b_0(0)$ and $b_1(0)$. Once the initial values of the smoothed statistics have been determined, the smoothing process can be begun and the simulation of historical data performed.

Example 4.3   We now consider double exponential smoothing as a forecasting tool to be used in forecasting the calculator sales time series of Example 4.1. The first step in implementing a double smoothing procedure to forecast future values of this time series is to choose a smoothing constant value. This choice will be made by simulating the historical time series in Table 4.1 using different values of α. Values

TABLE 4.2   *The First $n_1 = 6$ Observations*

Time	Calculator Sales
1	197
2	211
3	203
4	247
5	239
6	269

of α from .02 to .30 in increments of .02 will be used. The value of α that minimizes the sum of squared forecast errors will be chosen for actual forecasting. In order to initialize these simulations, the parameters β_0 and β_1 in the model

$$y_t = \beta_0 + \beta_1 t + \epsilon_t$$

must be estimated using the first $n_1 = 6$ observations of the time series (given in Table 4.2). Using regression analysis on these observations, we find that the least squares estimates of β_0 and β_1 are $b_0(0) = 178.87$ and $b_1(0) = 13.94$. These estimates may be used to find initial values for the smoothed statistics S_T and $S_T^{[2]}$. These initial values S_0 and $S_0^{[2]}$ will in turn be used to begin the simulations.

We illustrate the calculations involved in these simulations by considering the case in which $\alpha = .10$. We find the initial values S_0 and $S_0^{[2]}$ as follows.

$$S_0 = b_0(0) - \left(\frac{1-\alpha}{\alpha}\right) b_1(0)$$

$$= 178.87 - \left(\frac{1-.10}{.10}\right) 13.94$$

$$= 53.41.$$

$$S_0^{[2]} = b_0(0) - 2\left(\frac{1-\alpha}{\alpha}\right) b_1(0)$$

$$= 178.87 - 2\left(\frac{1-.10}{.10}\right) 13.94$$

$$= -72.05$$

Given these initial values we now generate a forecast for January of year 1 using the forecasting equation

$$\hat{y}_{T+\tau}(T) = \left(2 + \frac{\alpha\tau}{(1-\alpha)}\right) S_T - \left(1 + \frac{\alpha\tau}{(1-\alpha)}\right) S_T^{[2]}$$

$$\hat{y}_{0+1}(0) = \left(2 + \frac{\alpha(1)}{(1 - \alpha)}\right) S_0 - \left(1 + \frac{\alpha(1)}{(1 - \alpha)}\right) S_0^{[2]}$$

$$\hat{y}_1(0) = \left(2 + \frac{.10}{1 - .10}\right)(53.41) - \left(1 + \frac{.10}{1 - .10}\right)(-72.05)$$

$$= 193 \quad \text{(rounded)}$$

So the forecast for January of year 1 (period 1) made in period 0 is 193. In January of year 1 a sales figure of 197 was observed. This sales figure is now used to obtain new values of the smoothed statistics. We do this using the smoothing equations as follows.

$$S_1 = \alpha y_1 + (1 - \alpha)S_0$$
$$= .10(197) + (1 - .10)(53.41)$$
$$= 67.77$$
$$S_1^{[2]} = \alpha S_1 + (1 - \alpha)S_0^{[2]}$$
$$= .10(67.77) + (1 - .10)(-72.05)$$
$$= -58.07$$

We next obtain a forecast for February of year 1 (period 2) by using the forecasting equation

$$\hat{y}_{1+1}(1) = \left(2 + \frac{\alpha(1)}{(1 - \alpha)}\right) S_1 - \left(1 + \frac{\alpha(1)}{(1 - \alpha)}\right) S_1^{[2]}$$

$$\hat{y}_2(1) = \left(2 + \frac{.10}{1 - .10}\right)(67.77) - \left(1 + \frac{.10}{1 - .10}\right)(-58.07)$$

$$= 208 \quad \text{(rounded)}$$

The forecast for February of year 1 (period 2) made in period 1 is 208. In February of year 1 a sales figure of 211 was observed. In the same manner, we obtain a forecast for the next period, March of year 1:

$$S_2 = \alpha y_2 + (1 - \alpha)S_1$$
$$= .10(211) + (1 - .10)(67.77)$$
$$= 82.09$$
$$S_2^{[2]} = \alpha S_2 + (1 - \alpha)S_1^{[2]}$$
$$= .10(82.09) + (1 - .10)(-58.07)$$
$$= -44.05$$

and

$$\hat{y}_{2+1}(2) = \left(2 + \frac{\alpha(1)}{(1-\alpha)}\right) S_2 - \left(1 + \frac{\alpha(1)}{(1-\alpha)}\right) S_2^{[2]}$$

$$\hat{y}_3(2) = \left(2 + \frac{.10}{1-.10}\right)(82.09) - \left(1 + \frac{.10}{1-.10}\right)(-44.05)$$

$$= 222 \quad \text{(rounded)}$$

The forecast for March of year 1 (period 3) made in period 2 is 222. Continuing in this manner, we generate forecasts for the balance of the periods in our two years of historical data. For each period the forecast error and squared forecast error are calculated. Finally, the sum of squared forecast errors is calculated. These results are summarized in Table 4.3. The sum of squared forecast errors obtained for $\alpha = .10$ is 34,628.

TABLE 4.3   *Simulating One-Period-Ahead Forecasting of the Historical Calculator Sales Time Series Using Double Exponential Smoothing with $\alpha = .10$*

Year	Month	Calculator Sales y_T	S_T	$S_T^{[2]}$	Forecast	Forecast Made Last Period	Forecast Error	Squared Forecast Error
1	Jan.	197	67.77	−58.07	208	193	4	16
	Feb.	211	82.09	−44.05	222	208	3	9
	Mar.	203	94.18	−30.23	232	222	−19	361
	Apr.	247	109.46	−16.26	249	232	15	225
	May	239	122.42	−2.39	261	249	−10	100
	June	269	137.08	11.56	277	261	8	64
	July	308	154.17	25.82	297	277	31	961
	Aug.	262	164.95	39.73	304	297	−35	1,225
	Sept.	258	174.26	53.18	309	304	−46	2,116
	Oct.	256	182.43	66.11	312	309	−53	2,809
	Nov.	261	190.29	78.53	314	312	−51	2,601
	Dec.	288	200.06	90.68	322	314	−26	676
2	Jan.	296	209.65	102.58	329	322	−26	676
	Feb.	276	216.29	113.95	330	329	−53	2,809
	Mar.	305	225.16	125.07	336	330	−25	625
	Apr.	308	233.44	135.91	342	336	−28	784
	May	356	245.70	146.89	355	342	14	196
	June	393	260.43	158.24	374	355	38	1,444
	July	363	270.69	169.48	383	374	−11	121
	Aug.	386	282.22	180.76	395	383	3	9
	Sept.	443	298.30	192.51	416	395	48	2,304
	Oct.	308	299.27	203.19	406	416	−108	11,664
	Nov.	358	305.14	213.38	407	406	−48	2,304
	Dec.	384	313.02	223.35	413	407	−23	529

Sum of squared errors = 34,628

TABLE 4.4   *The Sums of Squared Forecast Errors for Different Values of α*

Smoothing Constant α	Sum of Squared Errors
.02	85,351
.04	59,946
.06	46,073
.08	38,440
.10	34,628
.12	32,518
.14	31,338
.16	30,725
.18	30,788
.20	30,822
.22	31,013
.24	31,258
.26	31,527
.28	31,609
.30	32,157

In a similar manner, a set of forecasts is simulated for each value of the smoothing constant α we wish to consider and the sum of squared forecast errors is calculated for each. Note that the same initial estimates $b_0(0) = 178.87$ and $b_1(0) = 13.94$ are used to begin each simulation, although the initial values of the smoothed statistics S_0 and $S_0^{[2]}$ will differ in each case since they depend on the value of α. The sums of squared errors obtained for all simulations are presented in Table 4.4. A smoothing constant of $\alpha = .16$ minimizes the sum of squared errors, and hence this value is chosen for use in actual forecasting.

In order to begin the actual forecasting procedure we must generate new estimates of β_0 and β_1 to be used to calculate the initial values of S_T and $S_T^{[2]}$. These initial estimates, $b_0(0)$ and $b_1(0)$, will now be based on the entire two-year sales history we have available and will be derived using regression. That is, we choose $n_2 = n = 24$. When regression analysis is applied to the sales history presented in Table 4.1, the least squares estimates of β_0 and β_1 are found to be $b_0(0) = 198.03$ and $b_1(0) = 8.07$. Next, these estimates will be used to obtain initial values of the smoothed statistics S_0 and $S_0^{[2]}$:

$$S_0 = b_0(0) - \left(\frac{1-\alpha}{\alpha}\right) b_1(0)$$

$$= 198.03 - \left(\frac{1-.16}{.16}\right)(8.07)$$

$$= 155.66$$

$$S_0^{[2]} = b_0(0) - 2\left(\frac{1-\alpha}{\alpha}\right) b_1(0)$$

$$= 198.03 - 2\left(\frac{1-.16}{.16}\right)(8.07)$$

$$= 113.295$$

Note that the "best" smoothing constant $\alpha = .16$ has been used in performing these calculations. These smoothed statistics will now be repeatedly updated using the observations in the two-year sales history. This process will insure that the forecasts we generate for future values of the time series will weight the most recent observations in the time series most heavily. The updating process begins with the observation from January of year 1 (period 1), which is $y_1 = 197$. We calculate updated smoothed statistics as follows.

$$S_1 = \alpha y_1 + (1 - \alpha)S_0$$

$$= .16(197) + (1 - .16)(155.66)$$

$$= 162.28$$

$$S_1^{[2]} = \alpha S_1 + (1 - \alpha)S_0^{[2]}$$

$$= .16(162.28) + (1 - .16)(113.295)$$

$$= 121.13$$

Using the observation for February of year 1 (period 2), which is $y_2 = 211$, we again calculate a pair of updated smoothed statistics.

$$S_2 = \alpha y_2 + (1 - \alpha)S_1$$

$$= .16(211) + (1 - .16)(162.28)$$

$$= 170.07$$

$$S_2^{[2]} = \alpha S_2 + (1 - \alpha)S_1^{[2]}$$

$$= .16(170.07) + (1 - .16)(121.13)$$

$$= 128.96$$

We continue calculating updated smoothed statistics for all other periods in the sales history in exactly the same way. The results of these calculations are shown in Table 4.5. So at the end of December of year 2 (period 24) we have updated single and double smoothed statistics of $S_{24} = 348.03$ and $S_{24}^{[2]} = 308.76$.

These quantities may now be used to generate forecasts of future values of the time series. For example, we can use the forecasting equation to obtain a forecast for January of year 3 (period 25) as follows.

TABLE 4.5   *Updating the Smoothed Statistics by Using Double Exponential Smoothing with the Optimal Smoothing Constant $\alpha = .16$ on the Historical Calculator Sales Time Series*

Year	Month	Time Period T	Calculator Sales y_T	Single Smoothed Statistic S_T	Double Smoothed Statistic $S_T^{[2]}$
1	Jan.	1	197	162.28	121.13
	Feb.	2	211	170.07	128.96
	Mar.	3	203	175.34	136.38
	Apr.	4	247	186.81	144.45
	May	5	239	195.16	152.56
	June	6	269	206.97	161.27
	July	7	308	223.14	171.17
	Aug.	8	262	229.35	180.48
	Sept.	9	258	233.94	189.03
	Oct.	10	256	237.47	196.78
	Nov.	11	261	241.23	203.89
	Dec.	12	288	248.72	211.06
2	Jan.	13	296	256.28	218.30
	Feb.	14	276	259.44	224.88
	Mar.	15	305	266.73	231.58
	Apr.	16	308	273.33	238.26
	May	17	356	286.56	245.98
	June	18	393	303.59	255.20
	July	19	363	313.09	264.46
	Aug.	20	386	324.76	274.11
	Sept.	21	443	343.68	285.24
	Oct.	22	308	337.97	293.68
	Nov.	23	358	341.17	301.28
	Dec.	24	384	348.03	308.76

$$\hat{y}_{T+\tau}(T) = \left(2 + \frac{\alpha\tau}{(1-\alpha)}\right) S_T - \left(1 + \frac{\alpha\tau}{(1-\alpha)}\right) S_T^{[2]}$$

$$\hat{y}_{24+1}(24) = \left(2 + \frac{\alpha(1)}{(1-\alpha)}\right) S_{24} - \left(1 + \frac{\alpha(1)}{(1-\alpha)}\right) S_{24}^{[2]}$$

$$\hat{y}_{25}(24) = \left(2 + \frac{.16}{(1-.16)}\right)(348.03) - \left(1 + \frac{.16}{(1-.16)}\right)(308.76)$$

$$= 395 \quad \text{(rounded)}$$

Thus the forecast for January of year 3 (period 25) made in December of year 2 (period 24) is 395. For comparison purposes it should be noticed that the regression approach of Section 4–2.1 made a forecast in December of year 2 (period 24) for January of year 3 (period 25) of 399.89 (or 400 rounded).

Now suppose that in January of year 3 (period 25) we observe a sales figure y_{25} = 358. In order to make a forecast for February of year 3 (period 26), we must first update the single and double smoothed statistics using the smoothing equations. We obtain

$$S_{25} = \alpha y_{25} + (1 - \alpha) S_{24}$$

$$= .16(358) + (1 - .16)(348.03)$$

$$= 349.62$$

$$S_{25}^{[2]} = \alpha S_{25} + (1 - \alpha) S_{24}^{[2]}$$

$$= .16(349.62) + (1 - .16)(308.76)$$

$$= 315.29$$

TABLE 4.6   *One-Period-Ahead Forecasts of Future Values of Calculator Sales Using Double Exponential Smoothing with $\alpha = .16$*

Year	Month	Period T	Calculator Sales y_T	S_T	$S_T^{[2]}$	Forecast Made Last Period	Forecast Error
3	Jan.	25	358	349.62	315.29	395	−37
	Feb.	26	325	345.68	320.16	390	−65
	Mar.	27	414	356.61	325.99	376	38
	Apr.	28	406	364.51	332.15	393	13
	May	29	425	374.19	338.88	403	22
	June	30	347	369.84	343.83	416	−69
	July	31	413	376.75	349.10	401	12
	Aug.	32	435	386.07	355.01	410	25
	Sept.	33	407	389.42	360.52	423	−16
	Oct.	34	438	397.19	366.38	424	14
	Nov.	35	479	410.28	373.41	434	45
	Dec.	36	373	404.31	378.35	454	−81
4	Jan.	37	467	414.34	384.11	435	32
	Feb.	38	500	428.05	391.14	450	50
	Mar.	39	535	445.16	399.78	472	63
	Apr.	40	525	457.93	409.09	499	26
	May	41	449	456.50	416.67	516	−67
	June	42	557	472.58	425.62	504	53
	July	43	543	483.85	434.94	528	15
	Aug.	44	433	475.71	441.46	542	−109
	Sept.	45	475	475.60	446.92	516	−41
	Oct.	46	592	494.22	454.49	510	82
	Nov.	47	548	502.83	462.22	542	6
	Dec.	48	520	505.57	469.16	551	−31

A forecast for February of year 3 (period 26) is now given by the forecasting equation

$$\hat{y}_{25+1}(25) = \left(2 + \frac{\alpha(1)}{(1-\alpha)}\right) S_{25} - \left(1 + \frac{\alpha(1)}{(1-\alpha)}\right) S_{25}^{[2]}$$

$$\hat{y}_{26}(25) = \left(2 + \frac{.16}{(1-.16)}\right)(349.62) - \left(1 + \frac{.16}{(1-.16)}\right)(315.29)$$

$$= 390 \text{ (rounded)}$$

So the forecast for February of year 3 (period 26) made in January of year 3 (period 25) is 390.

Forecasts for future months may be made in the same manner using newly observed values of the time series. Observed values of the time series for the next two years are given in Table 4.6. Also given in this table are updated values of the single and double smoothed statistics, one-period-ahead forecasts for each month, and the forecast error for each period. A plot of the actual sales and the forecasted sales is given in Figure 4.3.

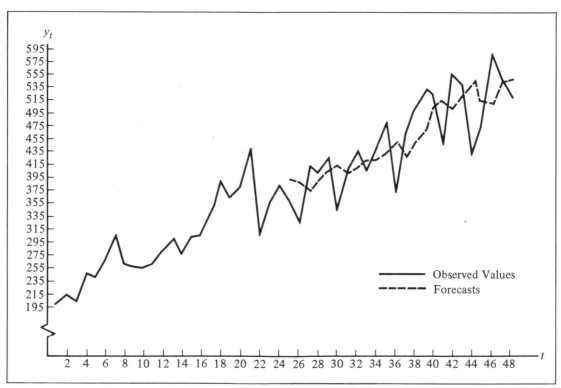

FIGURE 4.3   *One-Period-Ahead Forecasts of Future Values of Calculator Sales Using Double Exponential Smoothing with $\alpha = .16$*

4–3 GENERAL DIRECT SMOOTHING AND CONFIDENCE INTERVALS FOR SINGLE AND DOUBLE EXPONENTIAL SMOOTHING

Consider the following general time series regression model

$$y_t = \beta_0 + \beta_1 f_1(t) + \beta_2 f_2(t) + \cdots + \beta_p f_p(t) + \epsilon_t$$

where $f_1(t), f_2(t), \ldots, f_p(t)$ are mathematical functions of time and not functions of other random time series. This time series regression model is characterized by the fact that

$$\beta_0 + \beta_1 f_1(t) + \beta_2 f_2(t) + \cdots + \beta_p f_p(t)$$

is a *linear* function of the unknown parameters $\beta_0, \beta_1, \ldots, \beta_p$. Furthermore, there is a general method, called *general direct smoothing,* which updates the estimates of $\beta_0, \beta_1, \ldots, \beta_p$ in light of new observations obtained from time period to time period. We will not discuss the details of general direct smoothing in this book. The interested reader is referred to Brown [3] or Johnson and Montgomery [8]. It is important to know, however, that there is a theoretical, statistical basis behind this general method. This theoretical basis can be used to derive a $(100 - \alpha)$ percent confidence interval for a future observation, $y_{T+\tau}$, or a sum of future observations,

$$\sum_{t=1}^{\tau} y_{T+t}$$

In particular, the general updating method employs one smoothing constant. Assume that we are at time origin T and $b_0(T), b_1(T), \ldots, b_p(T)$ are the current estimates of $\beta_0, \beta_1, \ldots, \beta_p$. Then

$$\hat{y}_{T+\tau}(T) = b_0(T) + b_1(T)f_1(T + \tau) + b_2(T)f_2(T + \tau) + \cdots$$
$$+ b_p(T)f_p(T + \tau)$$

is the forecast of $y_{T+\tau}$. Moreover, let

$$\Delta(T) = \frac{\displaystyle\sum_{t=1}^{T} \left| y_t - \hat{y}_t(t - 1) \right|}{T}$$

and let $z_{\alpha/2}$ be the point on the scale of the normal curve having mean 0 and variance 1 such that there is an area of $(100 - \alpha)/100$ under this normal curve between $-z_{\alpha/2}$ and $z_{\alpha/2}$. Then, it can be shown that there exist constants d_τ and q_τ, which are independent of T, such that

1. the error $E_{T+\tau}^{(100-\alpha)}(T) = z_{\alpha/2} d_\tau \, \Delta(T)$ can be used to find an approximate $(100 - \alpha)$ percent confidence interval for $y_{T+\tau}$, which is given by

$$[\hat{y}_{T+\tau}(T) - E_{T+\tau}^{(100-\alpha)}(T), \, \hat{y}_{T+\tau}(T) + E_{T+\tau}^{(100-\alpha)}(T)]$$

2. the error $E_{[T+1,T+\tau]}^{(100-\alpha)}(T) = z_{\alpha/2} q_\tau \, \Delta(T)$ can be used to find an approximate $(100 - \alpha)$ percent confidence interval for $\Sigma_{t=1}^\tau y_{T+t}$, which is given by

$$\left[\sum_{t=1}^{\tau} \hat{y}_{T+t}(T) - E_{[T+1,T+\tau]}^{(100-\alpha)}(T), \, \sum_{t=1}^{\tau} \hat{y}_{T+t}(T) + E_{[T+1,T+\tau]}^{(100-\alpha)}(T)\right]$$

Note that these confidence intervals are calculated at time T.

In light of a new observation, y_{T+1}, $b_0(T)$, $b_1(T)$, . . . , $b_p(T)$ can be updated to $b_0(T + 1)$, $b_1(T + 1)$, . . . , $b_p(T + 1)$ using the general direct smoothing procedure. Also, $\Delta(T)$ can be updated to $\Delta(T + 1)$ using the formula

$$\Delta(T + 1) = \frac{T\Delta(T) + |y_{T+1} - \hat{y}_{T+1}(T)|}{T + 1}$$

Hence, letting

$$\hat{y}_{T+1+\tau}(T + 1) = b_0(T + 1) + b_1(T + 1)f_1(T + 1 + \tau)$$
$$+ b_2(T + 1)f_2(T + 1 + \tau) + \cdots + b_p(T + 1)f_p(T + 1 + \tau)$$

$$E_{T+1+\tau}^{(100-\alpha)}(T + 1) = z_{\alpha/2} d_\tau \, \Delta(T + 1)$$

and

$$E_{[T+2,T+1+\tau]}^{(100-\alpha)}(T + 1) = z_{\alpha/2} q_\tau \, \Delta(T + 1)$$

we can easily update the previous $(100 - \alpha)$ percent confidence intervals, which were made at time T for $y_{T+\tau}$ and $\Sigma_{t=1}^\tau y_{T+t}$. The updated $(100 - \alpha)$ percent confidence intervals for $y_{T+1+\tau}$ and $\Sigma_{t=1}^\tau y_{T+1+t}$ are, respectively,

$$[\hat{y}_{T+1+\tau}(T + 1) - E_{T+1+\tau}^{(100-\alpha)}(T + 1), \, \hat{y}_{T+1+\tau}(T + 1) + E_{T+1+\tau}^{(100-\alpha)}(T + 1)]$$

and

$$\left[\sum_{t=1}^{\tau} \hat{y}_{T+1+t}(T+1) - E^{(100-\alpha)}_{[T+2,\,T+1+\tau]}(T+1),\ \sum_{t=1}^{\tau} \hat{y}_{T+1+t}(T+1) + E^{(100-\alpha)}_{[T+2,\,T+1+\tau]}(T+1)\right]$$

Note that these confidence intervals are calculated at time $(T+1)$.

The constants d_τ and q_τ are functions of τ and the smoothing constant. The exact formulas for d_τ and q_τ depend upon the functions $f_1(t), f_2(t), \ldots, f_p(t)$. We will now discuss the values obtained for d_τ and q_τ when single and double exponential smoothing are being employed.

The model

$$y_t = \beta_0 + \epsilon_t$$

which is assumed to describe a time series that may be appropriately forecasted using simple exponential smoothing, is a special case of the general time series regression model

$$y_t = \beta_0 + \beta_1 f_1(t) + \beta_2 f_2(t) + \cdots + \beta_p f_p(t) + \epsilon_t$$

where $f_1(t) = f_2(t) = \cdots = f_p(t) = 0$.

If general direct smoothing is used to find an equation that successively updates the estimates of β_0, this equation is found to be identical to the updating equation given in Section 3–2.2, the section that discusses simple exponential smoothing. Then, using the statistical theory behind general direct smoothing, it can be shown that for simple exponential smoothing

$$d_\tau = 1.25$$

and

$$q_\tau = 1.25 \left[\frac{\alpha(\tau^2 - \tau) + 2\tau}{2}\right]^{1/2}$$

where α is the smoothing constant used in simple exponential smoothing.

The model

$$y_t = \beta_0 + \beta_1 t + \epsilon_t$$

which is assumed to describe a time series that may be appropriately forecasted using double exponential smoothing, is a special case of the general time series regression model

$$y_t = \beta_0 + \beta_1 f_1(t) + \beta_2 f_2(t) + \cdots + \beta_p f_p(t) + \epsilon_t$$

where $f_1(t) = t$ and $f_2(t) = \cdots = f_p(t) = 0$.

If general direct smoothing is used to find equations that successively update the estimates of β_0 and β_1, these equations are found to be identical to the updating equations given in Section 4–2.2, the section that discusses double exponential smoothing. Then, using the statistical theory behind general direct smoothing, it can be shown that for double exponential smoothing

$$d_\tau = 1.25 \left[\frac{1 + \dfrac{\alpha}{(1 + v)^3}\left[(1 + 4v + 5v^2) + 2\alpha(1 + 3v)\,\tau + 2\alpha^2\,\tau^2\right]}{1 + \dfrac{\alpha}{(1 + v)^3}\left[(1 + 4v + 5v^2) + 2\alpha(1 + 3v) + 2\alpha^2\right]} \right]^{1/2}$$

and

$$q_\tau = 1.25 \left[\frac{\tau + \dfrac{\alpha\tau^2}{2(1 + v)^3}\left[5(1 + 2v + v^2) + 4(1 - v^2)\tau + \alpha^2\tau^2\right]}{1 + \dfrac{\alpha}{(1 + v)^3}\left[(1 + 4v + 5v^2) + 2\alpha(1 + 3v) + 2\alpha^2\right]} \right]^{1/2}$$

where α is the smoothing constant employed in double exponential smoothing and $v = 1 - \alpha$.

Example 4.4   Let us reconsider Example 4.1. Assume that Smith's Department Stores, Inc. is considering purchasing the Bismark X-12 electronic calculator less frequently and in greater quantities when making a purchase in order to take advantage of quantity discounts offered by Bismark Electronics Company. Since such quantity purchasing will change Smith's inventory policy with respect to the Bismark X-12, Smith's not only needs a point forecast and a confidence interval forecast of next month's demand, but also needs point forecasts and confidence interval forecasts of demand in a single month three months in the future and of cumulative demand for the next three months.

Consider the two-year calculator sales history presented in Table 4.1. In Example 4.3 double exponential smoothing was used to forecast future values of this time series. We now consider calculating confidence intervals for future values of this time series. In particular, suppose that at time $T = 24$ we wish to find 95 percent confidence intervals for $y_{T+1} = y_{25}$, $y_{T+3} = y_{27}$, and $y_{25} + y_{26} + y_{27}$.

First, we note that if $\tau = 1$, then $d_\tau = 1.25$. Thus we have $d_1 = 1.25$. Second, recall that the optimal smoothing constant used in Example 4.3 was $\alpha = .16$. Thus $v = 1 - \alpha = 1 - .16 = .84$. Using $\tau = 3$, we now calculate d_3 and q_3 as follows.

$$d_\tau = 1.25 \left[\frac{1 + \dfrac{\alpha}{(1 + v)^3}[(1 + 4v + 5v^2) + 2\alpha(1 + 3v)\tau + 2\alpha^2\tau^2]}{1 + \dfrac{\alpha}{(1 + v)^3}[(1 + 4v + 5v^2) + 2\alpha(1 + 3v) + 2\alpha^2]} \right]^{1/2}$$

$$d_3 = 1.25 \left[\frac{1 + \dfrac{.16}{(1 + .84)^3}[(1 + 4(.84) + 5(.84)^2) + 2(.16)(1 + 3(.84))(3) + 2(.16)^2(3)^2]}{1 + \dfrac{.16}{(1 + .84)^3}[(1 + 4(.84) + 5(.84)^2) + 2(.16)(1 + 3(.84)) + 2(.16)^2]} \right]^{1/2}$$

$$= 1.2841986$$

$$q_\tau = 1.25 \left[\frac{\tau + \dfrac{\alpha\tau^2}{2(1 + v)^3}[5(1 + 2v + v^2) + 4(1 - v^2)\tau + \alpha^2\tau^2]}{1 + \dfrac{\alpha}{(1 + v)^3}[(1 + 4v + 5v^2) + 2\alpha(1 + 3v) + 2\alpha^2]} \right]^{1/2}$$

$$q_3 = 1.25 \left[\frac{3 + \dfrac{.16(3)^2}{2(1 + .84)^3}[5(1 + 2(.84) + (.84)^2) + 4(1 - (.84)^2)(3) + (.16)^2(3)^2]}{1 + \dfrac{.16}{(1 + .84)^3}[(1 + 4(.84) + 5(.84)^2) + 2(.16)(1 + 3(.84)) + 2(.16)^2]} \right]^{1/2}$$

$$= 2.6140252$$

Since 1.96 is the point on the scale of the normal curve having mean 0 and variance 1 such that there is an area of .95 under this normal curve between -1.96 and 1.96, we have $z_{5/2} = 1.96$. Thus the errors in 95 percent confidence intervals for y_{T+1}, y_{T+3}, and $y_{T+1} + y_{T+2} + y_{T+3}$ are, respectively,

$$E_{T+1}^{(95)}(T) = z_{5/2}d_1 \, \Delta(T) = 1.96(1.25) \, \Delta(T) = 2.45 \, \Delta(T)$$

$$E_{T+3}^{(95)}(T) = z_{5/2}d_3 \, \Delta(T) = 1.96(1.2841986) \, \Delta(T) = 2.5170292 \, \Delta(T)$$

$$E_{[T+1,T+3]}^{(95)}(T) = z_{5/2}q_3 \, \Delta(T) = 1.96(2.6140252) \, \Delta(T) = 5.1234893 \, \Delta(T)$$

Now, in order to calculate these errors we must determine $\Delta(24)$. We do this by considering the one-period-ahead forecasts generated by the double smoothing procedure in periods 1 through 24. Recall from Example 4.3 that when $\alpha = .16$ the initial values of the single and double smoothed statistics were $S_0 = 155.66$ and $S_0^{[2]} = 113.295$, and that the updated smoothed statistics in period 1 were $S_1 = 162.28$ and $S_1^{[2]} = 121.13$. Thus the one-period-ahead forecasts for y_1 and y_2 respectively are

$$\hat{y}_{T+\tau}(T) = \left(2 + \frac{\alpha\tau}{(1 - \alpha)} \right) S_T - \left(1 + \frac{\alpha\tau}{(1 - \alpha)} \right) S_T^{[2]}$$

$$\hat{y}_1(0) = \left(2 + \frac{.16(1)}{1 - .16}\right)(155.66) - \left(1 + \frac{.16(1)}{1 - .16}\right)(113.295)$$

$$= 2.1904761(155.66) - 1.1904761(113.295)$$

$$= 206.10$$

$$\hat{y}_2(1) = \left(2 + \frac{.16(1)}{1 - .16}\right)(162.28) - \left(1 + \frac{.16(1)}{1 - .16}\right)(121.13)$$

$$= 2.1904761(162.28) - 1.1904761(121.13)$$

$$= 211.26$$

In a similar manner, one-period-ahead forecasts are calculated for the remaining periods in the two-year sales history. The one-period-ahead forecasts and the forecast errors for periods 1 through 24 are given in Table 4.7.

We now use the forecast errors in Table 4.7 to calculate $\Delta(24)$. We have

$$\Delta(24) = \frac{\sum_{t=1}^{24} |y_t - \hat{y}_t(t - 1)|}{24} = \frac{582.74}{24} = 24.28$$

Thus the errors in 95 percent confidence intervals for y_{25}, y_{27}, and $y_{25} + y_{26} + y_{27}$ are, respectively,

$$E_{25}^{(95)}(24) = 2.45 \, \Delta(24) = 2.45(24.28) = 59.486$$

$$E_{27}^{(95)}(24) = 2.5170292 \, \Delta(24) = 2.5170292(24.28) = 61.113$$

$$E_{[25,27]}^{(95)}(24) = 5.1234893 \, \Delta(24) = 5.1234893(24.28) = 124.399$$

We next calculate one-, two-, and three-period-ahead forecasts, which are made in period 24 for y_t. Thus, since from Example 4.3

$$S_{24} = 348.03 \qquad \text{and} \qquad S_{24}^{[2]} = 308.76$$

we have

$$\hat{y}_{T+\tau}(T) = \left(2 + \frac{\alpha\tau}{(1 - \alpha)}\right) S_T - \left(1 + \frac{\alpha\tau}{(1 - \alpha)}\right) S_T^{[2]}$$

$$\hat{y}_{25}(24) = \left(2 + \frac{(.16)(1)}{1 - .16}\right) S_{24} - \left(1 + \frac{(.16)(1)}{1 - .16}\right) S_{24}^{[2]}$$

$$= 2.190476(348.03) - 1.190476(308.76)$$

$$= 394.77$$

TABLE 4.7   *Simulating One-Period-Ahead Forecasting of the Historical Calculator Sales Time Series Using Double Exponential Smoothing with the Optimal Smoothing Constant* $\alpha = .16$

Year	Month	Calculator Sales y_t	Forecast Made Last Period $\hat{y}_t(t-1)$	Forecast Error $y_t - \hat{y}_t(t-1)$
1	Jan.	197	206.10	− 9.10
	Feb.	211	211.26	− 0.26
	Mar.	203	219.01	−16.01
	Apr.	247	221.72	25.28
	May	239	237.23	1.77
	June	269	245.86	23.14
	July	308	261.38	46.62
	Aug.	262	285.00	−23.00
	Sept.	258	287.54	−29.54
	Oct.	256	287.40	−31.40
	Nov.	261	285.90	−24.90
	Dec.	288	285.68	2.32
2	Jan.	296	293.54	2.46
	Feb.	276	301.50	−25.50
	Mar.	305	300.57	4.43
	Apr.	308	308.57	− 0.57
	May	356	315.08	40.92
	June	393	334.86	58.14
	July	363	361.19	1.81
	Aug.	386	370.99	15.01
	Sept.	443	385.05	57.95
	Oct.	308	413.24	−105.24
	Nov.	358	390.70	−32.70
	Dec.	384	388.67	− 4.67

$$\hat{y}_{26}(24) = \left(2 + \frac{(.16)(2)}{1 - .16}\right) S_{24} - \left(1 + \frac{(.16)(2)}{1 - .16}\right) S_{24}^{[2]}$$

$$= 2.380952(348.03) - 1.380952(308.76)$$

$$= 402.26$$

$$\hat{y}_{27}(24) = \left(2 + \frac{(.16)(3)}{1 - .16}\right) S_{24} - \left(1 + \frac{(.16)(3)}{1 - .16}\right) S_{24}^{[2]}$$

$$= 2.571429(348.03) - 1.571429(308.76)$$

$$= 409.74$$

Given these forecasts and the errors previously calculated, we can now construct the desired confidence intervals. An approximate 95 percent confidence interval for y_{25} is

$$[\hat{y}_{25}(24) - E_{25}^{(95)}(24), \hat{y}_{25}(24) + E_{25}^{(95)}(24)]$$

$$[394.77 - 59.486, \quad 394.77 + 59.486]$$

$$[335.284, 454.256]$$

An approximate 95 percent confidence interval for y_{27} is

$$[\hat{y}_{27}(24) - E_{27}^{(95)}(24), \hat{y}_{27}(24) + E_{27}^{(95)}(24)]$$

$$[409.74 - 61.113, \quad 409.74 + 61.113]$$

$$[348.627, 470.853]$$

Since

$$\sum_{t=1}^{3} \hat{y}_{24+t}(24) = \hat{y}_{25}(24) + \hat{y}_{26}(24) + \hat{y}_{27}(24)$$

$$= 394.77 + 402.26 + 409.74$$

$$= 1206.77$$

an approximate 95 percent confidence interval for the cumulative sales in periods 25 through 27, $y_{25} + y_{26} + y_{27}$, is

$$\left[\sum_{t=1}^{3} \hat{y}_{24+t}(24) - E_{[25,27]}^{(95)}(24), \sum_{t=1}^{3} \hat{y}_{24+t}(24) + E_{[25,27]}^{(95)}(24) \right]$$

$$[1206.77 - 124.399, 1206.77 + 124.399]$$

$$[1082.371, 1331.169]$$

Now suppose that we observe a sales figure for period 25 of $y_{25} = 358$. We may update $\Delta(24)$ to $\Delta(25)$ as follows

$$\Delta(T + 1) = \frac{T\Delta(T) + |y_{T+1} - \hat{y}_{T+1}(T)|}{T + 1}$$

$$\Delta(25) = \frac{24\Delta(24) + |y_{25} - \hat{y}_{25}(24)|}{25}$$

$$= \frac{24(24.28) + |358 - 394.77|}{25}$$

$$= 24.78$$

Thus the updated errors in 95 percent confidence intervals for y_{26}, y_{28}, and $y_{26} + y_{27} + y_{28}$ are, respectively,

$$E_{26}^{(95)}(25) = 2.45 \, \Delta(25) = 2.45(24.78) = 60.711$$

$$E_{28}^{(95)}(25) = 2.5170292 \, \Delta(25) = 2.5170292(24.78) = 62.372$$

$$E_{[26,28]}^{(95)}(25) = 5.1234893 \, \Delta(25) = 5.1234893(24.78) = 126.960$$

We next calculate one-, two-, and three-period-ahead forecasts made in period 25 for y_t. These forecasts are calculated as follows, using $S_{25} = 349.62$ and $S_{25}^{[2]} = 315.29$ from Example 4.3.

$$\hat{y}_{T+\tau}(T) = \left(2 + \frac{\alpha\tau}{(1-\alpha)}\right) S_T - \left(1 + \frac{\alpha\tau}{(1-\alpha)}\right) S_T^{[2]}$$

$$\hat{y}_{26}(25) = \left(2 + \frac{(.16)(1)}{1-.16}\right) S_{25} - \left(1 + \frac{(.16)(1)}{1-.16}\right) S_{25}^{[2]}$$

$$= 2.190476(349.62) - 1.190476(315.29)$$

$$= 390.49$$

$$\hat{y}_{27}(25) = \left(2 + \frac{(.16)(2)}{1-.16}\right) S_{25} - \left(1 + \frac{(.16)(2)}{1-.16}\right) S_{25}^{[2]}$$

$$= 2.380952(349.62) - 1.380952(315.29)$$

$$= 397.03$$

$$\hat{y}_{28}(25) = \left(2 + \frac{(.16)(3)}{1-.16}\right) S_{25} - \left(1 + \frac{(.16)(3)}{1-.16}\right) S_{25}^{[2]}$$

$$= 2.571429(349.62) - 1.571429(315.29)$$

$$= 403.57$$

Given these forecasts and the updated errors previously calculated, we can calculate updated confidence intervals for y_{26}, y_{28}, and $y_{26} + y_{27} + y_{28}$. An approximate 95 percent confidence interval for y_{26} is

$$[\hat{y}_{26}(25) - E_{26}^{(95)}(25), \hat{y}_{26}(25) + E_{26}^{(95)}(25)]$$

$$[390.49 - 60.711, 390.49 + 60.711]$$

$$[329.779, 451.201]$$

An approximate 95 percent confidence interval for y_{28} is

$$[\hat{y}_{28}(25) - E_{28}^{(95)}(25), \hat{y}_{28}(25) + E_{28}^{(95)}(25)]$$

$$[403.57 - 62.372, 403.57 + 62.372]$$

$$[341.198, 465.942]$$

Since

$$\sum_{t=1}^{3} \hat{y}_{25+t}(25) = \hat{y}_{26}(25) + \hat{y}_{27}(25) + \hat{y}_{28}(25) = 390.49 + 397.03 + 403.57 = 1191.09$$

an approximate 95 percent confidence interval for the cumulative sales in periods 26 through 28, $y_{26} + y_{27} + y_{28}$, is

$$\left[\sum_{t=1}^{3} \hat{y}_{25+t}(25) - E_{[26,28]}^{(95)}(25), \sum_{t=1}^{3} \hat{y}_{25+t}(25) + E_{[26,28]}^{(95)}(25)\right]$$

$$[1191.09 - 126.960, 1191.09 + 126.960]$$

$$[1064.13, 1318.05]$$

This updating process, in which the errors for confidence intervals are updated by updating $\Delta(T)$, may be continued indefinitely as new sales figures are observed.

PROBLEMS

1. Consider Example 4.2 in which the regression approach is used to forecast the calculator sales time series of Table 4.1. Using this approach, compute calculator sales forecasts for periods 29 and 30.

2. Consider Example 4.3 in which the double exponential smoothing approach is used to forecast the calculator sales time series of Table 4.1.

    a. Verify that the single and double smoothed statistics given in Table 4.5 for March of year 1 (period 3) are

    $$S_3 = 175.34 \quad \text{and} \quad S_3^{[2]} = 136.38$$

    b. Verify that the single and double smoothed statistics given in Table 4.5 for April of year 1 (period 4) are

    $$S_4 = 186.81 \quad \text{and} \quad S_4^{[2]} = 144.45$$

3. Consider Example 4.3, in which the double exponential smoothing approach is used to forecast the calculator sales time series of Table 4.1.

a. Consider Table 4.6. Verify that the sales forecast made in February of year 3 (period 26) for March of year 3 (period 27) is 376 (rounded).

b. Again consider Table 4.6. Verify that the sales forecast made in March of year 3 (period 27) for April of year 3 (period 28) is 393 (rounded).

c. Calculate a sales forecast made in March of year 3 (period 27) for May of year 3 (period 29).

4. Find the single and double smoothed statistics for January of year 1 (period 1) that would be generated when simulating the calculator sales data in Table 4.1 using a smoothing constant of $\alpha = .20$.

---

*5. Consider the regression calculations in Section *4–2.1, which have been carried out for Example 4.2.

a. Determine the vector x'_{28} to be used in calculating a 95 percent confidence interval for calculator sales in period 28.

b. Determine the vector $\Sigma_{t=28}^{30}x'_t$ to be used in calculating a 95 percent confidence interval for cumulative calculator sales in periods 28 through 30.

---

6. Consider Example 4.2. Verify that the least squares estimates of β_1 and β_0 for the calculator sales data given in Table 4.1 are $b_1 = 8.07$ and $b_0 = 198.03$.

---

*7. Consider the regression calculations in Section *4–2.1, which are carried out for the calculator sales data presented in Table 4.1.

a. Verify that when a 95 percent confidence interval for calculator sales in period 25 is calculated, $f_{25} = 1.0851434$ (or 1.085 rounded).

b. Given that $t_{5/2}(22) = 2.074$ and $s = 31.67$, verify that a 95 percent confidence interval for calculator sales in period 25 is

$$[328.61, 471.17]$$

---

8. Consider Example 4.4, in which confidence intervals for future values of the calculator sales time series presented in Table 4.1 are calculated.

a. Suppose that we observe a sales figure for period 26 of $y_{26} = 325$. Update $\Delta(25)$ to $\Delta(26)$.

b. Find updated errors in approximate 95 percent confidence intervals for

$$y_{27}, y_{29}, \text{ and } \sum_{t=1}^{3} y_{26+t} = y_{27} + y_{28} + y_{29}$$

---

* Starred problems are only for students who have read Section *4–2.1.

c. Find the one-, two-, and three-period-ahead forecasts made in period 26 for y_{27}, y_{28}, and y_{29}.

d. Calculate updated approximate confidence intervals for

$$y_{27}, y_{29}, \text{ and } \sum_{t=1}^{3} y_{26+t} = y_{27} + y_{28} + y_{29}$$

9. Suppose that a forecaster wishes to calculate confidence intervals for future values of a time series that is being analyzed using simple exponential smoothing with $\alpha = .10$. Find expressions for $E_{T+1}^{(95)}(T)$, $E_{T+2}^{(95)}(T)$, and $E_{[T+1,T+2]}^{(95)}(T)$ in terms of $\Delta(T)$ in this situation.

CHAPTER 5

Triple Exponential Smoothing

5–1 INTRODUCTION

In this chapter we discuss the forecasting of time series with a quadratic trend. Both the regression and exponential smoothing approaches can be used to forecast such time series. In Section 5–2.1 we present the regression approach. In Section 5–2.2 we present the exponential smoothing approach that is called triple exponential smoothing.

Recall that the overall implementation of an exponential smoothing procedure is discussed in Section 3–4. As outlined in that section, this implementation involves a set of steps, several of which determine the smoothing constant to be used in forecasting future values of the time series. We conclude this chapter with a more complete discussion of several aspects of the initialization of single, double, and triple exponential smoothing procedures.

5–2 FORECASTING TIME SERIES WITH A QUADRATIC TREND

We next consider a situation in which the average level of the time series being forecast changes in a quadratic or curvilinear fashion over time. That is, we will now consider the situation in which the average level of the time series is either increasing at an increasing or decreasing rate or decreasing at an increasing or decreasing rate. Thus an appropriate model for the time series might be

$$y_t = \beta_0 + \beta_1 t + \beta_2 t^2 + \epsilon_t$$

This model implies that the time series can be represented by an average level that changes over time according to the quadratic function $\beta_0 + \beta_1 t + \beta_2 t^2$ combined with random fluctuations which cause the observations of the time series to deviate from the average level. The model is capable of representing any of the data patterns illustrated in Figure II.3.

Example 5.1   The State University Credit Union is a savings institution open to the faculty and staff of State University. The credit union, which handles savings accounts and makes loans to members, has been in operation for two years. In order to plan its investment strategies, the State University Credit Union needs both a point forecast and a confidence interval forecast of monthly loan requests (in hundreds of dollars) made by the faculty and staff. The point forecast is the best guess at monthly loan requests, while the confidence interval forecast gives the smallest and largest dollar values that monthly loan requests can reasonably be expected to be.

The State University Credit Union has recorded monthly loan requests (in hundreds of dollars) for its two years of operation, which we will call year 1 and year 2. The loan requests history is given in Table 5.1. If this data is plotted, it appears to randomly fluctuate around an average level that increases at a decreasing rate over time, as indicated in Figure 5.1. Since the credit union realizes that its memberships have been increasing at a decreasing rate for the previous two years, and since only members can apply for loans, the observed historical pattern of loan requests is consistent with the historical membership pattern. Also, the State University Credit Union subjectively believes that the observed quadratic trend will continue for at least the next two years, so it seems reasonable to use the model

$$y_t = \beta_0 + \beta_1 t + \beta_2 t^2 + \epsilon_t$$

to forecast monthly loan requests for the next two years.

TABLE 5.1   *Loan Requests*

Month	Year 1	Year 2
Jan.	297	808
Feb.	249	809
Mar.	340	867
Apr.	406	855
May	464	965
June	481	921
July	549	956
Aug.	553	990
Sept.	556	1019
Oct.	642	1021
Nov.	670	1033
Dec.	712	1127

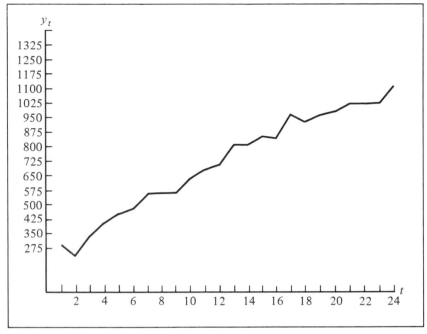

FIGURE 5.1   *Loan Requests*

5–2.1   THE REGRESSION APPROACH

Denoting the least squares estimates of β_0, β_1, and β_2 by b_0, b_1, and b_2, a forecast for a value y_t of the time series is

$$\hat{y}_t = b_0 + b_1 t + b_2 t^2$$

The computations for these least squares estimates are given in Section *5–2.1.

Example 5.2   Using the data of Table 5.1, the least squares estimates of β_0, β_1, and β_2 are found to be $b_0 = 199.62$, $b_1 = 50.94$, and $b_2 = -.57$. Moreover, it can be shown that $R^2 = .99$. The forecast of the value y_t of the time series is therefore

$$\hat{y}_t = b_0 + b_1 t + b_2 t^2$$

$$= 199.62 + 50.94t - .57t^2$$

For example, a prediction of loan requests in January of year 3 (period 25) is

$$\hat{y}_{25} = 199.62 + 50.94(25) - .57(25)^2$$

$$= 1118.21$$

Moreover, it can be shown that a 95 percent confidence interval for loan requests in January of year 3 is

$$[1118.21 - 78.14, 1118.21 + 78.14] \quad \text{or} \quad [1040.07, 1196.35]$$

As another example, a prediction of loan requests in February of year 3 (period 26) is

$$\hat{y}_{26} = 199.62 + 50.94(26) - .57(26)^2$$
$$= 1140.19$$

and it can be shown that a 95 percent confidence interval for loan requests in February of year 3 is

$$[1140.19 - 82.51, 1140.19 + 82.51] \quad \text{or} \quad [1057.68, 1222.70]$$

As yet another example, a prediction of loan requests in March of year 3 (period 27) is

$$\hat{y}_{27} = 199.62 + 50.94(27) - .57(27)^2$$
$$= 1161.04$$

and it can be shown that a 95 percent confidence interval for loan requests in March of year 3 is

$$[1161.04 - 87.81, 1161.04 + 87.81] \quad \text{or} \quad [1073.23, 1248.85]$$

Next, assume that to make longer range plans for its investments, the State University Credit Union needs a prediction of cumulative loan requests in January, February, and March of year 3. Such a prediction is

$$\hat{y}_{25} + \hat{y}_{26} + \hat{y}_{27} = 1118.21 + 1140.19 + 1161.04 = 3419.44$$

and it can be shown that a 95 percent confidence interval for these cumulative loan requests is

$$[3419.44 - 189.95, 3419.44 + 189.95] \quad \text{or} \quad [3229.49, 3609.39]$$

Note that the error, 189.95, in the above confidence interval for the cumulative loan requests is smaller than the sum of the errors of the forecasts for the individual loan request figures, which is

$$78.14 + 82.51 + 87.81 = 248.46$$

*5–2.1   THE REGRESSION CALCULATIONS

Using the data in Table 5.1, we see that if we wish to find the least squares estimates of β_0, β_1, and β_2 in the regression model,

$$y_t = \beta_0 + \beta_1 t + \beta_2 t^2 + \epsilon_t$$

we use

$$\mathbf{y} = \begin{bmatrix} 297 \\ 249 \\ 340 \\ \cdot \\ \cdot \\ \cdot \\ 1127 \end{bmatrix} \qquad \mathbf{X} = \begin{bmatrix} 1 & 1 & (1)^2 \\ 1 & 2 & (2)^2 \\ 1 & 3 & (3)^2 \\ \cdot & \cdot & \cdot \\ \cdot & \cdot & \cdot \\ \cdot & \cdot & \cdot \\ 1 & 24 & (24)^2 \end{bmatrix}$$

Thus, the least squares estimates of β_0, β_1, and β_2 are

$$\begin{bmatrix} b_0 \\ b_1 \\ b_2 \end{bmatrix} = (\mathbf{X}'\mathbf{X})^{-1}\mathbf{X}'\mathbf{y} = \begin{bmatrix} 199.62 \\ 50.94 \\ -.57 \end{bmatrix}$$

Moreover, a 95 percent confidence interval for y_t is

$$[\hat{y}_t - E_t(95), \; \hat{y}_t + E_t(95)]$$

for which

$$\hat{y}_t = b_0 + b_1 t + b_2 t^2 = 199.62 + 50.94t - .57t^2 = \begin{cases} 1118.21 & \text{if } t = 25 \\ 1140.19 & \text{if } t = 26 \\ 1161.04 & \text{if } t = 27 \end{cases}$$

$$s = \left(\frac{\displaystyle\sum_{t=1}^{24} (y_t - \hat{y}_t)^2}{24 - 3} \right)^{1/2} = 31.25$$

$$t_{5/2}(24 - 3) = t_{5/2}(21) = 2.080$$

$$\mathbf{x}'_t = [1 \; t \; t^2] = \begin{cases} [1 \; 25 \; (25)^2] & \text{if } t = 25 \\ [1 \; 26 \; (26)^2] & \text{if } t = 26 \\ [1 \; 27 \; (27)^2] & \text{if } t = 27 \end{cases}$$

$$f_t = (1 + \mathbf{x}'_t (\mathbf{X}'\mathbf{X})^{-1} \mathbf{x}_t)^{1/2} = \begin{cases} 1.20 & \text{if } t = 25 \\ 1.27 & \text{if } t = 26 \\ 1.35 & \text{if } t = 27 \end{cases}$$

and

$$E_t(95) = t_{5/2}(21) \cdot s \cdot f_t = \begin{cases} 78.14 \text{ if } & t = 25 \\ 82.51 \text{ if } & t = 26 \\ 87.81 \text{ if } & t = 27 \end{cases}$$

A 95 percent confidence interval for

$$\overset{t=l+q=25+2=27}{\underset{t=l=25}{\sum}} y_t$$

is

$$\left[\sum_{t=25}^{27} \hat{y}_t - E_{(25,25+2)}(95), \; \sum_{t=25}^{27} \hat{y}_t + E_{(25,25+2)}(95) \right]$$

where

$$E_{(25,25+2)}(95) = t_{5/2}(21) \cdot s \cdot \left\{ (2 + 1) + \left(\sum_{t=25}^{27} \mathbf{x}'_t \right) (\mathbf{X}'\mathbf{X})^{-1} \left(\sum_{t=25}^{27} \mathbf{x}_t \right) \right\}^{1/2}$$

$$= 189.95$$

since

$$\sum_{t=25}^{27} \mathbf{x}'_t = [1\ 25\ (25)^2] + [1\ 26\ (26)^2] + [1\ 27\ (27)^2]$$

$$= [3\ 78\ 2030]$$

5–2.2   THE EXPONENTIAL SMOOTHING APPROACH

To begin the exponential smoothing approach, let us rewrite our model as

$$\boxed{y_t = \beta_0 + \beta_1 t + \tfrac{1}{2}\beta_2 t^2 + \epsilon_t}$$

It should be noted that the term $\tfrac{1}{2}\beta_2 t^2$ is written in this rather unusual form in order to simplify the estimating equations that will be presented in the discussion of triple exponential smoothing to follow.

Suppose that at the end of period T we have a set of observations of the time series $y_1, y_2, \ldots, y_T$. We would like to find estimates of the model parameters β_0, β_1, and β_2 using these observations. Given these estimates, which we denote $b_0(T)$, $b_1(T)$, and $b_2(T)$, we obtain the following forecast for a future period $T + \tau$.

$$\hat{y}_{T+\tau}(T) = b_0(T) + b_1(T)(T + \tau) + \tfrac{1}{2}b_2(T)(T + \tau)^2$$

It should be noted that the estimates $b_0(T)$, $b_1(T)$, and $b_2(T)$ are defined in reference to an origin that is the time period immediately previous to the period corresponding to the first observation. We now redefine the origin to be the current period. That is, the origin will now be considered to be period T. This change of origin will alter the estimates of β_0, β_1, and β_2. We shall denote these new estimates $a_0(T)$, $a_1(T)$, and $a_2(T)$. Given these estimates of β_0, β_1, and β_2, we obtain the following forecast for period $T + \tau$.

$$\hat{y}_{T+\tau}(T) = a_0(T) + a_1(T)\tau + \tfrac{1}{2}a_2(T)\tau^2$$

Triple exponential smoothing involves the use of three smoothed statistics. These smoothed statistics are defined by the following equations.

$$S_T = \alpha y_T + (1 - \alpha)S_{T-1}$$

$$S_T^{[2]} = \alpha S_T + (1 - \alpha)S_{T-1}^{[2]}$$

$$S_T^{[3]} = \alpha S_T^{[2]} + (1 - \alpha)S_{T-1}^{[3]}$$

Here S_T and $S_T^{[2]}$ are the familiar single and double smoothed statistics used in the double exponential smoothing procedure. $S_T^{[3]}$ is known as the *triple smoothed statistic* and is obtained by applying the smoothing operation to the output of the double smoothing equation. That is, the series of $S_T^{[3]}$ values is obtained by smoothing the series of $S_T^{[2]}$ values using the triple smoothing equation

$$S_T^{[3]} = \alpha S_T^{[2]} + (1 - \alpha)S_{T-1}^{[3]}$$

It can be shown that the estimates $a_0(T)$, $a_1(T)$, and $a_2(T)$ are given by the following equations in any time period T.

$$a_0(T) = 3S_T - 3S_T^{[2]} + S_T^{[3]}$$

$$a_1(T) = \frac{\alpha}{2(1 - \alpha)^2}[(6 - 5\alpha)S_T - 2(5 - 4\alpha)S_T^{[2]} + (4 - 3\alpha)S_T^{[3]}]$$

$$a_2(T) = \left(\frac{\alpha}{1 - \alpha}\right)^2 (S_T - 2S_T^{[2]} + S_T^{[3]})$$

If we substitute these expressions for $a_0(T)$, $a_1(T)$, and $a_2(T)$ into the forecasting equation

$$\hat{y}_{T+\tau}(T) = a_0(T) + a_1(T)\tau + \tfrac{1}{2}a_2(T)\tau^2$$

we obtain the following expression, which can then be used as a forecasting equation.

$$\hat{y}_{T+\tau}(T) = [6(1 - \alpha)^2 + (6 - 5\alpha)\alpha\tau + \alpha^2\tau^2] \frac{S_T}{2(1 - \alpha)^2}$$

$$- [6(1 - \alpha)^2 + 2(5 - 4\alpha)\alpha\tau + 2\alpha^2\tau^2] \frac{S_T^{[2]}}{2(1 - \alpha)^2}$$

$$+ [2(1 - \alpha)^2 + (4 - 3\alpha)\alpha\tau + \alpha^2\tau^2] \frac{S_T^{[3]}}{2(1 - \alpha)^2}$$

Again, in order to begin the triple smoothing procedure, initial values for the smoothed statistics S_T, $S_T^{[2]}$, and $S_T^{[3]}$ must be supplied. These initial values can be found by solving the equations for the estimates $a_0(T)$, $a_1(T)$, and $a_2(T)$ for S_T, $S_T^{[2]}$, and $S_T^{[3]}$ with $T = 0$. Doing this, we obtain the following equations for the initial values S_0, $S_0^{[2]}$, and $S_0^{[3]}$.

$$S_0 = a_0(0) - \frac{(1 - \alpha)}{\alpha} a_1(0) + \frac{(1 - \alpha)(2 - \alpha)}{2\alpha^2} a_2(0)$$

$$S_0^{[2]} = a_0(0) - \frac{2(1 - \alpha)}{\alpha} a_1(0) + \frac{2(1 - \alpha)(3 - 2\alpha)}{2\alpha^2} a_2(0)$$

$$S_0^{[3]} = a_0(0) - \frac{3(1 - \alpha)}{\alpha} a_1(0) + \frac{3(1 - \alpha)(4 - 3\alpha)}{2\alpha^2} a_2(0)$$

Here, $a_0(0)$, $a_1(0)$, and $a_2(0)$ are initial estimates of the model parameters β_0, β_1, and β_2. These initial estimates might be obtained from historical data via least squares regression analysis. If such estimates are not available, it is common procedure to use the initial observation in the time series as an initial value for S_0, $S_0^{[2]}$, and $S_0^{[3]}$. This procedure is reasonable since the initial values of S_T, $S_T^{[2]}$, and $S_T^{[3]}$ contribute less and less to forecasts generated by the smoothing procedure as more observations are incorporated into the smoothed statistics.

Once the initial values of the smoothed statistics, S_0, $S_0^{[2]}$, and $S_0^{[3]}$, have been determined, the forecasting procedure is quite simple. In each time period, the single, double, and triple smoothed statistics are found using the three smoothing equations given previously. The smoothed statistics are then entered into the fore casting equation. This equation yields a τ-period-ahead forecast. These calculations will be illustrated in Example 5.3.

The model

$$y_t = \beta_0 + \beta_1 t + \tfrac{1}{2}\beta_2 t^2 + \epsilon_t$$

which is assumed to describe a time series that may be appropriately forecasted using triple exponential smoothing, is a special case of the general time series regression model

$$y_t = \beta_0 + \beta_1 f_1(t) + \beta_2 f_2(t) + \cdots + \beta_p f_p(t) + \epsilon_t$$

where $f_1(t) = t$, $f_2(t) = \tfrac{1}{2}t^2$, and $f_3(t) = \cdots = f_p(t) = 0$.

If general direct smoothing is used to find equations that successively update the estimates of β_0, β_1, and β_2, these equations are found to be identical to the updating equations for triple exponential smoothing. Then, using the statistical theory behind general direct smoothing, expressions for d_τ and q_τ can be found so that

$$E_{T+\tau}^{(100-\alpha)}(T) = z_{\alpha/2} d_\tau \Delta(T) \qquad \text{and} \qquad E_{[T+1,T+\tau]}^{(100-\alpha)}(T) = z_{\alpha/2} q_\tau \Delta(T)$$

are, respectively, the error in a $(100 - \alpha)$ percent confidence interval for $y_{T+\tau}$ and the error in a $(100 - \alpha)$ percent confidence interval for $\Sigma_{t=1}^{\tau} y_{T+t}$. Here $z_{\alpha/2}$ and $\Delta(T)$ are as defined in Section 4–3. Because the expressions for d_τ and q_τ are extremely complicated, we will not give them here. The interested reader is referred to Brown [3].

Example 5.3   We now consider triple exponential smoothing as a forecasting tool to be used in forecasting the loan requests time series of Example 5.1. First, the value of the smoothing constant α which minimizes the sum of squared forecast errors will be found by simulating forecasts for the historical data given in Table 5.1. This value will then be used for actual forecasting of future values of the time series.

In order to initialize these simulations, the parameters in the model

$$y_t = \beta_0 + \beta_1 t + \tfrac{1}{2}\beta_2 t^2 + \epsilon_t$$

must be estimated using the first n_1 observations of the time series. In this case, these estimates will be derived using the first $n_1 = 12$ observations of the time series. The reason for the use of twelve observations in this initialization rather than six observations (the number of observations used to initialize the simulations in our single and double exponential smoothing examples) will be explained in the next section. Regression analysis is used to fit the data given in Table 5.2 to the regression model

$$y_t = \beta_0 + \beta_1 t + \beta_2' t^2 + \epsilon_t \qquad (\beta_2' = \tfrac{1}{2}\beta_2)$$

TABLE 5.2   *The First $n_1 = 12$ Observations*

Loan Requests y_t	Time t	Time Squared t^2
297	1	1
249	2	4
340	3	9
406	4	16
464	5	25
481	6	36
549	7	49
553	8	64
556	9	81
642	10	100
670	11	121
712	12	144

This regression analysis yields an estimate of 212.16 for β_0, an estimate of 48.33 for β_1, and an estimate of $-.61$ for β_2'. Notice that the coefficient of the squared term in the regression model is $\beta_2' = \frac{1}{2}\beta_2$. Now, we desire an estimate of β_2, the parameter in the exponential smoothing model. Since regression analysis yields an estimate of $-.61$ for $\beta_2' = \frac{1}{2}\beta_2$, the estimate of β_2 must be twice the estimate of β_2' or -1.22. Thus we obtain the following initial estimates for the parameters in the triple exponential smoothing model.

$$a_0(0) = 212.16, \; a_1(0) = 48.33, \; a_2(0) = -1.22$$

These estimates may be used to find initial values for the smoothed statistics S_T, $S_T^{[2]}$, and $S_T^{[3]}$. The initial values S_0, $S_0^{[2]}$, and $S_0^{[3]}$ will in turn be used to begin the simulations.

We will illustrate the calculations involved in these simulations by considering the case in which $\alpha = .20$. We find the initial values S_0, $S_0^{[2]}$, and $S_0^{[3]}$ as follows.

$$S_0 = a_0(0) - \frac{(1 - \alpha)}{\alpha} a_1(0) + \frac{(1 - \alpha)(2 - \alpha)}{2\alpha^2} a_2(0)$$

$$= 212.16 - \frac{(1 - .20)}{.20}(48.33) + \frac{(1 - .20)(2 - .20)}{2(.20)^2}(-1.22)$$

$$= -3.12$$

$$S_0^{[2]} = a_0(0) - \frac{2(1-\alpha)}{\alpha} a_1(0) + \frac{2(1-\alpha)(3-2\alpha)}{2\alpha^2} a_2(0)$$

$$= 212.16 - \frac{2(1-.20)}{.20}(48.33) + \frac{2(1-.20)(3-2(.20))}{2(.20)^2}(-1.22)$$

$$= -237.92$$

$$S_0^{[3]} = a_0(0) - \frac{3(1-\alpha)}{\alpha} a_1(0) + \frac{3(1-\alpha)(4-3\alpha)}{2\alpha^2} a_2(0)$$

$$= 212.16 - \frac{3(1-.20)}{.20}(48.33) + \frac{3(1-.20)(4-3(.20))}{2(.20)^2}(-1.22)$$

$$= -492.24$$

Given these initial values, we can now generate a forecast for January of year 1 using the triple smoothing forecasting equation. Before this is done, it will be helpful to simplify the forecasting equation by letting the smoothing constant α be .20 and by setting τ equal to 1 since we desire a one-period-ahead forecast. Placing these values in the forecasting equation yields the following.

$$\hat{y}_{T+\tau}(T) = [6(1-\alpha)^2 + (6-5\alpha)\alpha\tau + \alpha^2\tau^2]\frac{S_T}{2(1-\alpha)^2}$$

$$- [6(1-\alpha)^2 + 2(5-4\alpha)\alpha\tau + 2\alpha^2\tau^2]\frac{S_T^{[2]}}{2(1-\alpha)^2}$$

$$+ [2(1-\alpha)^2 + (4-3\alpha)\alpha\tau + \alpha^2\tau^2]\frac{S_T^{[3]}}{2(1-\alpha)^2}$$

$$= [6(1-.2)^2 + (6-5(.2))(.2)(1) + (.2)^2(1)^2]\frac{S_T}{2(1-.2)^2}$$

$$- [6(1-.2)^2 + 2(5-4(.2))(.2)(1) + 2(.2)^2(1)^2]\frac{S_T^{[2]}}{2(1-.2)^2}$$

$$+ [2(1-.2)^2 + (4-3(.2))(.2)(1) + (.2)^2(1)^2]\frac{S_T^{[3]}}{2(1-.2)^2}$$

$$= 4.88\frac{S_T}{1.28} - 5.60\frac{S_T^{[2]}}{1.28} + 2.00\frac{S_T^{[3]}}{1.28}$$

or

$$\hat{y}_{T+1}(T) = 3.8125 S_T - 4.375 S_T^{[2]} + 1.5625 S_T^{[3]}$$

Thus the forecasting equation for $T = 0$ yielding a one-period-ahead forecast is

$$\hat{y}_1(0) = 3.8125 S_0 - 4.375 S_0^{[2]} + 1.5625 S_0^{[3]}$$

The forecast for January of year 1 (period 1) made in period 0 is

$$\hat{y}_1(0) = 3.8125(-3.12) - 4.375(-237.92) + 1.5625(-492.24)$$
$$= 260 \quad \text{(rounded)}$$

In January of year 1 loan requests of 297 were observed. This loan requests figure is now used to obtain new values of the smoothed statistics. We do this using the smoothing equations as follows.

$$S_1 = \alpha y_1 + (1 - \alpha)S_0$$
$$= .20(297) + (1 - .20)(-3.12)$$
$$= 56.90$$

$$S_1^{[2]} = \alpha S_1 + (1 - \alpha)S_0^{[2]}$$
$$= .20(56.90) + (1 - .20)(-237.92)$$
$$= -178.96$$

$$S_1^{[3]} = \alpha S_1^{[2]} + (1 - \alpha)S_0^{[3]}$$
$$= .20(-178.96) + (1 - .20)(-492.24)$$
$$= -429.58$$

We next obtain a forecast for February of year 1 (period 2) by using the forecasting equation

$$\hat{y}_{1+1}(1) = 3.8125\,S_1 - 4.375\,S_1^{[2]} + 1.5625\,S_1^{[3]}$$
$$\hat{y}_2(1) = 3.8125(56.90) - 4.375(-178.96) + 1.5625(-429.58)$$
$$= 329 \quad \text{(rounded)}$$

The forecast for February of year 1 (period 2) made in period 1 is 329. In February of year 1 a loan requests figure of 249 was observed. Again, we obtain a forecast for the next period, March of year 1, as shown below.

$$S_2 = \alpha y_2 + (1 - \alpha)S_1$$
$$= .20(249) + (1 - .20)(56.90)$$
$$= 95.32$$

$$S_2^{[2]} = \alpha S_2 + (1 - \alpha)S_1^{[2]}$$
$$= .20(95.32) + (1 - .20)(-178.96)$$
$$= -124.10$$

$$S_2^{[3]} = \alpha S_2^{[2]} + (1 - \alpha)S_1^{[3]}$$
$$= .20(-124.10) + (1 - .20)(-429.58)$$
$$= -368.49$$

$$\hat{y}_{2+1}(2) = 3.8125 S_2 - 4.375 S_2^{[2]} + 1.5625 S_2^{[3]}$$

$$\hat{y}_3(2) = 3.8125(95.32) - 4.375(-124.10) + 1.5625(-368.49)$$

$$= 331 \quad \text{(rounded)}$$

The forecast for March of year 1 (period 3) made in period 2 is 331.

Continuing in this manner, we generate forecasts for the balance of the periods in the two years of historical data. For each period the forecast error and squared forecast error are calculated. Finally, the sum of squared forecast errors is calculated. These results are summarized in Table 5.3. The sum of squared forecast errors obtained for $\alpha = .20$ is 33,965.

In a similar manner, a set of forecasts is simulated for each value of the smoothing constant α we wish to consider. The sum of squared forecast errors is calculated for each value. Remember that the same initial estimates $a_0(0) = 212.16$, $a_1(0) = 48.33$, and $a_2(0) = -1.22$ are used to begin each simulation, although the initial values of the smoothed statistics S_0, $S_0^{[2]}$, and $S_0^{[3]}$ will differ in each case since they depend on the value of α used. The sums of squared errors obtained for all simulations are presented in Table 5.4. A smoothing constant of $\alpha = .08$ minimizes the sum of squared errors and hence this value is chosen for use in actual forecasting.

In order to begin the actual forecasting procedure, we must generate new estimates of β_0, β_1, and β_2 which will be used to calculate initial values of S_T, $S_T^{[2]}$, and $S_T^{[3]}$. These initial estimates, $a_0(0)$, $a_1(0)$, and $a_2(0)$, will now be based on the entire two-year loan requests history we have available and will be derived using regression analysis. That is, we choose $n_2 = n = 24$. We use regression analysis to fit the loan requests history given in Table 5.1 to the model

$$y_t = \beta_0 + \beta_1 t + \beta_2' t^2 + \epsilon_t \qquad (\beta_2' = \tfrac{1}{2}\beta_2)$$

The least squares estimates of β_0, β_1, and β_2' are found to be 199.62, 50.94, and $-.568$. Since the estimate of $\beta_2' = \tfrac{1}{2}\beta_2$ is $-.568$, the estimate of β_2 in the exponential smoothing model is $2(-.568)$ or -1.14. Thus we obtain the following initial estimates.

$$a_0(0) = 199.62, \ a_1(0) = 50.94, \ a_2(0) = -1.14$$

Again, the estimates $a_0(0)$, $a_1(0)$, and $a_2(0)$ will be used to obtain initial values of the smoothed statistics. These initial values, S_0, $S_0^{[2]}$, and $S_0^{[3]}$, are calculated as follows.

TABLE 5.3   *Simulating One-Period-Ahead Forecasting of the Historical Loan Requests
Time Series Using Triple Exponential Smoothing with $\alpha = .20$*

Year	Month	Loan Requests y_T	S_T	$S_T^{[2]}$	$S_T^{[3]}$	Forecast Made Last Period	Forecast Error	Squared Forecast Error
1	Jan.	297	56.90	-178.95	-429.58	260	37	1369
	Feb.	249	95.32	-124.10	-368.49	329	-80	6400
	Mar.	340	144.26	-70.43	-308.87	331	9	81
	Apr.	406	196.61	-17.02	-250.50	375	31	961
	May	464	250.09	36.40	-193.12	433	31	961
	June	481	296.27	88.37	-136.82	492	-11	121
	July	549	346.81	140.06	-81.45	529	20	400
	Aug.	553	388.05	189.66	-27.22	582	-29	841
	Sept.	556	421.64	236.06	25.43	607	-51	2601
	Oct.	642	465.71	281.99	76.74	614	28	784
	Nov.	670	506.57	326.90	126.77	662	8	64
	Dec.	712	547.66	371.05	175.63	699	13	169
2	Jan.	808	599.72	416.79	223.86	739	69	4761
	Feb.	809	641.58	461.75	271.44	813	-4	16
	Mar.	867	686.66	506.73	318.50	850	17	289
	Apr.	855	720.33	549.45	364.69	899	-44	1936
	May	965	769.26	593.41	410.43	912	53	2809
	June	921	799.61	634.65	455.28	978	-57	3249
	July	956	830.89	673.90	499.00	983	-27	729
	Aug.	990	862.71	711.66	541.53	999	-9	81
	Sept.	1019	893.97	748.12	582.85	1022	-3	9
	Oct.	1021	919.37	782.37	622.75	1046	-25	625
	Nov.	1033	942.10	814.32	661.07	1055	-22	484
	Dec.	1127	979.08	847.27	698.31	1062	65	4225

Sum of squared errors = 33,965

$$S_0 = a_0(0) - \frac{(1 - \alpha)}{\alpha} a_1(0) + \frac{(1 - \alpha)(2 - \alpha)}{2\alpha^2} a_2(0)$$

$$= 199.62 - \frac{(1 - .08)}{.08} (50.94) + \frac{(1 - .08)(2 - .08)}{2(.08)^2} (-1.14)$$

$$= -543.51$$

$$S_0^{[2]} = a_0(0) - \frac{2(1 - \alpha)}{\alpha} a_1(0) + \frac{2(1 - \alpha)(3 - 2\alpha)}{2\alpha^2} a_2(0)$$

$$= 199.62 - \frac{2(1 - .08)}{.08} (50.94) + \frac{2(1 - .08)(3 - 2(.08))}{2(.08)^2} (-1.14)$$

$$= -1437.40$$

$$S_0^{[3]} = a_0(0) - \frac{3(1-\alpha)}{\alpha} a_1(0) + \frac{3(1-\alpha)(4-3\alpha)}{2\alpha^2} a_2(0)$$

$$= 199.62 - \frac{3(1-.08)}{.08}(50.94) + \frac{3(1-.08)(4-3(.08))}{2(.08)^2}(-1.14)$$

$$= -2482.06$$

Note that the "best" smoothing constant $\alpha = .08$ has been used in performing these calculations.

These smoothed statistics will now be repeatedly updated using the observations in the two-year loan requests history. This process will insure that the forecasts we generate for future values of the time series will weight the most recent observations in the time series most heavily. The updating process begins with the observation from January of year 1 (period 1) which is $y_1 = 297$. We calculate updated smoothed statistics as follows.

$$S_1 = \alpha y_1 + (1-\alpha)S_0$$

$$= .08(297) + (1-.08)(-543.51)$$

$$= -476.27$$

$$S_1^{[2]} = \alpha S_1 + (1-\alpha)S_0^{[2]}$$

$$= .08(-476.27) + (1-.08)(-1437.40)$$

$$= -1360.51$$

TABLE 5.4   *The Sums of Squared Forecast Errors for Different Values of α*

Smoothing Constant α	Sum of Squared Errors
.02	38307
.04	30669
.06	27838
.08	27251
.10	28022
.12	28604
.14	29792
.16	31113
.18	32462
.20	33965
.22	35430
.24	37082
.26	38725
.28	40733
.30	42454

$$S_1^{[3]} = \alpha S_1^{[2]} + (1 - \alpha)S_0^{[3]}$$
$$= .08(-1360.51) + (1 - .08)(-2482.06)$$
$$= -2392.34$$

Using the observation for February of year 1 (period 2) which is $y_2 = 249$, we again calculate three updated smoothed statistics.

$$S_2 = \alpha y_2 + (1 - \alpha)S_1$$
$$= .08(249) + (1 - .08)(-476.27)$$
$$= -418.25$$
$$S_2^{[2]} = \alpha S_2 + (1 - \alpha)S_1^{[2]}$$
$$= .08(-418.25) + (1 - .08)(-1360.51)$$
$$= -1285.13$$
$$S_2^{[3]} = \alpha S_2^{[2]} + (1 - \alpha)S_1^{[3]}$$
$$= .08(-1285.13) + (1 - .08)(-2392.34)$$
$$= -2303.76$$

We continue calculating updated smoothed statistics for all other periods in the loan requests history in exactly the same way. The results of these calculations are shown in Table 5.5. At the end of December of year 2 (period 24) we have updated single, double, and triple smoothed statistics of $S_{24} = 665.82$, $S_{24}^{[2]} = 86.52$, and $S_{24}^{[3]} = -643.85$.

    These quantities may now be used to generate forecasts for future values of the time series. In order to do this we will again simplify the forecasting equation. Since we desire one-period-ahead forecasts, we will set τ equal to 1. Since we wish to use the value of the "best" smoothing constant for forecasting purposes, we will set α equal to .08. Doing this, we obtain the following.

$$\hat{y}_{T+\tau}(T) = [6(1 - \alpha)^2 + (6 - 5\alpha)\alpha\tau + \alpha^2\tau^2] \frac{S_T}{2(1 - \alpha)^2}$$

$$- [6(1 - \alpha)^2 + 2(5 - 4\alpha)\alpha\tau + 2\alpha^2\tau^2] \frac{S_T^{[2]}}{2(1 - \alpha)^2}$$

$$+ [2(1 - \alpha)^2 + (4 - 3\alpha)\alpha\tau + \alpha^2\tau^2] \frac{S_T^{[3]}}{2(1 - \alpha)^2}$$

$$= [6(1 - .08)^2 + (6 - 5(.08))(.08)(1) + (.08)^2(1)^2] \frac{S_T}{2(1 - .08)^2}$$

TABLE 5.5   *Updating the Smoothed Statistics Using Triple Exponential Smoothing with the Optimal Smoothing Constant $\alpha = .08$ on the Historical Loan Requests Time Series*

Year	Month	Time Period T	Loan Requests y_T	Single Smoothed Statistic S_T	Double Smoothed Statistic $S_T^{[2]}$	Triple Smoothed Statistic $S_T^{[3]}$
1	Jan.	1	297	−476.25	−1360.51	−2392.34
	Feb.	2	249	−418.25	−1285.13	−2303.76
	Mar.	3	340	−357.59	−1210.93	−2216.33
	Apr.	4	406	−296.50	−1137.77	−2130.05
	May	5	464	−235.66	−1065.60	−2044.89
	June	6	481	−178.33	−994.62	−1960.87
	July	7	549	−120.14	−924.66	−1877.97
	Aug.	8	553	−66.29	−855.99	−1796.22
	Sept.	9	556	−16.51	−788.83	−1715.62
	Oct.	10	642	36.17	−722.83	−1636.20
	Nov.	11	670	86.88	−658.05	−1557.95
	Dec.	12	712	136.89	−594.46	−1480.87
2	Jan.	13	808	190.58	−531.66	−1404.93
	Feb.	14	809	240.05	−469.92	−1330.13
	Mar.	15	867	290.21	−409.11	−1256.45
	Apr.	16	855	335.39	−349.55	−1183.90
	May	17	965	385.76	−290.72	−1112.44
	June	18	921	428.58	−233.18	−1042.10
	July	19	956	470.77	−176.86	−972.88
	Aug.	20	990	512.31	−121.73	−904.79
	Sept.	21	1019	552.85	−67.76	−837.83
	Oct.	22	1021	590.30	−15.12	−772.01
	Nov.	23	1033	625.71	36.15	−707.36
	Dec.	24	1127	665.82	86.52	−643.85

$$- [6(1 - .08)^2 + 2(5 - 4(.08))(.08)(1) + 2(.08)^2(1)^2] \frac{S_T^{[2]}}{2(1 - .08)^2}$$

$$+ [2(1 - .08)^2 + (4 - 3(.08))(.08)(1) + (.08)^2(1)^2] \frac{S_T^{[3]}}{2(1 - .08)^2}$$

$$\hat{y}_{T+\tau}(T) = 5.5328 \frac{S_T}{1.6928} - 5.84 \frac{S_T^{[2]}}{1.6928} + 2.00 \frac{S_T^{[3]}}{1.6928}$$

$$= 3.26843 S_T - 3.4499 S_T^{[2]} + 1.18147 S_T^{[3]}$$

Thus the forecasting equation for $T = 24$ yielding a one-period-ahead forecast is

$$\hat{y}_{25}(24) = 3.26843 S_{24} - 3.4499 S_{24}^{[2]} + 1.18147 S_{24}^{[3]}$$

We place the values of the smoothed statistics in this equation and obtain a forecast for January of year 3 (period 25) as follows.

$$\hat{y}_{25}(24) = 3.26843(665.82) - 3.4499(86.52) + 1.18147(-643.85)$$

$$= 1117 \quad \text{(rounded)}$$

Thus the forecast for January of year 3 (period 25) made in December of year 2 (period 24) is 1117. For comparison purposes it should be noticed that the regression approach of Section 5–2.1 made a forecast in December of year 2 (period 24) for January of year 3 (period 25) of 1118.21 (or 1118 rounded). Now suppose that in January of year 3 (period 25) we observe a loan requests figure of $y_{25} = 1169$. In order to make a forecast for February of year 3 (period 26) we must first update the single, double, and triple smoothed statistics using the smoothing equations as shown next.

$$S_{25} = \alpha y_{25} + (1 - \alpha) S_{24}$$

$$= .08(1169) + (1 - .08)(665.82)$$

$$= 706.07$$

$$S_{25}^{[2]} = \alpha S_{25} + (1 - \alpha) S_{24}^{[2]}$$

$$= .08(706.07) + (1 - .08)(86.52) = 136.09$$

$$S_{25}^{[3]} = \alpha S_{25}^{[2]} + (1 - \alpha) S_{24}^{[3]}$$

$$= .08(136.09) + (1 - .08)(-643.85) = -581.45$$

A forecast for February of year 3 (period 26) is now given by the forecasting equation.

$$\hat{y}_{26}(25) = 3.26843 S_{25} - 3.4499 S_{25}^{[2]} + 1.18147 S_{25}^{[3]}$$

$$\hat{y}_{26}(25) = 3.26843(706.07) - 3.4499(136.09) + 1.18147(-581.45)$$

$$\hat{y}_{26}(25) = 1151 \quad \text{(rounded)}$$

So the forecast for February of year 3 (period 26) made in January of year 3 (period 25) is 1151. Subsequently, forecasts for future months may be made in the same manner using newly observed values of the time series. Observed values of the time series for the next two years are given in Table 5.6. Also given in this table are updated values of the single, double, and triple smoothed statistics, one-period-

TABLE 5.6   *One-Period-Ahead Forecasts of Future Values of Loan Requests Using Triple Exponential Smoothing with $\alpha = .08$*

Year	Month	Period T	Loan Requests y_T	S_T	$S_T^{[2]}$	$S_T^{[3]}$	Forecast Made Last Period	Forecast Error
3	Jan.	25	1169	706.07	136.09	−581.45	1117	52
	Feb.	26	1153	741.82	184.54	−520.17	1151	2
	Mar.	27	1193	777.92	232.01	−460.00	1173	20
	Apr.	28	1237	814.64	278.62	−400.91	1199	38
	May	29	1268	850.91	324.41	−342.88	1228	40
	June	30	1218	880.28	368.88	−285.94	1257	−39
	July	31	1288	912.90	412.40	−230.07	1267	21
	Aug.	32	1295	943.46	454.88	−175.28	1289	6
	Sept.	33	1305	972.39	496.28	−121.55	1307	−2
	Oct.	34	1367	1003.96	536.90	−68.88	1322	45
	Nov.	35	1343	1031.08	576.43	−17.25	1348	−5
	Dec.	36	1344	1056.11	614.81	33.31	1361	−17
4	Jan.	37	1340	1078.82	651.93	82.80	1370	−30
	Feb.	38	1361	1101.40	687.88	131.21	1375	−14
	Mar.	39	1384	1124.01	722.77	178.53	1382	2
	Apr.	40	1391	1145.36	756.58	224.78	1391	0
	May	41	1401	1165.82	789.32	269.94	1399	2
	June	42	1419	1186.07	821.06	314.03	1406	13
	July	43	1471	1208.86	852.08	357.07	1415	56
	Aug.	44	1375	1222.15	881.69	399.04	1433	−58
	Sept.	45	1442	1239.74	910.33	439.95	1424	18
	Oct.	46	1445	1256.16	938.00	479.79	1431	14
	Nov.	47	1420	1269.27	964.50	518.57	1437	−17
	Dec.	48	1444	1283.25	990.00	556.28	1434	10

ahead forecasts for each month, and the forecast error for each period. A plot of the observed and forecasted loan requests is given in Figure 5.2.

5–3   FURTHER ASPECTS OF THE INITIALIZATION OF EXPONENTIAL SMOOTHING PROCEDURES

The objective of this section is to further discuss the initialization of exponential smoothing procedures. In particular, we wish to consider the choice of n_1 and n_2. Recall from Section 3–4 that n_1 is the number of observations, chosen from the n total observations, that are used to calculate the values of the smoothed statistics

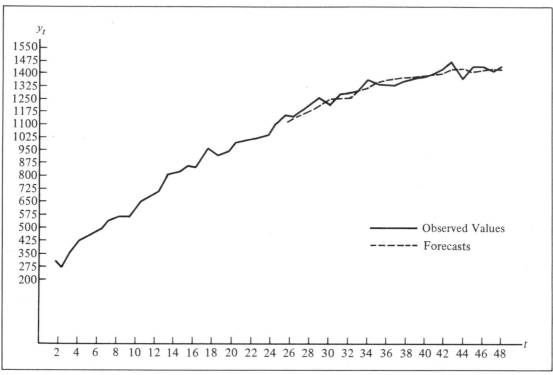

FIGURE 5.2   *One-Period-Ahead Forecasts of Future Values of Loan Requests Using Triple Exponential Smoothing with $\alpha = .08$*

initializing the simulations that determine an appropriate smoothing constant. Also remember that n_2 is the number of observations used to calculate the initial values of the smoothed statistics used to begin the actual forecasting procedure, after an appropriate smoothing constant has been determined. We will first discuss the choice of n_1 and n_2 in the context of triple exponential smoothing. We will then discuss the choice of n_1 and n_2 in the context of simple and double exponential smoothing.

5–3.1   CHOICE OF n_1 AND n_2 WHEN USING TRIPLE EXPONENTIAL SMOOTHING

In order to implement a triple exponential smoothing procedure, we know that the smoothing constant α must be chosen. As described previously, this choice is made by simulating forecasts for a set of historical data. Let us review this simulation process. We will assume that n observations of the time series are available. The choice of the smoothing constant is made in the following way. The first n_1 observations of the n observations available are chosen. These n_1 observations are used to obtain initial estimates, $a_0(0)$, $a_1(0)$, and $a_2(0)$, of the model parameters β_0, β_1, and β_2, by regression analysis. Initial values of the smoothed statistics $S_0, S_0^{[2]}$, and $S_0^{[3]}$ are found using the equations

$$S_0 = a_0(0) - \frac{(1 - \alpha)}{\alpha} a_1(0) + \frac{(1 - \alpha)(2 - \alpha)}{2\alpha^2} a_2(0)$$

$$S_0^{[2]} = a_0(0) - \frac{2(1 - \alpha)}{\alpha} a_1(0) + \frac{2(1 - \alpha)(3 - 2\alpha)}{2\alpha^2} a_2(0)$$

$$S_0^{[3]} = a_0(0) - \frac{3(1 - \alpha)}{\alpha} a_1(0) + \frac{3(1 - \alpha)(4 - 3\alpha)}{2\alpha^2} a_2(0)$$

These initial values are then used to begin forecast simulations of the historical data for different values (usually between .01 and .30) of the smoothing constant α. The sum of squared forecast errors is calculated for each forecast simulation and the simulation yielding the minimum sum of squared forecast errors is determined. The smoothing constant α that was used in the forecast simulation yielding the minimum sum of squared forecast errors is the smoothing constant chosen for actual forecasting.

Johnson and Montgomery [8] suggest that n_1 be set equal to six. That is, they suggest that regression analysis be applied to the first six observations of the available historical data, and that the estimates of β_0, β_1, and β_2 so obtained be used for $a_0(0)$, $a_1(0)$, and $a_2(0)$. Procedures such as this, with $n_1 = 6$, seem to work quite well when used to determine the smoothing constant to be used in single and double exponential smoothing. However, when this procedure is used to determine the smoothing constant to be used in triple exponential smoothing, the forecasts obtained may not be at all satisfactory. This fact will be illustrated in the development that follows. It should be realized that we are not criticizing Johnson and Montgomery [8]. Rather, we are attempting to illustrate how sensitive the forecasts generated by triple exponential smoothing are to the number of observations used to determine the values of the smoothed statistics initializing the simulations that determine the appropriate smoothing constant.

Consider the time series presented in Table 5.7. This time series represents four years of monthly loan requests and was generated by adding a random error term with mean zero and standard deviation 35 to a quadratic trend given by $200 + 50t - .5t^2$.

We shall consider the first 24 observations in this time series to be 2 years of historical data. Note that these observations are the same as those considered in Example 5.3. Triple exponential smoothing has been used to generate one-period-ahead forecasts for the last 24 observations in the time series using different values for n_1 in order to determine the smoothing constant to be used. The accuracy of the forecasts obtained will be compared for the various values of n_1.

We will illustrate our calculations by considering the case in which $n_1 = 12$. Regression analysis on the first 12 observations of the time series yields estimates of β_0, β_1, and β_2 as follows.

$$a_0(0) = 212.16, \ a_1(0) = 48.33, \ a_2(0) = -1.22$$

TABLE 5.7   *Loan Requests*

Year 1		Year 2		Year 3		Year 4	
Period t	y_t	Period t	y_t	Period t	y_t	Period t	y_t
1	297	13	808	25	1169	37	1340
2	249	14	809	26	1153	38	1361
3	340	15	867	27	1193	39	1384
4	406	16	855	28	1237	40	1391
5	464	17	965	29	1268	41	1401
6	481	18	921	30	1218	42	1419
7	549	19	956	31	1288	43	1471
8	553	20	990	32	1295	44	1375
9	556	21	1019	33	1305	45	1442
10	642	22	1021	34	1367	46	1445
11	670	23	1033	35	1343	47	1420
12	712	24	1127	36	1344	48	1444

We first let $\alpha = .08$. Initial values of the smoothed statistics are obtained as follows.

$$S_0 = a_0(0) - \frac{(1 - \alpha)}{\alpha} a_1(0) + \frac{(1 - \alpha)(2 - \alpha)}{2\alpha^2} a_2(0)$$

$$= 212.16 - \frac{(1 - .08)}{.08} (48.33) + \frac{(1 - .08)(2 - .08)}{2(.08)^2} (-1.22)$$

$$= -511.99$$

$$S_0^{[2]} = a_0(0) - \frac{2(1 - \alpha)}{\alpha} a_1(0) + \frac{2(1 - \alpha)(3 - 2\alpha)}{2\alpha^2} a_2(0)$$

$$= 212.16 - \frac{2(1 - .08)}{.08} (48.33) + \frac{2(1 - .08)(3 - .16)}{2(.08)^2} (-1.22)$$

$$= -1397.49$$

$$S_0^{[3]} = a_0(0) - \frac{3(1 - \alpha)}{\alpha} a_1(0) + \frac{3(1 - \alpha)(4 - 3\alpha)}{2\alpha^2} a_2(0)$$

$$= 212.16 - \frac{3(1 - .08)}{.08} (48.33) + \frac{3(1 - .08)(4 - .24)}{2(.08)^2} (-1.22)$$

$$= -2444.34$$

The triple smoothing forecasting equation yields a forecast for period 1 of 260 (rounded) as follows.

$$\hat{y}_{T+\tau}(T) = [6(1 - \alpha)^2 + (6 - 5\alpha)\alpha\tau + \alpha^2\tau^2] \frac{S_T}{2(1 - \alpha)^2}$$

$$- [6(1 - \alpha)^2 + 2(5 - 4\alpha)\alpha\tau + 2\alpha^2\tau^2] \frac{S_T^{[2]}}{2(1 - \alpha)^2}$$

$$+ [2(1 - \alpha)^2 + (4 - 3\alpha)\alpha\tau + \alpha^2\tau^2] \frac{S_T^{[3]}}{2(1 - \alpha)^2}$$

$$\hat{y}_{T+1}(T) = [6(1 - .08)^2 + (6 - 5(.08))(.08)(1) + (.08)^2(1)^2] \frac{S_T}{2(1 - .08)^2}$$

$$- [6(1 - .08)^2 + 2(5 - 4(.08))(.08)(1) + 2(.08)^2(1)^2] \frac{S_T^{[2]}}{2(1 - .08)^2}$$

$$+ [2(1 - .08)^2 + (4 - 3(.08))(.08)(1) + (.08)^2(1)^2] \frac{S_T^{[3]}}{2(1 - .08)^2}$$

$$= 3.26843 S_T - 3.4499 S_T^{[2]} + 1.181474 S_T^{[3]}$$

Thus

$$\hat{y}_{0+1}(0) = 3.26843 S_0 - 3.4499 S_0^{[2]} + 1.181474 S_0^{[3]}$$

$$= 3.26843(-511.99) - 3.4499(-1397.49)$$

$$+ 1.181474(-2444.34)$$

$$= 260 \quad \text{(rounded)}$$

Given the first observation of the time series, $y_1 = 297$, updated values of the smoothed statistics are found using the exponential smoothing equations as follows.

$$S_1 = \alpha y_1 + (1 - \alpha)S_0$$

$$= .08(297) + (1 - .08)(-511.99)$$

$$= -447.27$$

$$S_1^{[2]} = \alpha S_1 + (1 - \alpha)S_0^{[2]}$$

$$= .08(-447.27) + (1 - .08)(-1397.49)$$

$$= -1321.48$$

$$S_1^{[3]} = \alpha S_1^{[2]} + (1 - \alpha)S_0^{[3]}$$

$$= .08(-1321.48) + (1 - .08)(-2444.34)$$

$$= -2354.51$$

The triple smoothing forecasting equation yields a forecast for period 2 of

$$\hat{y}_2(1) = 3.26843S_1 - 3.4499S_1^{[2]} + 1.181474S_1^{[3]}$$

$$= 3.26843(-447.27) - 3.4499(-1321.48)$$

$$+ 1.181474(-2354.51)$$

$$= 315 \quad \text{(rounded)}$$

Similarly, these calculations are performed for the balance of periods 1 through 24. The results of these calculations are summarized in Table 5.8, which gives the observed value of the time series, values of the smoothed statistics, the one-period-ahead forecast, and the forecast error for each period. The sum of the squared forecast errors for this simulation with $\alpha = .08$ is 27,251.

In a similar manner, forecast simulations have been generated using values of α from .02 to .30 in increments of .02. In each simulation, the initial values of the

TABLE 5.8   *Simulating One-Period-Ahead Forecasting of the Historical Loan Requests Time Series Using Triple Exponential Smoothing with $\alpha = .08$. Obtained from $n_1 = 12$ Initial Observations*

Period T	Observation y_T	S_T	$S_T^{[2]}$	$S_T^{[3]}$	Forecast Made Last Period	Forecast Error
1	297	−447.27	−1321.48	−2354.51	260	37
2	249	−391.57	−1247.08	−2265.91	315	−66
3	340	−333.05	−1173.96	−2178.56	345	−5
4	406	−273.92	−1101.96	−2092.43	388	18
5	464	−214.89	−1030.99	−2007.51	434	30
6	481	−159.22	−961.25	−1923.81	483	−2
7	549	−102.56	−892.55	−1841.31	523	26
8	553	−50.12	−825.16	−1760.02	569	−16
9	556	−1.63	−759.28	−1679.96	604	−48
10	642	49.86	−694.54	−1601.13	629	13
11	670	99.47	−631.02	−1523.52	667	3
12	712	148.48	−568.66	−1447.13	702	10
13	808	201.24	−507.07	−1371.92	737	71
14	809	249.86	−446.52	−1297.89	786	23
15	867	299.23	−386.86	−1225.01	824	43
16	855	343.69	−328.41	−1153.28	865	−10
17	965	393.40	−270.67	−1082.67	894	71
18	921	435.60	−214.17	−1013.19	940	−19
19	956	477.24	−158.85	−944.84	966	−10
20	990	518.26	−104.68	−877.63	992	−2
21	1019	558.32	−51.64	−811.55	1018	1
22	1021	595.33	0.11	−746.62	1044	−23
23	1033	630.34	50.53	−682.84	1063	−30
24	1127	670.08	100.10	−620.21	1079	48

smoothed statistics were obtained using $a_0(0) = 212.16$, $a_1(0) = 48.33$, and $a_2(0) = -1.22$, which are the initial estimates of β_0, β_1, and β_2 obtained from regression analysis performed on the first $n_1 = 12$ observations in the historical time series. The sums of squared forecast errors obtained for these simulations are given in Table 5.4. A smoothing constant value of $\alpha = .08$ minimizes the sum of squared forecast errors. Thus this value of α would be used to obtain forecasts of future values of the time series when $n_1 = 12$.

One-period-ahead forecasts for values of the time series in periods 25 through 48 were generated as follows. Initial estimates of β_0, β_1, and β_2 were found using regression analysis on the entire 24-period historical time series. In general, remember that we denote the nummber of observations used to obtain these estimates in the actual forecasting process as n_2. Hence, in this case, $n_2 = 24$. The estimates obtained using the 24 observations were $a_0(0) = 199.62$, $a_1(0) = 50.94$, and $a_2(0) = -1.14$. As shown in Example 5.3, these estimates yielded initial values for the smoothed statistics of $S_0 = -543.51$, $S_0^{[2]} = -1437.40$, and $S_0^{[3]} = -2482.06$. The triple exponential smoothing equations were used to repeatedly update the smoothed statistics through period 24. This updating process yielded values of $S_{24} = 665.82$, $S_{24}^{[2]} = 86.52$, and $S_{24}^{[3]} = -643.85$ for the smoothed statistics in period 24. The forecasting equation yielded a one-period-ahead forecast for period 25 of 1117. One-period-ahead forecasts for periods 26 through 48 were obtained by repeated use of the smoothing equations, which yield updated smoothed statistics, and the forecasting equation, which yields a forecast in each period. The results of these calculations are summarized in Table 5.6, which gives the observed value of the time series, values of the smoothed statistics, the one-period-ahead forecast, and the forecast error for each period.

We now consider using other values for n_1, the number of observations used to obtain the initial estimates of β_0, β_1, and β_2, which are used to initialize the simulations determining the optimal value of α to be used in forecasting the time series. We will consider initializing these simulations using values of $n_1 = 1, 6, 12, 18$, and 24. When $n_1 = 1$, the initial values of the smoothed statistics used in the simulations were assigned the first value in the time series. When $n_1 = 6, 12, 18$, and 24, the initial estimates $a_0(0)$, $a_1(0)$, and $a_2(0)$ were obtained using regression analysis on the first n_1 observations in the time series. These estimates were then used to obtain the initial values of the smoothed statistics used in the simulations. For each value of n_1, simulations were performed for values of α from .02 to .30 in increments of .02 and the value of the smoothing constant α which minimized the sum of squared forecast errors was found as shown previously. The results of these calculations are summarized in Table 5.9.

We now wish to investigate the use of $n_1 = 6$. Table 5.10 summarizes the results of the simulation initialized using $n_1 = 6$ observations and a smoothing constant of $\alpha = .30$ (the optimal value of α obtained using $n_1 = 6$). The table presents the observed value of the time series, the forecast generated by the simulation, and the forecast error for each period. It should be noted that after period 6, "most" of the forecast errors are negative. This indicates that the triple smoothing procedure quite consistently produced forecasts larger than the observed values of the time series. This occurred because the average level of the time series

TABLE 5.9   *Minimum Sums of Squares for Different Values of n_1 in Triple Exponential Smoothing*

n_1	Initial Estimates of β_0, β_1, and β_2	Initial Values of Smoothed Statistics	Optimal Value of α	Minimum Sum of Squares (Given by Optimal α)
$n_1 = 1$	Not available	$S_0 = 297$ $S_0^{[2]} = 297$ $S_0^{[3]} = 297$	$\alpha = .28$	52,635
$n_1 = 6$	$a_0(0) = 241.90$ $a_1(0) = 22.48$ $a_2(0) = 6.90$	$S_0 = 250.83$ $S_0^{[2]} = 328.95$ $S_0^{[3]} = 476.26$	$\alpha = .30$	47,361
$n_1 = 12$	$a_0(0) = 212.16$ $a_1(0) = 48.33$ $a_2(0) = -1.22$	$S_0 = -511.99$ $S_0^{[2]} = -1,397.49$ $S_0^{[3]} = -2,444.34$	$\alpha = .08$	27,251
$n_1 = 18$	$a_0(0) = 215.99$ $a_1(0) = 45.03$ $a_2(0) = -.42$	$S_0 = -3,009.19$ $S_0^{[2]} = -7,242.78$ $S_0^{[3]} = -12,484.80$	$\alpha = .12$	30,193
$n_1 = 24$	$a_0(0) = 199.62$ $a_1(0) = 50.94$ $a_2(0) = -1.14$	$S_0 = -5,061.51$ $S_0^{[2]} = -13,059.77$ $S_0^{[3]} = -23,795.18$	$\alpha = .02$	21,648

increases at a decreasing rate while the initial estimates of β_0, β_1, and β_2 were obtained using the first 6 observations of the time series. During the first 6 periods, the average level of the time series is increasing rapidly and the initial estimates of β_0, β_1, and β_2 based on these 6 observations reflect this rapid increase. However, during the next 18 periods, the average level of the time series is not increasing as rapidly. In order to compensate for this, the triple exponential smoothing procedure requires the use of a large smoothing constant, and even with $\alpha = .30$ most of the forecasts generated are too large. It should be noted that when smoothing constants less than .30 are used in these forecast simulations, the forecast errors obtained are larger in the negative direction.

Now let us consider how the use of this large smoothing constant influences the performance of the triple smoothing procedure when it is used to forecast values of the time series for periods 25 through 48. Forecasts were generated for these periods using the optimal value of $\alpha = .30$ and using initial estimates of β_0, β_1, and β_2 based on the first $n_2 = 24$ periods of the historical time series. The results of these calculations are given in Table 5.11. It will be shown that the use of this large smoothing constant produced forecasts that were influenced by random fluctuations in the time series to a greater degree than were forecasts generated using the optimal value of α obtained when n_1 was assigned values other than 6. Because of this, the performance of the smoothing procedure when $n_1 = 6$ was not as good as its performance for other values of n_1.

We measure the performance of the triple smoothing procedure by considering the sum of squared forecast errors obtained for periods 25 through 48. Table 5.12

gives the sum of squared forecast errors obtained for these periods using several initialization procedures. In particular, we consider two aspects of the initialization. First, we consider how n_1, the number of periods used to obtain initial estimates of β_0, β_1, and β_2 that initialize the forecast simulations determining α, affects the accuracy of the forecasts generated by the smoothing procedure. We also examine how n_2, the number of periods used to obtain initial estimates of β_0, β_1, and β_2 that initialize the actual forecasting process, influences the accuracy of the forecasts.

Notice, in Table 5.12, that the combination ($n_1 = 6$, $n_2 = 24$) does not perform nearly as well as either of the combinations ($n_1 = 12$, $n_2 = 24$) or ($n_1 = 18$, $n_2 = 24$). As described previously, this is because the large smoothing constant obtained by using $n_1 = 6$ allows the forecasts for periods 25 through 48 to be unduly influenced by random fluctuations in the time series. It should also be noted that the combination ($n_1 = 24$, $n_2 = 24$) performs very poorly. In this case, since the esti-

TABLE 5.10   *Simulating One-Period-Ahead Forecasting of the Historical Loan Requests Time Series Using Triple Exponential Smoothing with $\alpha = .30$, Obtained from $n_1 = 6$ Initial Observations*

Period T	Observation y_T	Forecast Made Last Period	Forecast Error
1	297	268	29
2	249	327	−78
3	340	304	36
4	406	371	35
5	464	452	12
6	481	528	−47
7	549	562	−13
8	553	622	−69
9	556	634	−78
10	642	626	16
11	670	686	−16
12	712	723	−11
13	808	763	45
14	809	853	−44
15	867	877	−10
16	855	923	−68
17	965	916	49
18	921	997	−76
19	956	979	−23
20	990	989	1
21	1019	1014	5
22	1021	1041	−20
23	1033	1046	−13
24	1127	1050	77

TABLE 5.11   *One-Period-Ahead Forecasts of Future Values of Loan Requests Using Triple Exponential Smoothing with $\alpha = .30$, Obtained from $n_1 = 6$ Initial Observations*

Period T	y_T	Forecast Made Last Period	Forecast Error
25	1169	1128	41
26	1153	1193	−40
27	1193	1196	−3
28	1237	1222	15
29	1268	1263	5
30	1218	1299	−81
31	1288	1259	29
32	1295	1296	−1
33	1305	1312	−7
34	1367	1321	46
35	1343	1375	−32
36	1344	1369	−25
37	1340	1361	−21
38	1361	1349	12
39	1384	1359	25
40	1391	1382	9
41	1401	1395	6
42	1419	1407	12
43	1471	1425	46
44	1375	1476	−101
45	1442	1406	36
46	1445	1433	12
47	1420	1446	−26
48	1444	1427	17

TABLE 5.12   *Sum of Squares for Periods 25–48*

		$n_1 = $ # Periods Used to Initialize Simulations Determining α				
		1	6	12	18	24
$n_2 = $ # Periods Used	1	29,380				
to Initialize	6		30,840			
Actual Forecasting	12			20,381		
	18				20,642	
	24	29,579	30,950	19,175	20,278	44,938

mates of β_0, β_1, and β_2 initializing the simulation procedure to determine α are based on all 24 observations, there are no historical periods, other than those used to obtain these initial estimates, which remain to be simulated. Hence, a small smoothing constant ($\alpha = .02$) is obtained. This small smoothing constant does not allow the forecasts generated in periods 25 through 48 to compensate for the increasing average level of the time series.

In general, examination of Table 5.12 indicates that the performance of the triple smoothing procedure is very sensitive to the choice of n_1. On the other hand, given the choice of α as determined by n_1, the choice of n_2 does not seem to influence the accuracy of the forecasts nearly as much.

Because the choice of n_1 can be very important, we now consider how this choice might be made. One possible way in which this choice can be made is suggested by comparing Tables 5.9 and 5.12. *Consider determining the optimal value of α for each of several values of* n_1*, where each value of* n_1 *is sufficiently smaller than* n*, the number of observations in the historical time series, so that an adequate number of periods, other than those used to obtain the initial estimates, remain to be simulated.* For example, then, 24 is not a reasonable candidate for n_1, because, as explained previously, setting n_1 equal to 24 leaves no periods, other than those used to obtain the initial estimates, which remain to be simulated. *Next, consider the minimum sum of squared forecast errors obtained for each of these optimal values of α. One might choose the value of* n_1 *which yields the smallest minimum sum of squared forecast errors for an optimal value of α.* For example, examination of Table 5.9 shows that the smallest minimum sum of squared forecast errors obtained for an optimal value of α is 27,251, which corresponds to the optimal value of $\alpha = .08$. These values were obtained using $n_1 = 12$ and thus a value of $n_1 = 12$ and a smoothing constant of $\alpha = .08$ were used for actual forecasting in Example 5.3. Examination of Table 5.12 shows that the choice of $n_1 = 12$ does yield more accurate forecasts for periods 25 through 48 than does any other value of n_1.

5–3.2   CHOICE OF n_1 AND n_2 WHEN USING SIMPLE OR DOUBLE EXPONENTIAL SMOOTHING

The time series presented in Tables 4.1 and 4.6 of Example 4.1, the double exponential smoothing example, has also been analyzed in a similar manner. Simulations determining the optimal value of the smoothing constant α were initialized using values of $n_1 = 1, 6, 12, 18,$ and 24. When $n_1 = 1$, the initial values of the smoothed statistics used in the simulations were assigned the first value in the time series. When $n_1 = 6, 12, 18,$ and 24, the initial estimates of β_0 and β_1 were calculated using regression analysis on the first n_1 observations in the time series. These estimates were then used to find the initial values of the single and double smoothed statistics that initialized the simulations. For each value of n_1, simulations were performed for values of α from .02 to .30 in increments of .02 and the value of the smoothing constant α that minimized the sum of squared forecast errors was

TABLE 5.13   *Minimum Sums of Squares for Different Values of n_1 in Double Exponential Smoothing*

n_1	Initial Estimates of β_0 and β_1	Initial Values of Smoothed Statistics	Optimal Value of α	Minimum Sum of Squares (Given by Optimal α)
$n_1 = 1$	Not available	$S_0 = 197$ $S_0^{[2]} = 197$	.20	31,886
$n_1 = 6$	$b_0(0) = 178.87$ $b_1(0) = 13.94$	$S_0 = 155.66$ $S_0^{[2]} = 113.295$	.16	30,725
$n_1 = 12$	$b_0(0) = 204.80$ $b_1(0) = 6.94$	$S_0 = -135.26$ $S_0^{[2]} = -475.32$	.02	24,755
$n_1 = 18$	$b_0(0) = 196.34$ $b_1(0) = 8.18$	$S_0 = -204.48$ $S_0^{[2]} = -605.30$	.02	23,038
$n_1 = 24$	$b_0(0) = 198.03$ $b_1(0) = 8.07$	$S_0 = -197.40$ $S_0^{[2]} = -592.83$	.02	22,942

found. The results of these calculations are summarized in Table 5.13. The performance of the double smoothing procedure was measured by considering the sum of squared forecast errors obtained for periods 25 through 48. Table 5.14 gives the sum of squared forecast errors obtained for these periods using initialization procedures that employ the values of n_1 specified above as well as various values of n_2, the number of periods used to obtain the initial estimates of β_0 and β_1 used to begin the actual forecasting process. Examination of Table 5.14 indicates that when $n_2 = 24$, the performance of the double smoothing procedure is quite insensitive to the choice of n_1. The use of the values $n_1 = 12$, 18, and 24 yield somewhat better results than the use of the values $n_1 = 1$ and $n_1 = 6$ when $n_2 = 24$. However, the use of $n_1 = 6$, although not the best possible choice when $n_2 = 24$, seems adequate. It should also be noticed that the use of $n_1 = 12$ and $n_2 = 12$ yields a sum of squared forecast errors that is "substantially" smaller than the sum obtained using any other

TABLE 5.14   *Sum of Squares for Periods 25-48*

		$n_1 =$ # Periods Used to Initialize Simulations Determining α				
		1	6	12	18	24
$n_2 =$ # Periods Used	1	62,748				
to Initialize	6		59,642			
Actual Forecasting	12			46,349		
	18				59,021	
	24	62,926	59,234	57,283	57,283	57,283

combination of n_1 and n_2. This might have been expected by having inspected Table 5.13, which indicates that, in simulating the first 24 observations, $n_1 = 12$ yields a minimum sum of squared forecast errors that is nearly as small as the sum obtained by using any other value of n_1. However, it seems strange, since we have generally found that given a choice of n_1 the performance of the smoothing procedure is quite insensitive to the choice of n_2, that $n_1 = 12$ and $n_2 = 24$ does substantially worse than $n_1 = 12$ and $n_2 = 12$. Consequently, one might guess that the fact that $n_1 = 12$ and $n_2 = 12$ does substantially better is due to sampling variation and does not have any long-run implications. Certainly, more simulations, similar to the ones we have done, need to be carried out to better understand the appropriate choice of n_1 and n_2.

A similar analysis can be carried out using data for which simple exponential smoothing is an appropriate forecasting technique. It has been found that, for the cod catch data of Example 3.3, the performance of a simple exponential smoothing procedure is quite insensitive to the choice of n_1 and n_2. Hence, the use of $n_1 = 6$ and $n_2 = 24$ seems adequate when simple exponential smoothing is being applied to the data of Example 3.3.

5–3.3   A SUMMARY RELATING TO CHOOSING n_1 AND n_2

We will now summarize our discussion of choosing n_1 and n_2. It seems that the usual procedure of setting $n_2 = n$ is generally reasonable. However, the procedure of setting $n_1 = 6$, while being fairly reasonable for single and double exponential smoothing, can lead to very inaccurate forecasts when using triple exponential smoothing, if only a moderate number (for example, $n = 24$) of historical observations exist for developing a forecasting model. One method to choose n_1 has been previously discussed and can be summarized as follows: *Consider determining the optimal value of α for each of several values of* n_1, *where each value of* n_1 *is sufficiently smaller than* n, *the number of observations in the historical time series, so that an adequate number of periods, other than those used to obtain the initial estimates, remain to be simulated. Next, consider the minimum sum of squared forecast errors obtained for each of these optimal values of α. One might choose the value of* n_1 *which yields the smallest minimum sum of squared forecast errors for an optimal value of α.* This method to choose n_1 seems intuitively reasonable and has been suggested by the simulations we have carried out. It is interesting to note that these same simulations suggest that there is a very simple method to choose n_1 that gives reasonable results. This method is to *choose* n_1 *equal to one half of the number of historical observations available to develop a forecasting model.*

Of course, the choice of n_1 is most important when a limited amount of historical data is available. The larger the amount of historical data available, the less important is the choice of n_1, because the simulation procedure used to determine the smoothing constant has more time periods in which to correct for inaccurate initial estimates of the model parameters. However, the smaller the number of observations available in the historical time series, the greater is the danger that inaccurate initial estimates of the parameters of the time series will adversely influence the results obtained, and the more critical is the value of n_1 chosen. Clearly, if the number of historical observations is extremely small, there may be no method for choosing n_1 (or n_2) that will give accurate forecasts of future time series values. However, if there exist a moderate number of historical observations, simulations indicate that both of the above suggested methods for choosing n_1 will give substantially more accurate forecasts when using triple exponential smoothing and equivalent or somewhat better forecasts when using single or double exponential smoothing than will the procedure of setting $n_1 = 6$.

PROBLEMS

1. Consider Example 5.2 in which the regression approach is used to forecast the loan requests time series of Table 5.1. Using this approach, calculate loan request forecasts for periods 30 and 35.

2. Consider Example 5.3 in which the triple exponential smoothing approach is used to forecast the loan requests time series of Table 5.1.

   a. Verify that the single, double, and triple smoothed statistics given in Table 5.3 for March of year 1 (period 3) are

$$S_3 = 144.26, \ S_3^{[2]} = -70.43, \text{ and } S_3^{[3]} = -308.87$$

   b. Verify that the single, double, and triple smoothed statistics given in Table 5.3 for April of year 1 (period 4) are

$$S_4 = 196.61, \ S_4^{[2]} = -17.02, \text{ and } S_4^{[3]} = -250.50$$

3. Consider Example 5.3 in which the triple exponential smoothing approach is used to forecast the loan requests time series of Table 5.1.

   a. Consider Table 5.6. Verify that the loan request forecast made in February of year 3 (period 26) for March of year 3 (period 27) is 1173 (rounded).
   b. Again consider Table 5.6. Verify that the loan request forecast made in March of year 3 (period 27) for April of year 3 (period 28) is 1199 (rounded).

4. Find the initial values of the single, double, and triple smoothed statistics to be used when simulating the historical loan request data in Table 5.1 with a smoothing constant of $\alpha = .10$.

*5. Consider the regression calculations in Section *5–2.1, which have been carried out for Example 5.2.

  a. Determine the vector $\mathbf{x}'_{30}$ to be used in calculating a 95 percent confidence interval for loan requests in period 30.

  b. Determine the vector $\sum_{t=30}^{31} \mathbf{x}'_t$ to be used in calculating a 95 percent confidence interval for cumulative loan requests in periods 30 and 31.

---

* Starred problems are only for students who have read Section *5–2.1.

PART II SUMMARY

In Part II you have learned how to analyze and forecast time series that can be described by trend and irregular components. In particular, you have studied time series that display no trend, a linear trend, and a quadratic trend. You have seen that the regression approach and the exponential smoothing approach can both be used to forecast these time series. The exponential smoothing approach achieves unequal weighting of the observed values of the time series by using a smoothing constant. This unequal weighting insures that recent observations contribute more heavily than remote observations when a forecast is made.

In Chapter 3 you learned how to use both regression and simple exponential smoothing to forecast time series with no trend. When using exponential smoothing, the choice of a smoothing constant is important. You learned that this choice can be made through simulation of historical data. The choice of a smoothing constant is one of several steps involved in the overall implementation of an exponential smoothing procedure. You have learned these steps and have also seen that forecast errors must be analyzed so that the smoothing constant in an exponential smoothing procedure can be changed when necessary. Procedures that automatically change the smoothing constant when necessary are called adaptive control processes. You have studied one adaptive control process, called Chow's Method, in detail.

In Chapter 4 you learned how to forecast time series with a linear trend. Both regression and double exponential smoothing can be used to forecast such time series. You have seen that both methods can be used to obtain point forecasts and confidence interval forecasts. In particular, you have learned that a method called general direct smoothing can be used to calculate confidence interval forecasts when exponential smoothing is being used to forecast a time series.

In Chapter 5 you learned how to forecast time series with a quadratic trend. Both regression and triple exponential smoothing can be used to forecast such time series. You have seen that the initialization of a triple exponential smoothing procedure can be quite complicated and can affect the accuracy of the forecasts that are generated by the procedure. Specifically, it is important to consider the choice of n_1, the number of observations used to obtain initial estimates of the model parameters that determine initial values of the smoothed statistics that begin the simulations which find the best smoothing constant. It is also important to consider n_2, the number of observations used to obtain initial estimates of the model parameters used to begin actual forecasting.

EXERCISES

1. Describe the behavior of a time series that exhibits no trend.

2. Describe the behavior of a time series that exhibits a linear trend.

3. Describe the behavior of four different data patterns that can be described by a quadratic trend.

4. What are the two basic approaches used to forecast time series described by trend and irregular components?

5. Explain why the name "simple exponential smoothing" was given to the procedure described in Section 2–2.2.

6. In simple exponential smoothing, which observations make the largest contributions to the current estimate of the average level of the time series being forecasted?

7. Explain how simulation of historical data is used to determine an appropriate smoothing constant.

8. In the context of the overall implementation of an exponential smoothing procedure, explain the meanings of n_1 and n_2.

9. Explain the purpose of adaptive control procedures in exponential smoothing.

10. Describe the historical data pattern that would indicate that simple exponential smoothing might be an appropriate method to be used in forecasting a time series.

11. Describe the historical data patterns that would indicate that double exponential smoothing might be an appropriate method to be used in forecasting a time series.

12. Describe the historical data patterns that would indicate that triple exponential smoothing might be an appropriate method to be used in forecasting a time series.

13. Why is the value of n_1 chosen to be smaller than n (the number of observations in the historical time series)?

14. Explain how an appropriate value of n_1 might be chosen.

15. When are the choices of n_1 and n_2 the most critical?

16. Why is the theoretical, statistical basis behind general direct smoothing important?

VOCABULARY LIST

No trend	Adaptive control procedures
Linear trend	Chow's procedure
Quadratic trend	Double exponential smoothing
Simple exponential smoothing	Double smoothed statistic
Smoothed statistic	Triple exponential smoothing
Smoothing constant	Triple smoothed statistic
Tracking signal	General direct smoothing

QUIZ

Answer TRUE if the statement is always true. If the statement is not true, replace the underlined word(s) with a word(s) that makes the statement true.

1. A time series with no trend is randomly fluctuating around an average level that is <u>changing rapidly</u> over time.

2. A time series with a linear trend is randomly fluctuating around an average level that changes in a linear or <u>curvilinear</u> fashion over time.

3. A time series with a quadratic trend could be randomly fluctuating around an average level that is <u>decreasing</u> <u>at</u> <u>an</u> <u>increasing rate</u>.

4. <u>Simple</u> exponential smoothing is used when the historical data pattern of a time series exhibits a linear trend.

5. In simple exponential smoothing, the smoothed estimate or statistic can be shown to be a <u>linear combination</u> of all <u>past</u> <u>observations</u> of the time series.

6. In simple exponential smoothing, the most recent observation makes the <u>smallest</u> contribution to the estimate of β_0, while older observations make <u>larger</u> <u>and</u> <u>larger</u> contributions at each successive time point.

7. In simple exponential smoothing, large values of the smoothing constant dampen out remote observations in the time series <u>very</u> <u>slowly</u>.

8. Simulation of historical data may be used to determine an appropriate smoothing constant in <u>single</u>, <u>double</u>, <u>and</u> <u>triple</u> exponential smoothing.

9. One arbitrary aspect of the implementation of an exponential smoothing procedure is the choice of n_1, the number of observations used to calculate the initial values of the smoothed statistics used to begin actual forecasting of future values of a time series.

10. A small value of the tracking signal implies that the forecasting system is producing errors that are either consistently positive or consistently negative.

11. The regression approach can be used to obtain point and confidence interval forecasts for time series that exhibit no trend, a linear trend, or a quadratic trend.

12. Double exponential smoothing is used when the historical data pattern of a time series exhibits a quadratic trend.

13. The double smoothed statistic is found by applying the smoothing operation to the output of the single smoothing equation.

14. Initial values of the single and double smoothed statistics are often obtained from estimates of the coefficients β_0 and β_1 generated through simple exponential smoothing of historical data.

15. Triple exponential smoothing is used when the historical data pattern of a time series exhibits a quadratic trend.

16. Triple exponential smoothing involves the use of three smoothing constants.

17. The triple smoothed statistic is found by applying the smoothing operation to the cube of the single smoothed statistic.

18. General direct smoothing can be used to calculate approximate confidence intervals for future values of time series being forecasted by single, double, or triple exponential smoothing.

19. Confidence intervals for sums of future values of time series exhibiting no trend, a linear trend, or a quadratic trend can be obtained using the regression approach, but not the exponential smoothing approach.

20. General direct smoothing can be used to calculate confidence intervals that can be updated in light of new observations of the time series being forecasted.

REFERENCES

1. Box, G. E. P., and G. M. Jenkins, *Time Series Analysis, Forecasting and Control,* Holden-Day, Inc., San Francisco, 1970.

2. Brown, R. G., *Statistical Forecasting for Inventory Control,* McGraw-Hill Book Company, New York, 1959.

3. Brown, R. G., *Smoothing, Forecasting and Prediction of Discrete Time Series,* Prentice-Hall, Inc., Englewood Cliffs, New Jersey, 1962.

4. Brown, R. G., *Decision Rules for Inventory Management,* Holt, Rinehart and Winston, Inc., New York, 1967.

5. Chow, W. M., "Adaptive Control of the Exponential Smoothing Constant," *Journal of Industrial Engineering,* vol. 16, no. 5, pp. 314–317, 1965.

6. Draper, N. R., and H. Smith, *Applied Regression Analysis,* John Wiley & Sons, Inc., New York, 1976.

7. Fuller, Wayne A., *Introduction to Statistical Time Series,* John Wiley & Sons, Inc., New York, 1976.

8. Johnson, L. A., and D. C. Montgomery, *Forecasting and Time Series Analysis,* McGraw-Hill Book Company, New York, 1976.

9. Nelson, C. R., *Applied Time Series Analysis for Managerial Forecasting,* Holden-Day, Inc., San Francisco, 1973.

10. Neter, John, and William Wasserman, *Applied Linear Statistical Models,* R. D. Irwin, Homewood, Illinois, 1974.

11. Wheelright, S. C., and S. Makridakis, *Forecasting Methods for Management,* John Wiley & Sons, Inc., New York, 1973.

12. Winters, P. R., "Forecasting Sales by Exponentially Weighted Moving Averages," *Management Science,* vol. 6, no. 3, pp. 324–342, 1960.

PART III

Forecasting Seasonal Time Series

CONTENTS

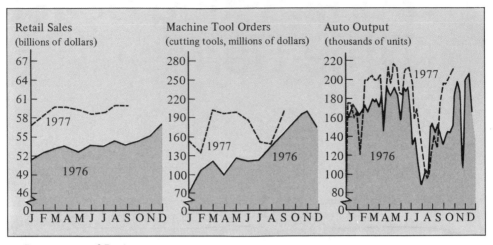

Barometers of Business

Source: *Industry Week,* November 7, 1977. Data for retail sales from U.S. Department of Commerce; data for machine tool orders from National Machine Tool Builders' Association; data for auto output from Ward's Automotive Reports. Reprinted by permission of NMTBA, Ward's Automotive Reports, and *Industry Week*.

OBJECTIVES FOR PART III

As pointed out in Chapter 1, many time series possess seasonal variation. The objective of Part III is to study the analysis and forecasting of time series that can be adequately described by trend, seasonal, and irregular components. The multiplicative decomposition method, Winters' Method, and several regression approaches to forecasting seasonal time series will be presented. As examples of the techniques to be studied, we will forecast demand for nutrition seminars by elementary school teachers, sales of Rola-Cola by the Pop-A-Top Pop Shop, Inc., and sales of the Bass Grabber fishing lure by Alluring Tackle, Inc.

INTRODUCTION TO PART III

III–1 ADDITIVE AND MULTIPLICATIVE SEASONAL VARIATION

In Part III we study time series that are seasonal in nature. That is, we will consider time series that include seasonal variation. We will define and concentrate on two types of seasonal variation. The first type is *additive seasonal variation*. If a time series displays additive seasonal variation, the magnitude of the "seasonal swing" of the time series is *independent* of the average level as determined by the trend. Additive seasonal variation is illustrated in Figure III.1. The second type is *multiplicative seasonal variation*. If a time series displays multiplicative seasonal variation, the magnitude of the "seasonal swing" of the time series is *proportional* to the average level as determined by the trend. Thus, if the average level of the time series is increasing, so is the magnitude of the seasonal swing, while if the average

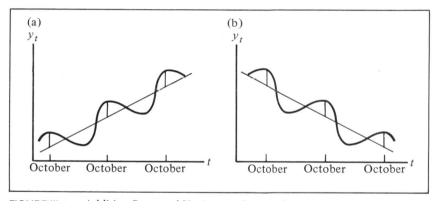

FIGURE III.1   *Additive Seasonal Variation: Seasonal Swing Is the Same as the Time Series Increases or Decreases*

level of the time series is decreasing, the magnitude of the seasonal swing is also decreasing. Multiplicative seasonal variation is illustrated in Figure III.2.

Very few real-world time series possess seasonal variation that is precisely additive or precisely multiplicative in nature. However, in Part III we will attempt to classify each seasonal time series we study as either approximately possessing additive seasonal variation or approximately possessing multiplicative seasonal variation. It is important to do this so that we may choose an appropriate model to use in analyzing the time series. Some models are appropriate for analyzing additive seasonal variation, while others are appropriate for analyzing multiplicative seasonal variation. Many of the time series models studied in Part III are special cases of the model

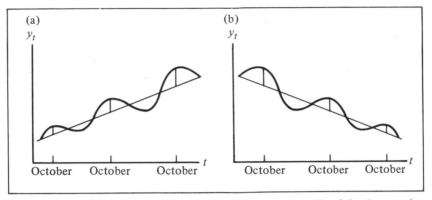

FIGURE III.2 *Multiplicative Seasonal Variation: Magnitude of the Seasonal Swing Is Proportional to the Average Level of the Time Series*

$$y_t = f(TR_t, SN_t) + \epsilon_t$$

where y_t = the observed value of the time series in time period t
  TR_t = the trend factor of the time series in time period t
  SN_t = the seasonal factor of the time series in time period t
    f = a function relating the observed value of the time series to the trend and seasonal factors
    ϵ_t = the irregular factor of the time series in time period t

In Part III we will study various methods that yield estimates of TR_t and SN_t. These estimates will be based on previously observed historical data. Denoting the estimate of TR_t by the symbol tr_t and the estimate of SN_t by the symbol sn_t, the estimate of the value of the time series in period t is

$$\hat{y}_t = f(tr_t, sn_t)$$

When analyzing a time series having additive seasonal variation, it is generally assumed that

$$f(TR_t, SN_t) = TR_t + SN_t$$

which implies that the magnitude of the "seasonal swing" of the time series is independent of the average level as determined by the trend. In this case the estimate of the value y_t is

$$\hat{y}_t = tr_t + sn_t$$

When analyzing a time series having multiplicative seasonal variation, it is generally assumed that

$$f(TR_t, SN_t) = TR_t \times SN_t$$

which implies that the "seasonal swing" of the time series is proportional to the average level as determined by the trend. In this case the estimate of the value y_t is

$$\hat{y}_t = tr_t \times sn_t$$

The trend factor of the time series in time period t, TR_t, is frequently assumed to be given by one of the following equations.

1. $TR_t = \beta_0$, which implies that there is no trend,
2. $TR_t = \beta_0 + \beta_1 t$, which implies that there is a linear trend, or
3. $TR_t = \beta_0 + \beta_1 t + \beta_2 t^2$, which implies that there is a quadratic trend.

We will find that most of the time series analyzed in Part III exhibit a linear trend. We will, therefore, assume that $TR_t = \beta_0 + \beta_1 t$. The reader should realize, however, that many of the methods discussed in Part III can be easily modified to deal with other types of trend.

III–2 THE SEASONAL FACTOR

We will now discuss the meaning of the seasonal factor SN_t. This factor is a correction factor that accounts or adjusts for the seasonality in the time series. For example, suppose that sales of outboard motors by Power Drive Corporation are seasonal in nature, with sales in the first quarter (January–March) being the lowest, sales in the second quarter (April–June) being the highest because of the beginning of the boating season, sales in the third quarter (July–September) being somewhat higher than normal, and sales in the fourth quarter (October–December) being somewhat lower than normal. Further, suppose that the trend factor for outboard motor sales by Power Drive Corporation is

$$TR_t = 500 + 50t$$

where the origin is considered to be the fourth quarter of 1978. Then if trend alone is considered, outboard motor sales for the four quarters of 1979 (periods 1, 2, 3, and 4) are expected to be

$$TR_1 = 500 + 50(1) = 550 \quad \text{(January–March)}$$
$$TR_2 = 500 + 50(2) = 600 \quad \text{(April–June)}$$
$$TR_3 = 500 + 50(3) = 650 \quad \text{(July–September)}$$
$$TR_4 = 500 + 50(4) = 700 \quad \text{(October–December)}$$

However, since sales are seasonal in this situation, we must correct these trend values to account for the seasonal nature of the time series. Assume that the four seasonal factors (one for each quarter in the year) describing the seasonal pattern of outboard motor sales are

$$SN_1 = .4 \quad \text{(January–March)}$$
$$SN_2 = 1.6 \quad \text{(April–June)}$$
$$SN_3 = 1.2 \quad \text{(July–September)}$$
$$SN_4 = .8 \quad \text{(October–December)}$$

Thus, if we assume that outboard motor sales possess multiplicative seasonal variation, and if trend and seasonal variation alone are considered, sales in period t are expected to be

$$TR_t \times SN_t$$

So, in the first quarter (January–March) of 1979, sales are expected to be

$$TR_1 \times SN_1 = (500 + 50(1)) \times .4 = 220$$

Thus the seasonal factor $SN_1 = .4$ corrects for the seasonality in the time series and reduces sales below the level that would be expected if trend alone is considered. In the second quarter (April–June) of 1979 sales are expected to be

$$TR_2 \times SN_2 = (500 + 50(2)) \times 1.6 = 960$$

Again, the seasonal factor $SN_2 = 1.6$ corrects for the seasonality in the time series, this time increasing sales above the level that would be expected if trend alone is considered. Similarly, sales for the third quarter (July–September) and fourth quarter (October–December) of 1979 are expected to be

$$TR_3 \times SN_3 = (500 + 50(3)) \times 1.2 = 780$$

$$TR_4 \times SN_4 = (500 + 50(4)) \times .8 = 560$$

In Figure III.3a, we plot expected outboard motor sales for the four quarters of 1979 when trend alone is considered. In Figure III.3b, we plot expected outboard motor sales for the four quarters of 1979 when both trend and seasonal variation are considered. Note that multiplication of the trend values by the appropriate seasonal factors has resulted in expected sales displaying a seasonal pattern.

In the foregoing discussion we assumed that outboard motor sales possess multiplicative seasonal variation. If, instead, we assume that outboard motor sales possess additive seasonal variation, seasonal factors can be defined that will generate the same seasonal pattern for the four quarters of 1979. Assume that the four seasonal factors describing the seasonal pattern of outboard motors are, in this case,

$$SN_1 = -330 \qquad \text{(January–March)}$$

$$SN_2 = 360 \qquad \text{(April–June)}$$

$$SN_3 = 130 \qquad \text{(July–September)}$$

$$SN_4 = -140 \qquad \text{(October–December)}$$

Thus, if we assume that outboard motor sales possess additive seasonal variation, and if both trend and seasonal variation are considered, sales in period t are expected to be

$$TR_t + SN_t$$

So sales for the four quarters of 1979 are expected to be

$$TR_1 + SN_1 = (500 + 50(1)) - 330 = 220$$

$$TR_2 + SN_2 = (500 + 50(2)) + 360 = 960$$

$$TR_3 + SN_3 = (500 + 50(3)) + 130 = 780$$

$$TR_4 + SN_4 = (500 + 50(4)) - 140 = 560$$

In this case, the seasonal factors add the appropriate correction factor to trend levels to account for the seasonality of the time series.

In most of this book we will assume that the seasonal factors for all time periods in the same season remain constant. That is, for example, we will assume that the seasonal factor for outboard motor sales in quarter 2 of 1979 is the same as the seasonal factor for outboard motor sales in quarter 2 of 1980, 1981, 1982, and so

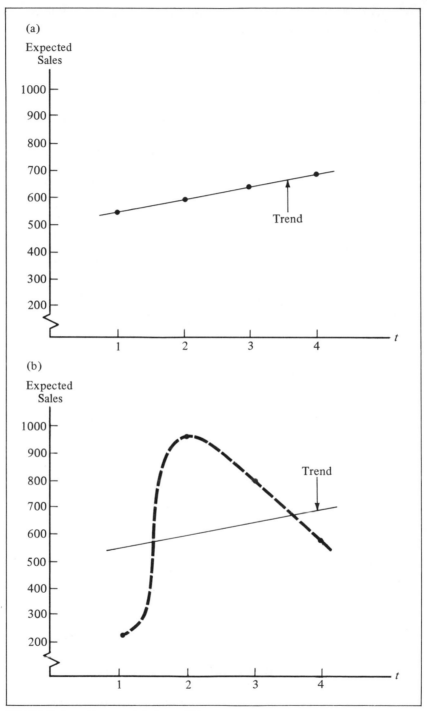

FIGURE III.3a   *Expected Sales* $= TR_t$
FIGURE III.3b   *Expected Sales* $= TR_t \times SN_t$

on. If a time series possesses additive seasonal variation, this assumption implies that the magnitude of the seasonal swing in the time series remains constant as time advances, as illustrated in Figure III.1. For example, if we assume outboard motor sales possess additive seasonal variation, in quarter 2 of 1979 (period 2),

$$TR_2 = 500 + 50(2) = 600$$

while sales are expected to be

$$TR_2 + SN_2 = (500 + 50(2)) + 360 = 960$$

Thus, for quarter 2 of 1979, sales are expected to be 360 units above the trend value for that time period. In quarter 2 of 1980 (period 6),

$$TR_6 = 500 + 50(6) = 800$$

while sales are expected to be

$$TR_6 + SN_6 = (500 + 50(6)) + 360 = 1160$$

Here, $SN_6 = SN_2$ since time periods 2 and 6 are in the same season (quarter 2). Again, for quarter 2 of 1980, sales are expected to be 360 units above the trend value for the time period, and thus the magnitude of the seasonal swing remains constant.

However, if a time series possesses multiplicative seasonal variation, the assumption that the seasonal factors for all time periods in the same season remain constant does not imply that the seasonal swing in the time series remains the same as time advances. Rather, the seasonal swing in the time series is proportional to the trend as illustrated in Figure III.2. For example, if we assume outboard motor sales possess multiplicative seasonal variation, in quarter 2 of 1979 (period 2),

$$TR_2 = 500 + 50(2) = 600$$

while sales are expected to be

$$TR_2 \times SN_2 = (500 + 50(2)) \times 1.6 = 960$$

So, in quarter 2 of 1979, sales are expected to be 360 units above trend. However, in quarter 2 of 1980 (period 6),

$$TR_6 = 500 + 50(6) = 800$$

while sales are expected to be

$$TR_6 \times SN_6 = (500 + 50(6)) \times 1.6 = 1280$$

Again $SN_6 = SN_2$ since time periods 2 and 6 are in the same season (quarter 2). But, notice that in quarter 2 of 1980, sales are expected to be 480 units above trend. Thus, although the seasonal factor for quarter 2 remains constant from 1979 to 1980, the magnitude of the seasonal swing has increased from 360 to 480 because it is now proportional to the trend due to the multiplicative nature of the seasonal variation. In some situations, however, the seasonal factors for all time periods in the same season do not remain constant. We will discuss a method that can be used to analyze shifting seasonal factors in Section 6–4.

III–3 THE CYCLICAL FACTOR

Recall from Chapter 1 that seasonal variation in a time series, and hence the seasonal factor SN_t, reflects cyclical patterns in a time series that are completed within one calendar year. If a time series is also influenced by a cyclical pattern that has a duration of more than one year, a cyclical factor can also be defined. This factor, denoted CL_t, adjusts for the influence of the cyclical pattern that exists. For example, suppose that all four quarters of 1979 are included in a "boom period" of the business cycle. This would lead to an increase in outboard motor sales above sales that would be expected if economic activity was normal. Assume that the cyclical factors describing the "boom" in economic activity during the four quarters in 1979 are

$$CL_1 = 1.08 \qquad \text{(January–March)}$$
$$CL_2 = 1.09 \qquad \text{(April–June)}$$
$$CL_3 = 1.09 \qquad \text{(July–September)}$$
$$CL_4 = 1.10 \qquad \text{(October–December)}$$

Thus, if we assume a multiplicative model, and if trend, seasonal variation, and cyclical variation are considered, sales in period t are expected to be

$$TR_t \times SN_t \times CL_t$$

So sales for the four quarters of 1979 are expected to be

$$TR_1 \times SN_1 \times CL_1 = 220(1.08) = 238$$
$$TR_2 \times SN_2 \times CL_2 = 960(1.09) = 1046$$
$$TR_3 \times SN_3 \times CL_3 = 780(1.09) = 850$$
$$TR_4 \times SN_4 \times CL_4 = 560(1.10) = 616$$

Thus the cyclical factors have increased expected sales above the level that would be expected if trend and seasonal variation alone are considered. The increased expected sales reflect the "boom" in economic activity.

Note that in the foregoing discussion we have assumed that the trend, seasonal, and cyclical factors, TR_t, SN_t, and CL_t, are known. Of course, when analyzing a real time series these factors are not known to the forecaster. In the chapters that follow we will present methods that can be used to estimate these factors.

III–4 TWO EXAMPLES

Now, returning to our discussion of multiplicative and additive seasonal variation, using a model appropriate for one type of seasonal variation to analyze and forecast a time series possessing the other type of seasonal variation can lead to inaccurate forecasts. To demonstrate the above points, we will concentrate on analyzing and forecasting two time series. In the following paragraphs we will describe the two time series and give hypothetical, "real life" situations in which it would be necessary to forecast future values of these time series.

The first time series we will analyze and forecast in Part III describes demand for nutrition seminars by elementary school teachers. During the 1970s the importance of proper nutrition has become increasingly apparent. One of the duties of the State Nutrition Center, which is located in State Capitol, is to give seminars on nutrition to interested groups from around the state. Since it is felt that it is important for elementary school children to be exposed to the ideas of proper nutrition, the State Nutrition Center has made available to elementary school teachers a two-day seminar on proper nutrition. Any individual elementary school teacher or any group of elementary school teachers may attend one of these seminars. The State Nutrition Center uses attendance requests from elementary school teachers to schedule seminars accommodating from 50 to 100 teachers. Applying teachers specify the dates that they would prefer to attend a seminar. In the past, the State Nutrition Center has not only honored all requests for seminars but also has been successful in honoring date preferences. In order to plan for future seminars, the State Nutrition Center needs to forecast "seminar demand" for certain periods of time. "Seminar demand" for a given period of time is defined to be the number of two-day seminars that are needed to accommodate attendance requests by elementary school teachers during that period of time. For forecasting purposes, the State Nutrition Center has divided the year into four quarters, winter (January through March), spring (April through June), summer (July through September), and fall (October through December). At the end of each quarter, the State Nutrition Center desires to have point forecasts and confidence interval forecasts of both seminar demand in individual future quarters and total seminar demand for the next three quarters. The point forecast is the best guess at seminar demand for the particular period, while the confidence interval forecast states the smallest and largest that seminar demand can reasonably be expected to be in the future period.

The State Nutrition Center has recorded quarterly seminar demand for the previous four years, which we will call year 1, year 2, year 3, and year 4. The seminar demand time series is given in Table III.1 and is plotted in Figure III.4.

TABLE III.1 *Quarterly Seminar Demand*

Year	Quarter	t	Demand y_t	Year	Quarter	t	Demand y_t
1	1 (Winter)	1	10	3	1	9	13
	2 (Spring)	2	31		2	10	34
	3 (Summer)	3	43		3	11	48
	4 (Fall)	4	16		4	12	19
2	1	5	11	4	1	13	15
	2	6	33		2	14	37
	3	7	45		3	15	51
	4	8	17		4	16	21

Notice that, in addition to having a linear trend, this time series certainly possesses seasonal variation, with seminar demand being lowest in the winter, increasing in the spring and summer, and then decreasing in the fall. The State Nutrition Center realizes that the reasons for this seasonal pattern are as follows. First, it is generally true that seminars are held at the State Nutrition Center in State Capitol, and hence teachers have to arrange transportation to State Capitol. Due to the fact that in the winter quarter the teachers have to deal with bad weather, "recovery" from the holiday period, and the beginning of a new semester, seminar demand is lowest in

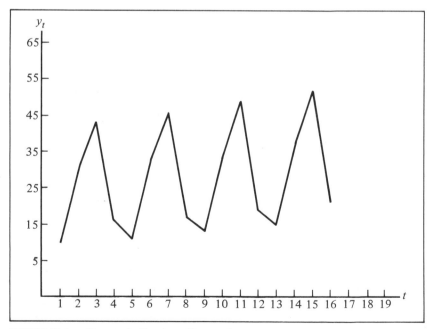

FIGURE III.4 *Quarterly Seminar Demand*

the winter quarter. In the spring quarter, seminar demand increases because the weather is better and teachers have the current semester better under control. Seminar demand is the greatest in the summer quarter because the weather is the best and because, since there is no school, teachers have more time to attend seminars. In the fall quarter, seminar demand is lower than in the spring and summer and greater than in the winter because teachers have to deal with the beginning of the school year and because the weather becomes worse as the fall quarter progresses. Although the successive differences between summer and winter seminar demand, which are 33 in year 1, 34 in year 2, 35 in year 3, and 36 in year 4, are increasing as time passes, the successive differences between seminar demand in each of the other seasons and winter seminar demand are not increasing as time passes. Consequently, the seminar demand time series is perhaps best described as possessing additive seasonal variation.

The second time series we will analyze and forecast in Part III describes sales of a particular soft drink[1] by a chain of soft drink stores. The Pop-A-Top Pop Shop, Inc., owns and operates ten drive-in soft drink stores in York City, which has a population of roughly one million. This chain of soft drink stores, often referred to as Pop-A-Top, has been selling Rola-Cola, a soft drink which was introduced on the market just three years ago and has been gaining in popularity with the inhabitants of York City. Periodically, Pop-A-Top orders a supply of Rola-Cola from the regional distributor of Rola-Cola. Pop-A-Top uses an inventory policy that attempts to insure that its stores will have enough Rola-Cola to meet practically all of the demand for Rola-Cola, while at the same time insuring that the company does not needlessly tie up its money by ordering much more Rola-Cola than can be reasonably be expected to be sold. As previously discussed, it is true that ''sales'' and ''demand'' are not the same. However, since Pop-A-Top has a policy of having enough Rola-Cola in stock to insure that it very seldom loses a potential sale, we will in this example consider sales and demand to be the same. In order to implement its inventory policy, Pop-A-Top needs to forecast Rola-Cola sales. ''Rola-Cola sales'' for a particular period is defined to be the total sales in hundreds of cases of Rola-Cola to patrons shopping in Pop-A-Top's ten York City stores during that period of time. At the end of each month, Pop-A-Top desires point forecasts and confidence interval forecasts of total Rola-Cola sales in individual future months, total Rola-Cola sales in the next three months, and total Rola-Cola sales in the next four months. The point forecast is the best guess at Rola-Cola sales for the particular period, while the confidence interval forecast states the smallest and largest that Rola-Cola sales can reasonably be expected to be in the future period.

Pop-A-Top has recorded monthly Rola-Cola sales for the previous three years, which we will call year 1, year 2, and year 3. This time series, which we will henceforth refer to as either the Rola-Cola sales time series or the soft drink sales time series, is given in Table III.2 and plotted in Figure III.5. Notice that, in addition to having a linear trend, the Rola-Cola sales time series possesses seasonal variation, with sales of the soft drink being greatest in the summer and early fall months and smallest in the winter months. Notice, also, the plot of this time series

---

[1] This example was motivated by a similar example in Johnson and Montgomery [8].

TABLE III.2   *Monthly Sales of Rola-Cola (in hundreds of cases)*

Year	Month	t	Sales y_t	Year	Month	t	Sales y_t
1	1 (Jan.)	1	189	2	7	19	831
	2 (Feb.)	2	229		8	20	960
	3 (Mar.)	3	249		9	21	1152
	4 (Apr.)	4	289		10	22	759
	5 (May)	5	260		11	23	607
	6 (June)	6	431		12	24	371
	7 (July)	7	660	3	1	25	298
	8 (Aug.)	8	777		2	26	378
	9 (Sept.)	9	915		3	27	373
	10 (Oct.)	10	613		4	28	443
	11 (Nov.)	11	485		5	29	374
	12 (Dec.)	12	277		6	30	660
2	1	13	244		7	31	1004
	2	14	296		8	32	1153
	3	15	319		9	33	1388
	4	16	370		10	34	904
	5	17	313		11	35	715
	6	18	556		12	36	441

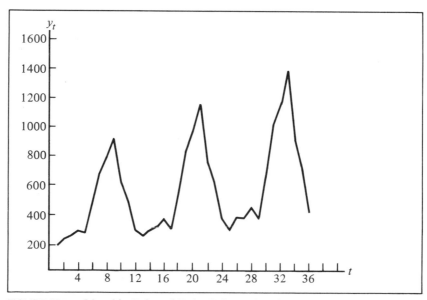

FIGURE III.5   *Monthly Sales of Rola-Cola (in hundreds of cases)*

has the appearance of Figure III.2a rather than the appearance of Figure III.1a. Hence, this time series is perhaps best described as possessing multiplicative seasonal variation. In general, plotting a time series is of great help in classifying it as possessing either additive or multiplicative seasonal variation.

The seminar demand and Rola-Cola time series were in fact artificially generated. The seminar demand time series, which appears to approximately possess a linear trend and additive seasonal variation, was generated by *adding* seasonal factors to a linear trend and then adding error terms generated so that they would be independent and identically distributed according to a normal distribution. The Rola-Cola time series, which appears to approximately possess a linear trend and multiplicative seasonal variation, was generated by *multiplying* a linear trend by seasonal factors and then adding error terms generated so that they would be independent and identically distributed according to a normal distribution. We discuss this artificial generation because, for purposes of comparison, we will analyze and forecast each of these time series using both a model appropriate for additive seasonal variation and a model appropriate for multiplicative seasonal variation. Then, in Chapter 9 we will summarize our findings. It will be found, as we would expect, that a multiplicative model forecasts future values of the generated Rola-Cola sales time series, which possesses multiplicative seasonal variation, more accurately than does an additive model, and an additive model forecasts future values of the generated seminar demand time series, which possesses additive seasonal variation, more accurately than does a multiplicative model.

It should be noticed that the seminar demand time series consists of quarterly seasonal data, where there are four "seasons," and the Rola-Cola sales time series consists of monthly seasonal data, where there are twelve "seasons." Many time series fall into one of these two categories. It should, of course, also be realized that although both of our time series exhibit seasonal variation such that the largest values of the time series are in the summer, this is not generally the case. For example, jewelry sales are generally greatest in December because of Christmas and are much smaller in the summer. The methods to be developed in Part III will apply to any number of "seasons" and to any seasonal pattern.

III–5 METHODS TO BE DISCUSSED

In Chapter 6 we will study the multiplicative decomposition method of analyzing and forecasting time series with multiplicative seasonal variation. In Chapter 7 we will discuss Winters' Method, which is an extension of exponential smoothing to forecasting time series with multiplicative seasonal variation. In Chapter 8 we will discuss forecasting time series with additive seasonal variation. Both a regression approach and Winters' Method, this time as an extension of exponential smoothing to forecasting time series with additive seasonal variation, will be presented. In Chapter 9 we will discuss some additional topics concerning seasonal variation. First, we will summarize our previous modeling of time series possessing additive and multiplicative seasonal variation. Next, we will discuss two more regression approaches to forecasting time series with seasonal variation—regression models having trigonometric terms and regression models involving causal variables and mathematical functions of time.

The methods discussed in Part III assume that the error terms ϵ_1, ϵ_2, . . . in the general seasonal model

$$y_t = f(TR_t, SN_t) + \epsilon_t$$

adhere to the assumptions discussed in Section 1–5. One of these assumptions basically states that the observations are not related to each other. If the observations are only weakly related, then the methods discussed in Part III will still probably produce fairly accurate forecasts. But, if the observations are strongly related, then one should remember that the Box-Jenkins methodology uses the dependency to produce forecasts that are likely to be more accurate than the forecasts produced by the methods of Part III.

CHAPTER 6

Forecasting Using The Multiplicative Decomposition Method

6–1 INTRODUCTION

In this chapter we discuss analyzing and forecasting a time series that has multiplicative seasonal variation. Since multiplicative seasonal variation implies that the magnitude of the "seasonal swing" of the time series is proportional to the average level as determined by the trend, it follows that a reasonable model to analyze a time series with multiplicative seasonal variation is

$$y_t = TR_t \times SN_t + \epsilon_t$$

where the meaning of the notation above has been previously discussed.

This chapter will discuss the use of the multiplicative decomposition method in estimating TR_t and SN_t in the above model. In Section 6–2 we will show how a multiplicative time series can be "decomposed" into trend, seasonal, cyclical, and irregular components. Section 6–3 describes how to use the estimates obtained through the decomposition process to forecast future values of the time series. In Section 6–3 we also present an intuitive method that can be used to calculate fairly accurate confidence interval forecasts when using the multiplicative decomposition method. Section 6–4 presents a procedure that can be used to analyze and forecast a time series that displays a shifting seasonal pattern.

6-2 MULTIPLICATIVE DECOMPOSITION

The model used in the *multiplicative decomposition method* is somewhat different from the model

$$y_t = TR_t \times SN_t + \epsilon_t$$

It is usually written as

$$\boxed{y_t = TR_t \times SN_t \times CL_t \times IR_t}$$

where y_t, TR_t, and SN_t are as defined previously. Here, CL_t is the cyclical factor of the time series in period t, and IR_t is the irregular factor of the time series in period t. In this section we will describe how to use the above model to "decompose" a time series and thus obtain estimates tr_t, sn_t, cl_t, and ir_t of the components TR_t, SN_t, CL_t, and IR_t. In the next section we will use the estimates of TR_t and SN_t so obtained to forecast values of the time series, using the model

$$y_t = TR_t \times SN_t + \epsilon_t$$

A forecast of y_t will then be

$$\boxed{\hat{y}_t = tr_t \times sn_t}$$

In the next section we will explain why the estimate of the cyclical factor often is not used in forecasting the time series.

We will present the multiplicative decomposition method in the following example, which makes use of the time series concerning seminar demand given in Table III.1. It should be remembered that this time series approximately possesses additive seasonal variation. Hence, it is somewhat inappropriate to analyze and forecast this time series using a multiplicative model. We do this, however, because the multiplicative decomposition procedure is easier to initially understand if it is explained using quarterly data, rather than monthly data. In Chapter 8 we will analyze and forecast the seminar demand time series using an additive model. We will then compare the results obtained using the two models.

Example 6.1   Let us consider the time series on seminar demand that was given in Table III.1.

Quarter	Year 1	Year 2	Year 3	Year 4
1	10	11	13	15
2	31	33	34	37
3	43	45	48	51
4	16	17	19	21

The first step in the analysis of this time series will be the estimation of seasonal factors for each quarter. This will be done by first calculating a *moving average,* which will remove the seasonal variation from the time series. A moving average is calculated by adding the observations for a number of periods in the time series and dividing the result by the number of periods included in the average. Here the number of periods included in the moving average is determined by the number of periods in a year. In analyzing the seminar demand time series we shall use a four-period moving average since we have quarterly data. If the time series consisted of monthly data, a twelve-period moving average would be used. In this case the first average calculated is the average for the four observations in year 1. We obtain

$$\frac{10 + 31 + 43 + 16}{4} = \frac{100}{4} = 25$$

The second average is obtained by eliminating the first observation in year 1 from the average and including the first observation for year 2 in the new average. Thus we obtain

$$\frac{31 + 43 + 16 + 11}{4} = \frac{101}{4} = 25.25$$

The third average is obtained by dropping the second observation in year 1 and adding the second observation in year 2 for a new average. This yields

$$\frac{43 + 16 + 11 + 33}{4} = \frac{103}{4} = 25.75$$

Each successive average is obtained by dropping the first observation included in the previous average and adding the next observation in the time series to obtain the new average. The four-quarter moving totals and four-quarter moving averages for the seminar demand time series are given in Table 6.1.

Since the first average is the average of the observations in quarters 1, 2, 3, and 4, the average corresponds to a point in time midway between the second and third quarters. In order to obtain an average corresponding to one of the time periods in the original time series, we calculate a *centered moving average.* This is done by computing a two-period moving average of the moving averages we calculated previously. Thus our first centered moving average is

$$\frac{25.00 + 25.25}{2} = 25.12$$

This centered moving average corresponds to the third quarter of year 1. The centered moving average corresponding to the fourth quarter of year 1 is

$$\frac{25.25 + 25.75}{2} = 25.5$$

TABLE 6.1 *Calculation of Moving Averages*

Year	Quarter	Observation y_t	Moving Total	Moving Average	Centered Moving Average $tr_t \times cl_t$	$sn_t \times ir_t = \dfrac{y_t}{tr_t \times cl_t}$
1	1	10				
	2	31				
			100	25.00		
	3	43			25.12	1.71
			101	25.25		
	4	16			25.50	.63
			103	25.75		
2	1	11			26.00	.42
			105	26.25		
	2	33			26.38	1.25
			106	26.50		
	3	45			26.75	1.68
			108	27.00		
	4	17			27.12	.63
			109	27.25		
3	1	13			27.62	.47
			112	28.00		
	2	34			28.25	1.20
			114	28.50		
	3	48			28.75	1.67
			116	29.00		
	4	19			29.38	.65
			119	29.75		
4	1	15			30.12	.50
			122	30.50		
	2	37			30.75	1.20
			124	31.00		
	3	51				
	4	21				

The balance of the centered moving averages for the time series are given in Table 6.1. Of course, if the original moving averages had been based on an odd number of observations, the centering procedure would not have been necessary. For example, if the moving average had been a three-period moving average, the first average would correspond to period 2, the second average would correspond to period 3, and so on. Unfortunately, since most business time series are quarterly or monthly, the centering procedure is necessary in most cases.

Let us now consider these centered moving averages. First, note that the averaging has removed seasonal variations from the data, since each moving average is computed using exactly one observation from each season. Second, the averaging has hopefully cancelled the effects of irregular factors as well. Hence, these centered moving averages represent a combination of trend and cyclical factors. We consider the centered moving average corresponding to period t as $tr_t \times cl_t$, the estimate of $TR_t \times CL_t$. Since

$$y_t = TR_t \times SN_t \times CL_t \times IR_t$$

and

$$sn_t \times ir_t = \frac{tr_t \times sn_t \times cl_t \times ir_t}{tr_t \times cl_t}$$

we obtain the estimate of $SN_t \times IR_t$ using the formula

$$sn_t \times ir_t = \frac{y_t}{tr_t \times cl_t}$$

These estimates are also given in Table 6.1. Next, to find the estimate of SN_t, we will group the values of $sn_t \times ir_t$ by quarters and average these values for each quarter. This averaging procedure hopefully removes the influence of the irregular factors. Thus, the averages that are calculated represent the effects of seasonal factors alone. We shall denote these averages as $\overline{sn}_t$, for $t = 1, 2, 3$, and 4. They are listed below.

Quarter 1	Quarter 2	Quarter 3	Quarter 4
0.42	1.25	1.71	0.63
0.47	1.20	1.68	0.63
0.50	1.20	1.67	0.65
$\overline{sn}_1 = 0.46$	$\overline{sn}_2 = 1.22$	$\overline{sn}_3 = 1.69$	$\overline{sn}_4 = 0.64$

These seasonal factors are now normalized so that they add to the number of periods in a year, L. Here the factors are forced to add to $L = 4$ since the time series consists of quarterly data. This normalization is accomplished by multiplying each value of $\overline{sn}_t$ by the quantity

$$\frac{L}{\sum\limits_{t=1}^{L} \overline{sn}_t} = \frac{4}{\sum\limits_{t=1}^{4} \overline{sn}_t} = \frac{4}{4.01} = .9975$$

Thus, the estimate of SN_t, sn_t, is given by

$$sn_t = \overline{sn}_t \left[\frac{L}{\sum\limits_{t=1}^{L} \overline{sn}_t} \right]$$

$$= \overline{sn}_t[.9975] \qquad \text{for } t = 1, \ldots, L$$

This normalization process results in the estimates $sn_1 = .46$, $sn_2 = 1.22$, $sn_3 = 1.68$, and $sn_4 = .64$. It should be noticed that we are assured that the above normalization process will make the seasonal factors add to L. This is because

$$\sum_{t=1}^{L} sn_t = \sum_{t=1}^{L} \left(\overline{sn}_t \left[\frac{L}{\sum\limits_{t=1}^{L} \overline{sn}_t} \right] \right) = L \left[\frac{\sum\limits_{t=1}^{L} \overline{sn}_t}{\sum\limits_{t=1}^{L} \overline{sn}_t} \right] = L$$

Once the estimates of the seasonal factors have been calculated, we may obtain an estimate of the trend for the time series. This is done by first computing the *deseasonalized data*. The deseasonalized observation for period t, d_t, is found by dividing the observation y_t in period t by the seasonal factor for period t:

$$d_t = \frac{y_t}{sn_t}$$

Calculating the deseasonalized data is an attempt to eliminate seasonal effects. Thus, if an observation corresponds to a season that has a seasonal factor of less than one, the deseasonalized observation, d_t, is larger than the actual observation, y_t; and, therefore, the deseasonalized observation is closer to the trend value. If an observation corresponds to a season with a seasonal factor greater than one, the deseasonalized observation is less than the actual observation, and again, the deseasonalized observation is closer to the trend value. The deseasonalized data for the seminar demand time series are given in Table 6.2 and plotted in Figure 6.1.

We find the estimate of TR_t, tr_t, by using the deseasonalized data. Since Figure 6.1 indicates that the deseasonalized data increase over time in a linear fashion, it seems reasonable to assume that

$$TR_t = \beta_0 + \beta_1 t$$

Hence, we will estimate TR_t by fitting a straight line to the deseasonalized data.

TABLE 6.2   *Deseasonalized Seminar Demand*

Year	Quarter	Observation y_t	sn_t	$d_t = \dfrac{y_t}{sn_t}$
1	1	10	0.46	21.74
	2	31	1.22	25.41
	3	43	1.68	25.60
	4	16	0.64	25.00
2	1	11	0.46	23.91
	2	33	1.22	27.05
	3	45	1.68	26.78
	4	17	0.64	26.56
3	1	13	0.46	28.26
	2	34	1.22	27.87
	3	48	1.68	28.57
	4	19	0.64	29.69
4	1	15	0.46	32.61
	2	37	1.22	30.33
	3	51	1.68	30.36
	4	21	0.64	32.81

Computing

$$b_1 = \frac{16 \sum_{t=1}^{16} t d_t - \left(\sum_{t=1}^{16} t\right)\left(\sum_{t=1}^{16} d_t\right)}{16 \sum_{t=1}^{16} t^2 - \left(\sum_{t=1}^{16} t\right)^2} = .59$$

and

$$b_0 = \frac{\sum_{t=1}^{16} d_t}{16} - b_1 \left(\frac{\sum_{t=1}^{16} t}{16}\right) = 22.61$$

we find that the estimate of TR_t, tr_t, is given by

$$tr_t = b_0 + b_1 t = 22.61 + .59t$$

It should be noted that this regression could be performed using the original time series data for the dependent variable. However, this procedure may result in the estimate of the trend component being distorted by the seasonal nature of the

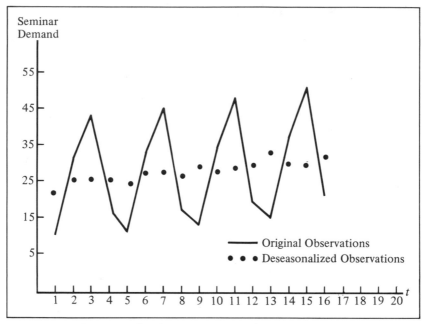

FIGURE 6.1   *Deseasonalized Seminar Demand*

time series. For example, suppose that the seasonal factor for the first period in the year is substantially less than 1 while the seasonal factor for the last period in the year is substantially greater than 1. Then the observation for the first period in the year will be below the trend value for that period, while the observation for the last period in the year will be above the trend value for the period. It is easy to see that such a situation could lead to overestimation of the slope if the original time series data are used for the dependent variable. Use of the deseasonalized data should result in a more accurate estimate of the trend component.

The calculation of cyclical factors is possible after the trend and seasonal factors have been derived. Since

$$y_t = TR_t \times SN_t \times CL_t \times IR_t$$

and

$$cl_t \times ir_t = \frac{tr_t \times sn_t \times cl_t \times ir_t}{tr_t \times sn_t}$$

we obtain the estimate $cl_t \times ir_t$ using the equation

$$cl_t \times ir_t = \frac{y_t}{tr_t \times sn_t}$$

where sn_t has been calculated previously and where

$$tr_t = b_0 + b_1 t$$

$$= 22.61 + .59t$$

These calculations are summarized in Table 6.3.

Next we will calculate the estimates of the cyclical factors. This may be done by averaging the values of $cl_t \times ir_t$. We will use a three-quarter moving average in this case. This moving average is appropriate since irregular movements rarely last

TABLE 6.3   *The Estimates of $CL_t \times IR_t$*

Year	Quarter	t	y_t	$tr_t = 22.61 + .59t$	sn_t	$tr_t \times sn_t$	$cl_t \times ir_t = \dfrac{y_t}{tr_t \times sn_t}$
1	1	1	10	23.20	0.46	10.67	0.94
	2	2	31	23.79	1.22	29.02	1.07
	3	3	43	24.38	1.68	40.96	1.05
	4	4	16	24.97	0.64	15.98	1.00
2	1	5	11	25.56	0.46	11.76	0.94
	2	6	33	26.15	1.22	31.90	1.03
	3	7	45	26.74	1.68	44.92	1.00
	4	8	17	27.33	0.64	17.49	0.97
3	1	9	13	27.92	0.46	12.84	1.01
	2	10	34	28.51	1.22	34.78	0.98
	3	11	48	29.10	1.68	48.89	0.98
	4	12	19	29.69	0.64	19.00	1.00
4	1	13	15	30.28	0.46	13.93	1.08
	2	14	37	30.87	1.22	37.66	0.98
	3	15	51	31.46	1.68	52.85	0.96
	4	16	21	32.05	0.64	20.51	1.02

longer than two or three months and hence a three-quarter moving average will hopefully remove their influence. A three-quarter moving average is also convenient since the average need not be centered. These moving averages for the seminar demand time series are given in Table 6.4. As an example of the calculations involved, the first average, that corresponding to the second period, is

$$\frac{0.94 + 1.07 + 1.05}{3} = 1.02$$

The second average, corresponding to the third period, is

$$\frac{1.07 + 1.05 + 1.00}{3} = 1.04$$

The balance of the averages are calculated in a similar manner. These averages represent the effects of cyclical variations alone, and so we use these averages as the estimates of the cyclical factors.

Once the estimates of the cyclical factors have been derived, the estimates of the irregular movements in the time series may be obtained. This is done using the formula

$$ir_t = \frac{(cl_t \times ir_t)}{cl_t}$$

The estimates of the irregular movements for the seminar demand time series are given in the last column of Table 6.4.

Although we have illustrated the multiplicative decomposition method using quarterly data, the general procedure also applies to data having an arbitrary number of seasonal periods. In the next section we will apply the multiplicative decomposition method to monthly data.

Having decomposed the time series into the estimates of the four model

TABLE 6.4   *The Estimates of CL_t and IR_t*

Year	Quarter	$cl_t \times ir_t$	3-Period Moving Average cl_t	$ir_t = \dfrac{cl_t \times ir_t}{cl_t}$
1	1	0.94		
	2	1.07	1.02	1.05
	3	1.05	1.04	1.01
	4	1.00	1.00	1.00
2	1	0.94	0.99	0.95
	2	1.03	0.99	1.04
	3	1.00	1.00	1.00
	4	0.97	0.99	0.98
3	1	1.01	0.99	1.02
	2	0.98	0.99	0.99
	3	0.98	0.99	0.99
	4	1.00	1.02	0.98
4	1	1.08	1.02	1.06
	2	0.98	1.01	0.97
	3	0.96	0.99	0.97
	4	1.02		

components—trend, seasonal variations, cyclical variations, and irregular movements—we can use this information to forecast future values of the time series. It should be noted that the forecasting procedure to be described is based on the fundamental assumption used in the other quantitative forecasting procedures discussed in this book that the time series will behave in the future in the same way as it has behaved in the past.

6–3 FORECASTING

The most common use of the information provided by the decomposition method involves generation of forecasts using the trend and seasonal factors only. Forecasters often restrict their attention to these two components because the other components, cyclical variations and irregular movements, cannot be accurately predicted for future time periods. Obviously, irregular movements, by their very nature, cannot be predicted for future time periods. However, theoretically the cyclical variations in the time series ought to be predictable. That is, the cyclical factors we have derived could be treated in the same manner as the seasonal factors we derived. The cyclical factors calculated for time periods in the same portion of different cycles could be averaged, just as the seasonal factors for the various quarters in different years were averaged. These averages could then be normalized so that the cyclical factors add to the number of periods in a cycle. This procedure, although theoretically possible to implement, is often difficult if not impossible to use. First, it can be used only if a well-defined, repeating cycle of a reasonably constant duration can be identified. In many situations this is not the case. Second, even if such a cycle does exist, it may not be possible to analyze the cycle because of lack of data. In order to obtain accurate estimates of cyclical factors, data for several cycles must exist. Since cyclical fluctuations by definition have a duration of from 2 to 10 years or longer, more than 30 years of data may be required for the calculation of estimates of cyclical factors that can be used in forecasting. Many times these data requirements cannot be met. For these reasons, the cyclical variations in a time series often cannot be predicted accurately for future time periods. When such is the case, forecasts are made using the trend and seasonal factors only. Note that when a limited number of historical observations is available, the cyclical behavior of a time series is often reflected in the estimate of the trend. Hence, using only the trend and seasonal factors does not necessarily mean that the cyclical behavior of the time series is being ignored.

Having obtained the estimates tr_t and sn_t of, respectively, TR_t and SN_t, and assuming that the seasonal factors for all time periods in the same season remain constant, the forecast of a value y_t of the time series is

$$\hat{y}_t = tr_t \times sn_t$$

We next consider finding a confidence interval for a value y_t of the time series. Since there is no "statistical theory" behind the time series decomposition method, the method by which we obtain the estimates tr_t and sn_t, there is no "statistically correct" confidence interval for y_t. There is an intuitive method, however, for finding a confidence interval for y_t, which works reasonably well. This method yields an approximate $(100 - \alpha)$ percent confidence interval for y_t of

$$[\hat{y}_t - E_t(100 - \alpha), \hat{y}_t + E_t(100 - \alpha)]$$

where $E_t(100 - \alpha)$ is the error obtained in finding a $(100 - \alpha)$ percent confidence interval for the deseasonalized observation, d_t. Here we assume that the regression model

$$d_t = TR_t + \epsilon_t$$

describes d_t. This intuitive method may be extended to the determination of an approximate confidence interval for

$$\sum_{t=l}^{l+q} y_t$$

the sum of the values of the observations in the time series from time l to time $l + q$:

$$\left[\sum_{t=l}^{l+q} \hat{y}_t - E_{(l,l+q)}(100 - \alpha), \sum_{t=l}^{l+q} \hat{y}_t + E_{(l,l+q)}(100 - \alpha) \right]$$

where $E_{(l,l+q)}(100 - \alpha)$ equals the error obtained in finding a $(100 - \alpha)$ percent confidence interval for the corresponding sum of values of the deseasonalized observations,

$$\sum_{t=l}^{l+q} d_t$$

Again we assume that the regression model

$$d_t = TR_t + \epsilon_t$$

describes d_t. This methodology will be illustrated in the next two examples.

Example 6.2   Let us consider forecasting quarterly seminar demand using the data given in Table III.1. In Example 6.1 we found that the estimates of SN_1, SN_2, SN_3, and SN_4 are, respectively, $sn_1 = .46$, $sn_2 = 1.22$, $sn_3 = 1.69$, and $sn_4 = .64$. We also computed the deseasonalized data, which is given in Table 6.2 and is repeated in Table 6.5. Moreover, assuming that

$$TR_t = \beta_0 + \beta_1 t$$

TABLE 6.5   *Forecasts of Historical Seminar Demand Calculated Using the Multiplicative Decomposition Method*

t	y_t	sn_t	$d_t = y_t/sn_t$	$tr_t = 22.61 + .59t$	$\hat{y}_t = tr_t \times sn_t$	$(y_t - \hat{y}_t)^2$
1	10	.46	21.74	23.20	10.67	$(-.67)^2$
2	31	1.22	25.41	23.79	29.02	$(1.98)^2$
3	43	1.68	25.60	24.38	40.96	$(2.04)^2$
4	16	.64	25.00	24.97	15.98	$(.02)^2$
5	11	.46	23.91	25.56	11.76	$(-.76)^2$
6	33	1.22	27.05	26.15	31.90	$(1.10)^2$
7	45	1.68	26.78	26.74	44.92	$(.08)^2$
8	17	.64	26.56	27.33	17.49	$(-.49)^2$
9	13	.46	28.26	27.92	12.84	$(.16)^2$
10	34	1.22	27.87	28.51	34.78	$(-.78)^2$
11	48	1.68	28.57	29.10	48.89	$(-.89)^2$
12	19	.64	29.69	29.69	19.00	$(0)^2$
13	15	.46	32.61	30.28	13.93	$(1.07)^2$
14	37	1.22	30.33	30.87	37.66	$(-.66)^2$
15	51	1.68	30.36	31.46	52.85	$(-1.85)^2$
16	21	.64	32.81	32.05	20.51	$(.49)^2$

$$\sum_{t=1}^{16} (y_t - \hat{y}_t)^2 = 17.234$$

we fitted a straight line to the deseasonalized data using d_t as the dependent variable and t as the independent variable and found the estimate of TR_t to be

$$tr_t = 22.61 + .59t$$

A forecast of y_t, seminar demand in period t, is then given by

$$\hat{y}_t = tr_t \times sn_t$$

Moreover, assuming that

$$d_t = TR_t + \epsilon_t = \beta_0 + \beta_1 t + \epsilon_t$$

we can compute a $(100 - \alpha)$ percent confidence interval for d_t. This interval is

$$[tr_t - E_t(100 - \alpha), \, tr_t + E_t(100 - \alpha)]$$

Then, we can use the error $E_t(100 - \alpha)$ in this confidence interval to compute an approximate $(100 - \alpha)$ percent confidence interval for y_t:

$$[\hat{y}_t - E_t(100 - \alpha), \, \hat{y}_t + E_t(100 - \alpha)]$$

For example, a forecast of y_{17} is

$$\hat{y}_{17} = tr_{17} \times sn_{17}$$
$$= (22.61 + .59(17))(.46)$$
$$= 15.01$$

Moreover, the error in a 95 percent confidence interval for d_{17} is shown to be E_{17} (95) = 2.80 in Section *6–3. So we have an approximate 95 percent confidence interval for y_{17} of

$$[15.01 - 2.80, 15.01 + 2.80] \quad \text{or} \quad [12.21, 17.81]$$

Continuing in this manner, we can obtain a forecast and an approximate confidence interval for any y_t. In Table 6.6 we give forecasts and confidence intervals for y_{17}, y_{18}, y_{19}, and y_{20}.

It should be emphasized that the forecasts for periods 17, 18, 19, and 20 are made using the first 16 periods of data. Next, assume that the actual seminar demands in these four periods are observed to be $y_{17} = 16$, $y_{18} = 39$, $y_{19} = 53$, and $y_{20} = 22$. The observed value, y_t, and the squared forecast error, $(y_t - \hat{y}_t)^2$, are listed in Table 6.6 for periods $t = 17, 18, 19,$ and 20. These quantities are also listed in Table 6.5 for periods 1 through 16. Moreover, in Figure 6.2 we have plotted the observed value y_t and the rounded prediction $\hat{y}_t$ for periods 1 through 20. Notice that in year 1 and year 2 the predictions of winter and fall seminar demand tend to be too large, and the predictions of spring and summer seminar demand tend to be too small, while in year 3 and year 4 the predictions of winter and fall seminar demand tend to be too small, and the predictions of spring and summer seminar demand tend to be too large. This "systematic" bias is partially caused by the fact that wc arc using a *multiplicative* model to forecast a time series that approximately possesses *additive* seasonal variation. Note, also (Table 6.6), that 3 out of 4 confidence intervals contain the observed value y_t. We can gain further information

TABLE 6.6   *Forecasts of Future Values of Seminar Demand Calculated Using the Multiplicative Decomposition Method*

t	sn_t	$tr_t = 22.61 + .59t$	$\hat{y}_t = tr_t \times sn_t$	$E_t(95)$	$[\hat{y}_t - E_t(95), \hat{y}_t + E_t(95)]$	y_t	$(y_t - \hat{y}_t)^2$
17	.46	32.64	15.01	2.80	[12.21, 17.81]	16	$(.99)^2$
18	1.22	33.23	40.54	2.85	[37.69, 43.39]	39	$(-1.54)^2$
19	1.68	33.82	56.82	2.92	[53.90, 59.74]	53	$(-3.82)^2$
20	.64	34.41	22.02	2.98	[19.04, 25.00]	22	$(-.08)^2$

$$\sum_{t=17}^{20}(y_t - \hat{y}_t)^2 = 17.945$$

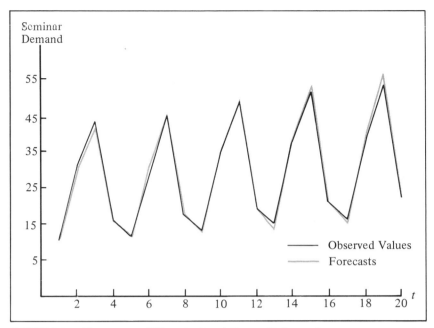

FIGURE 6.2   *Forecasts of Historical and Future Values of Seminar Demand
Calculated Using the Multiplicative Decomposition Method*

concerning the fit of our model to the observed data by computing the sum of the squared forecast errors. For periods 1 through 16 this quantity is

$$\sum_{t=1}^{16} (y_t - \hat{y}_t)^2 = 17.234$$

while for periods 17 through 20 this quantity is

$$\sum_{t=17}^{20} (y_t - \hat{y}_t)^2 = 17.945$$

In Section 9–2 we will use these quantities to summarize our previous modeling of time series possessing additive and multiplicative seasonal variation.

Example 6.3   Let us next consider forecasting monthly sales of Rola-Cola using the data given in Table III.2. Since this time series approximately possesses multiplicative seasonal variation, it is appropriate to decompose this time series using the multiplicative decomposition method. The results of this decomposition are summarized in Table 6.7. Let us consider, for example, how the centered moving average $tr_9 \times cl_9$ = 460.9 has been computed. This centered moving average equals the average of two moving averages, the first moving average being the average of the observa-

TABLE 6.7   *Analysis of the Historical Rola-Cola Sales Time Series Using the Multiplicative Decomposition Method*

t	y_t	Centered Moving Average $tr_t \times cl_t$	$sn_t \times ir_t = \dfrac{y_t}{tr_t \times cl_t}$	sn_t	$d_t = \dfrac{y_t}{sn_t}$	$tr_t = 380.163 + 9.489t$	$\hat{y}_t = tr_t \times sn_t$	$(y_t - \hat{y}_t)^2$
1	189			.493	383.37	389.652	192.10	$(-3.1)^2$
2	229			.596	384.23	399.141	237.89	$(-8.89)^2$
3	249			.595	418.49	408.630	243.13	$(5.87)^2$
4	289			.680	425	418.119	284.32	$(4.68)^2$
5	260			.564	460.99	427.608	241.17	$(18.83)^2$
6	431			.986	437.12	437.097	430.98	$(.02)^2$
7	660	450.1	1.466	1.467	449.9	446.586	655.14	$(4.86)^2$
8	777	455.2	1.707	1.693	458.95	456.075	772.13	$(4.87)^2$
9	915	460.9	1.985	1.990	459.79	465.564	926.47	$(-11.47)^2$
10	613	467.2	1.312	1.307	469.01	475.053	620.89	$(-7.89)^2$
11	485	472.8	1.026	1.029	471.33	489.542	498.59	$(-13.59)^2$
12	277	480.2	.577	.600	461.67	494.031	296.42	$(-19.42)^2$
13	244	492.5	.495	.493	494.97	503.520	248.24	$(-4.24)^2$
14	296	507.3	.583	.596	496.64	513.009	305.75	$(-9.75)^2$
15	319	524.8	.608	.595	536.13	522.498	310.89	$(8.11)^2$
16	370	540.7	.684	.680	544.12	531.987	361.75	$(8.25)^2$
17	313	551.9	.567	.564	554.97	541.476	305.39	$(7.61)^2$
18	556	560.9	.991	.986	563.89	550.965	543.25	$(12.75)^2$
19	831	567.1	1.465	1.467	566.46	560.454	822.19	$(8.81)^2$
20	960	572.7	1.676	1.693	567.04	569.943	964.91	$(-4.91)^2$
21	1152	578.4	1.992	1.990	578.89	579.432	1153.07	$(-1.07)^2$
22	759	583.7	1.300	1.307	580.72	588.921	769.72	$(-10.72)^2$
23	607	589.3	1.030	1.029	589.89	598.410	615.76	$(-8.76)^2$
24	371	596.2	.622	.600	618.33	607.899	364.74	$(6.26)^2$
25	298	607.7	.490	.493	604.46	617.388	304.37	$(-6.37)^2$
26	378	623.0	.607	.596	634.23	626.877	373.62	$(4.38)^2$
27	373	640.8	.582	.595	626.89	636.366	378.64	$(-5.64)^2$
28	443	656.7	.675	.680	651.47	645.855	439.18	$(3.82)^2$
29	374	667.3	.561	.564	663.12	655.344	369.61	$(4.39)^2$
30	660	674.7	.978	.986	669.37	664.833	655.53	$(4.47)^2$
31	1004			1.467	684.39	674.322	989.23	$(14.77)^2$
32	1153			1.693	681.04	683.811	1157.69	$(-4.69)^2$
33	1388			1.990	697.49	693.300	1379.67	$(8.33)^2$
34	904			1.307	691.66	702.789	918.55	$(-14.55)^2$
35	715			1.029	694.85	712.278	732.93	$(-17.93)^2$
36	441			.600	735	721.707	433.06	$(7.94)^2$

$$\sum_{t=1}^{36} (y_t - \hat{y}_t)^2 = 3154.7919$$

tions in periods 3 through 14, which is 458, and the second moving average being the average of the observations in periods 4 through 15, which is 463.833. That is,

$$tr_9 \times cl_9 = \frac{458 + 463.833}{2} = 460.9$$

where

458

$$= \frac{(249 + 289 + 260 + 431 + 660 + 777 + 915 + 613 + 485 + 277 + 244 + 296)}{12}$$

and

463.833

$$= \frac{(289 + 260 + 431 + 660 + 777 + 915 + 613 + 485 + 277 + 244 + 296 + 319)}{12}$$

The other centered moving averages are computed in a similar fashion. To find sn_t, the estimate of SN_t, we group the values of $sn_t \times ir_t$ by months and calculate an average, $\overline{sn}_t$, for each month. These seasonal factors are then normalized so that they add to $L = 12$, the number of periods in a year. This normalization is accomplished by multiplying each value of $\overline{sn}_t$ by the quantity

$$\frac{L}{\sum\limits_{t=1}^{L} \overline{sn}_t} = \frac{12}{11.9895} = 1.0008758$$

This normalization process results in the estimate $sn_t = 1.0008758(\overline{sn}_t)$, which is the estimate of SN_t. These calculations are summarized in Table 6.8. Since the plot of the deseasonalized data given in Figure 6.3 has a "straight line appearance," it seems reasonable to assume that

$$TR_t = \beta_0 + \beta_1 t$$

So we will estimate TR_t by fitting a straight line to the deseasonalized data using d_t as the dependent variable and t as the independent variable. We obtain tr_t, the estimate of TR_t, by computing

$$b_1 = \frac{36 \sum\limits_{t=1}^{36} td_t - \left(\sum\limits_{t=1}^{36} t\right)\left(\sum\limits_{t=1}^{36} d_t\right)}{36 \sum\limits_{t=1}^{36} t^2 - \left(\sum\limits_{t=1}^{36} t\right)^2} = 9.489$$

TABLE 6.8   *Estimates of the Seasonal Factors of the Rola-Cola Sales Time Series*

		$sn_t \times ir_t = y_t/(tr_t \times cl_t)$			
		Year 1	Year 2	$\overline{sn}_t$	sn_t
1	Jan.	.495	.490	.4925	.493
2	Feb.	.583	.607	.595	.596
3	Mar.	.608	.582	.595	.595
4	Apr.	.684	.675	.6795	.680
5	May	.567	.561	.564	.564
6	June	.991	.978	.9845	.986
7	July	1.466	1.465	1.4655	1.467
8	Aug.	1.707	1.676	1.6915	1.693
9	Sept.	1.985	1.992	1.9885	1.990
10	Oct.	1.312	1.300	1.306	1.307
11	Nov.	1.026	1.030	1.028	1.029
12	Dec.	.577	.622	.5995	.600

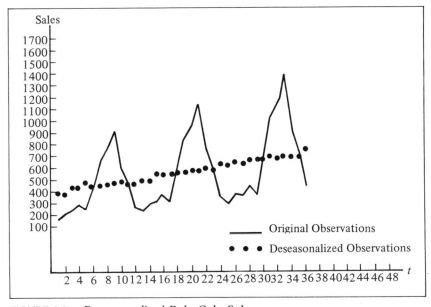

FIGURE 6.3   *Deseasonalized Rola-Cola Sales*

$$b_0 = \frac{\sum\limits_{t=1}^{36} d_t}{36} - b_1 \left(\frac{\sum\limits_{t=1}^{36} t}{36} \right) = 380.163$$

which gives us,

$$tr_t = b_0 + b_1 t = 380.163 + 9.489t$$

A forecast of y_t, sales of Rola-Cola in period t, is then given by

$$\hat{y}_t = tr_t \times sn_t$$

Furthermore, assuming that

$$d_t = TR_t + \epsilon_t = \beta_0 + \beta_1 t + \epsilon_t$$

we can compute a $(100 - \alpha)$ percent confidence interval for d_t of

$$[tr_t - E_t(100 - \alpha), \ tr_t + E_t(100 - \alpha)]$$

Then, we can use the error $E_t(100 - \alpha)$ in the above confidence interval to compute an approximate $(100 - \alpha)$ percent confidence interval for y_t. This interval is given by

$$[\hat{y}_t - E_t(100 - \alpha), \ \hat{y}_t + E_t(100 - \alpha)]$$

For example, a forecast of y_{37} is

$$\hat{y}_{37} = tr_{37} \times sn_{37}$$
$$= (380.163 + 9.489(37))(.493)$$
$$= 360.52$$

The error in a 95 percent confidence interval for d_{37} is shown in Section *6–3 to be $E_{37}(95) = 26.80$. So we have an approximate 95 percent confidence interval for y_{37} of

$$[360.52 - 26.80, 360.52 + 26.80] \qquad \text{or} \qquad [333.72, 387.32]$$

Continuing in this manner, we can obtain a forecast and an approximate confidence interval for any y_t. In Table 6.9 we give forecasts and confidence intervals for y_{37} through y_{48}. It should be emphasized that these forecasts were made using the first 36 periods of data. Assume, now, that the actual Rola-Cola sales in periods 37 through 48 are observed to be those given in Table 6.9, under the column headed "y_t." The squared forecast errors, $(y_t - \hat{y}_t)^2$, for periods 37 through 48 are

TABLE 6.9   *Forecasts of Future Values of Rola-Cola Sales Calculated Using the Multiplicative Decomposition Method*

t	sn_t	$tr_t = 380.163 + 9.489t$	$\hat{y}_t = tr_t \times sn_t$	$E_t(95)$	$[\hat{y}_t - E_t(95), \hat{y}_t + E_t(95)]$	y_t	$(y_t - \hat{y}_t)^2$
37	.493	731.273	360.52	26.80	[333.72, 387.32]	352	$(-8.52)^2$
38	.596	740.762	441.48	26.92	[414.56, 468.40]	445	$(3.51)^2$
39	.595	750.252	446.40	27.04	[419.36, 473.44]	453	$(6.6)^2$
40	.680	759.741	516.62	27.17	[489.45, 543.79]	541	$(24.38)^2$
41	.564	769.231	433.85	27.30	[406.55, 461.15]	457	$(23.15)^2$
42	.986	778.720	767.82	27.44	[740.38, 795.26]	762	$(-5.82)^2$
43	1.467	788.209	1156.30	27.59	[1128.71, 1183.89]	1194	$(37.7)^2$
44	1.693	797.699	1350.50	27.74	[1322.76, 1378.24]	1361	$(10.5)^2$
45	1.990	807.188	1606.30	27.89	[1578.41, 1634.19]	1615	$(8.7)^2$
46	1.307	816.678	1067.40	28.05	[1039.35, 1095.45]	1059	$(-8.4)^2$
47	1.029	826.167	850.12	28.22	[821.90, 878.34]	824	$(-26.12)^2$
48	.600	835.657	501.39	28.39	[473, 529.78]	495	$(-6.39)^2$

$$\sum_{t=37}^{48}(y_t - \hat{y}_t)^2 = 3693.5263$$

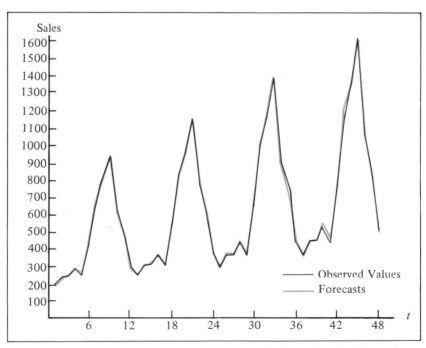

FIGURE 6.4   *Forecasts of Historical and Future Values of Rola-Cola Sales Calculated Using the Multiplicative Decomposition Method*

also listed in Table 6.9. The quantities y_t and $(y_t - \hat{y}_t)^2$ for periods 1 through 36 are listed in Table 6.7. In Figure 6.4 we plot the observed value y_t and the rounded prediction $\hat{y}_t$ for periods 1 through 48. Note that the multiplicative model seems to give fairly accurate predictions of the observed data. This is reasonable since the observed data seems to approximately possess multiplicative seasonal variation. Note, also, that 11 out of 12 confidence intervals contain the observed value y_t. We can gain further information concerning the fit of the model to the observed data by computing the sum of the squared forecast errors. For periods 1 through 36 this quantity is

$$\sum_{t=1}^{36} (y_t - \hat{y}_t)^2 = 3154.7919$$

while for periods 37 through 48, this quantity is

$$\sum_{t=37}^{48} (y_t - \hat{y}_t)^2 = 3693.5263$$

In Section 9–2 we will use these quantities to summarize our previous modeling of time series possessing additive and multiplicative seasonal variation.

A forecast of $\sum_{t=l}^{l+q} y_t$ is

$$\sum_{t=l}^{l+q} \hat{y}_t = \sum_{t=l}^{l+q} tr_t \times sn_t$$

Assuming that

$$d_t = TR_t + \epsilon_t = \beta_0 + \beta_1 t + \epsilon_t$$

we can compute a $(100 - \alpha)$ percent confidence interval for $\sum_{t=l}^{l+q} d_t$ of

$$\left[\sum_{t=l}^{l+q} tr_t - E_{(l,l+q)}(100 - \alpha), \sum_{t=l}^{l+q} tr_t + E_{(l,l+q)}(100 - \alpha) \right]$$

Then, we can use the error $E_{(l,l+q)}(100 - \alpha)$ in the above confidence interval to give an approximate $(100 - \alpha)$ percent confidence interval for $\sum_{t=l}^{l+q} y_t$ of

$$\left[\sum_{t=l}^{l+q} \hat{y}_t - E_{(l,l+q)}(100 - \alpha), \sum_{t=l}^{l+q} \hat{y}_t + E_{(l,l+q)}(100 - \alpha) \right]$$

For example, using the results in Table 6.9, a forecast of the sum of the observations for periods 38 through 41, based on the first 36 observations, is

$$\sum_{t=38}^{41} \hat{y}_t = 441.49 + 446.40 + 516.62 + 433.85 = 1838.36$$

Next, the error in a 95 percent confidence interval of the sum of the deseasonalized observations for periods 38 through 41 is shown in Section *6–3 to be $E_{(38,41)}(95) = 63.49$. So an approximate 95 percent confidence interval for $\sum_{t=38}^{41} y_t$, based on the first 36 observations, is

$$[1838.36 - 63.49, \ 1838.36 + 63.49] \qquad \text{or} \qquad [1774.87, 1901.85]$$

From Table 6.9 we find that

$$\sum_{t=38}^{41} y_t = 1896$$

is contained in this interval. Notice that $E_{(38,41)}(95) = 63.49$ is substantially smaller than

$$\sum_{t=38}^{41} E_t(95) = 26.92 + 27.04 + 27.17 + 27.30 = 108.43$$

which is the sum of the "errors of the individual confidence intervals." This illustrates that the "error for the sum is not the sum of the errors."

*6–3   THE REGRESSION CALCULATIONS

In this starred subsection we demonstrate how $E_{17}(95) = 2.80$ for Example 6.2 and $E_{37}(95) = 26.80$ and $E_{(38,41)}(95) = 63.49$ for Example 6.3 are calculated.

First, referring to Example 6.2 and the data of Table 6.5, we have found that

$$tr_t = 22.61 + .59t$$

Hence, assuming that

$$d_t = \beta_0 + \beta_1 t + \epsilon_t$$

the error in a 95 percent confidence interval for d_t is

$$E_t(95) = t_{5/2}(14) \cdot s \cdot f_t = 2.80 \qquad \text{if } t = 17$$

This is true since

$$s = \left(\frac{\sum_{t=1}^{16} (d_t - tr_t)^2}{16 - 2} \right)^{1/2} = 1.15$$

$$t_{5/2}(16 - 2) = t_{5/2}(14) = 2.145$$

and

$$f_t = \left(1 + \frac{1}{n} + \frac{(t - \bar{x})^2}{\displaystyle\sum_{i=1}^{} (i - \bar{x})^2}\right)^{1/2} = \left(1 + \frac{1}{16} + \frac{(t - \bar{x})^2}{\displaystyle\sum_{i=1}^{16} (i - \bar{x})^2}\right)^{1/2}$$

$$= 1.135 \qquad \text{if } t = 17$$

Here, $\bar{x} = \sum_{i=1}^{16} i/16$.

Next, referring to Example 6.3 and the data of Table 6.7, we have found that

$$tr_t = 380.163 + 9.489t$$

Hence, assuming that

$$d_t = \beta_0 + \beta_1 t + \epsilon_t$$

the error in a 95 percent confidence interval for d_t is

$$E_t(95) = t_{5/2}(34) \cdot s \cdot f_t = 26.80 \qquad \text{if } t = 37$$

This is true since

$$s = \left(\frac{\displaystyle\sum_{t=1}^{36} (d_t - tr_t)^2}{36 - 2}\right)^{1/2} = 12.487$$

$$t_{5/2}(36 - 2) = t_{5/2}(34) = 2.032$$

and

$$f_t = \left(1 + \frac{1}{n} + \frac{(t - \bar{x})^2}{\displaystyle\sum_{i=1}^{} (i - \bar{x})^2}\right)^{1/2} = \left(1 + \frac{1}{36} + \frac{(t - \bar{x})^2}{\displaystyle\sum_{i=1}^{36} (i - \bar{x})^2}\right)^{1/2}$$

$$= 1.056 \qquad \text{if } t = 37$$

Here $\bar{x} = \sum_{i=1}^{36} i/36$. Moreover, the error in a 95 percent confidence interval for

$$\sum_{t=l=38}^{l+q=38+3=41} d_t$$

is

$$E_{(l,l+q)}(95) = t_{5/2}(34) \cdot s \cdot f_{(l,l+q)} = 63.49$$

This is true since

$$f_{(l,l+q)} = \left\{ (q + 1) + \left(\sum_{t=l}^{l+q} \mathbf{x}'_t \right) (\mathbf{X}'\mathbf{X})^{-1} \left(\sum_{t=l}^{l+q} \mathbf{x}_t \right) \right\}^{1/2}$$

$$= \sqrt{(3 + 1) + 2.2606}$$

$$= 2.5021$$

where

$$\mathbf{x}'_{38} = [1 \ 38]$$

$$\mathbf{x}'_{39} = [1 \ 39]$$

$$\mathbf{x}'_{40} = [1 \ 40]$$

$$\mathbf{x}'_{41} = [1 \ 41]$$

$$\sum_{t=l}^{l+q} \mathbf{x}'_t = \sum_{t=38}^{41} \mathbf{x}'_t = [4,158]$$

$$\mathbf{X} = \begin{bmatrix} 1 & 1 \\ 1 & 2 \\ 1 & 3 \\ \cdot & \cdot \\ \cdot & \cdot \\ \cdot & \cdot \\ 1 & 36 \end{bmatrix}$$

6–4 SHIFTING SEASONAL PATTERNS

The previous sections in this chapter deal with a situation in which the seasonal factors for all time periods in the same season remain constant. It is possible, however, that the seasonal factors for time periods in the same season may change over time, resulting in a shifting seasonal pattern. In this section we present an example that illustrates a method that can be used to analyze a shifting seasonal pattern.

Example 6.4   The I.C. Winters Corporation produces and markets the Blizzard snow blower. Sales of the Blizzard exhibit a seasonal pattern with highest sales in the fourth quarter (October–December) and lowest sales in the second quarter (April–June). In order to set up more uniform production schedules and stabilize employment levels at its assembly facilities, I.C. Winters Corporation has carried out a series of special promotional campaigns that are designed to dampen the seasonal pattern of Blizzard sales. These promotional campaigns have been run for the last

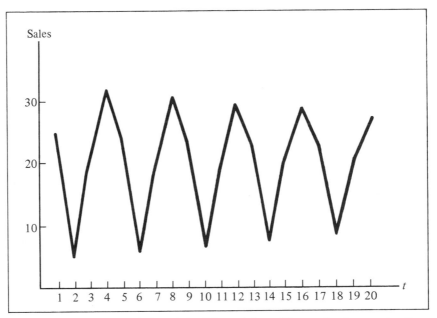

FIGURE 6.5   *Quarterly Sales of the Blizzard Snow Blower (in thousands of units)*

few years and have been successful in reducing the seasonal fluctuation in Blizzard sales. Quarterly sales of the Blizzard for the last five years are plotted in Figure 6.5. As indicated in this figure, the average level of sales has remained nearly constant over this five-year period. Although the seasonal pattern of Blizzard sales has persisted over this time period, the seasonal fluctuations have been gradually dampened. Thus Blizzard sales might be said to display a shifting seasonal pattern. In order to gain additional insight into the nature of the shifting seasonal pattern, the I.C. Winters Corporation uses a four-period centered moving average to derive an estimate of $SN_t \times IR_t$ for each time period in the five years of historical data. Since a four-period moving average is being used, estimates $sn_t \times ir_t$ are obtained for all time periods except the first two quarters of the first year and the last two quarters of the fifth year (for example, refer to Table 6.1) in the historical data. The estimates $sn_t \times ir_t$ are plotted by quarters in Figure 6.6. That is, estimates $sn_t \times ir_t$ have been calculated for each of the first quarters in years 2, 3, 4, and 5. These estimates are plotted against the year in which they were observed, in Figure 6.6a. Similarly, the estimates $sn_t \times ir_t$ for each of the second, third, and fourth quarters are plotted against the year in which they were observed, in Figures 6.6b, 6.6c, and 6.6d. Notice that, as time passes, these estimates are gradually decreasing in the first and fourth quarters and are gradually increasing in the second and third quarters. These estimates appear to be increasing or decreasing in a linear fashion for each of the first, second, third, and fourth quarters.

In order to calculate sales forecasts for future time periods it will be necessary to predict the values of the seasonal factors sn_t in future time periods. Suppose that the I.C. Winters Corporation wishes to forecast sales for the four quarters in year 6.

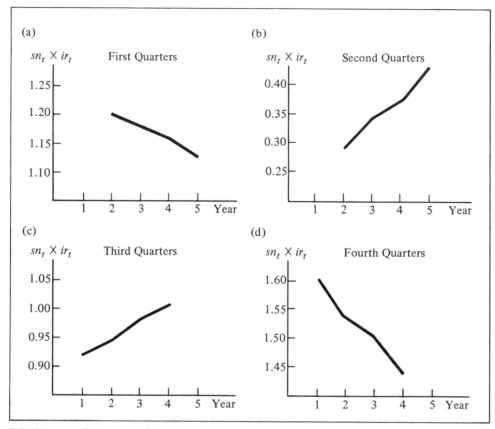

FIGURE 6.6   *Estimates of* $SN_t \times IR_t$ *by Quarters*

To do this, values of the seasonal factors for each of these quarters must be predicted. This can be done by using linear regression to fit lines to the data depicted in Figures 6.6a, b, c, and d, and then extrapolating these lines into year 6 to obtain estimates of $SN_t \times IR_t$ for the four quarters in year 6. When regression analysis is used to fit lines to the data (Figure 6.6), the following results are obtained.

For first quarters: $sn_t \times ir_t = 1.241 - .0195t$

For second quarters: $sn_t \times ir_t = 0.218 + .0402t$

For third quarters: $sn_t \times ir_t = 0.888 + .030t$

For fourth quarters: $sn_t \times ir_t = 1.652 - .051t$

These fitted lines yield estimates of $SN_t \times IR_t$ for the four quarters in year 6 as follows.

For first quarter of year 6: $sn_t \times ir_t = 1.241 - .0195(6) = 1.124$

For second quarter of year 6: $sn_t \times ir_t = 0.218 + .0402(6) = 0.4592$

For third quarter of year 6: $sn_t \times ir_t = 0.888 + .030(6) = 1.068$

For fourth quarter of year 6: $sn_t \times ir_t = 1.652 - .051(6) = 1.346$

These estimates are now normalized so that they add to 4 (the number of seasons in a year). The normalized estimates will then be used as the seasonal factors for the four quarters in year 6. The normalization procedure is carried out as follows.

$$\frac{4}{1.124 + .4592 + 1.068 + 1.346} = \frac{4}{3.9972} = 1.0007$$

For first quarter of year 6: $sn_1 = 1.0007(1.124) = 1.1247868$

For second quarter of year 6: $sn_2 = 1.0007(0.4592) = 0.4595214$

For third quarter of year 6: $sn_3 = 1.0007(1.068) = 1.0687476$

For fourth quarter of year 6: $sn_4 = 1.0007(1.346) = 1.3469422$

Next, the trend factor, TR_t, must be estimated for the Blizzard sales time series. Since average sales of the Blizzard have remained nearly constant for the last five years, we will assume that $TR_t = \beta_0$. That is, we will assume that Blizzard sales exhibit no trend. In order to estimate β_0, the Blizzard sales data must be deseasonalized. Since the seasonal factors for this data are gradually changing, the data is deseasonalized by dividing each sales observation by its corresponding estimate of $SN_t \times IR_t$. That is, in this case

$$d_t = \frac{y_t}{sn_t \times ir_t}$$

Regression analysis is then applied to the deseasonalized observations to obtain an estimate of β_0. In this case, since we are assuming that $TR_t = \beta_0$, the estimate of β_0 is $\bar{d}_t$, the average of the deseasonalized observations. For the data depicted in Figure 6.5, we obtain an estimate of

$$b_0 = \bar{d}_t = 20,004.76$$

Thus, the estimate of TR_t is

$$tr_t = 20,004.76$$

Forecasts of future Blizzard sales can now be calculated. A forecast of y_t, Blizzard sales in time period t, is

$$\hat{y}_t = tr_t \times sn_t$$

Forecasts of Blizzard sales for the four quarters of year 6 are as follows.

For first quarter: $\hat{y}_t = (20{,}004.76)(1.1247868) = 22{,}501.09$

For second quarter: $\hat{y}_t = (20{,}004.76)(0.4595214) = 9{,}192.62$

For third quarter: $\hat{y}_t = (20{,}004.76)(1.0687476) = 21{,}380.04$

For fourth quarter: $\hat{y}_t = (20{,}004.76)(1.3469422) = 26{,}945.255$

Before leaving this example, it is important to realize that, in general, when using the multiplicative decomposition method the estimates $sn_t \times ir_t$ for like seasons should always be plotted against time as in Figure 6.6. If these estimates appear to be relatively constant, the methods presented in Sections 6–2 and 6–3 may be appropriately used to forecast the time series in question. However, if the estimates $sn_t \times ir_t$ appear to be shifting over time, then the procedures given in this section should be used. Of course, if the estimates $sn_t \times ir_t$ are shifting over time, they need not shift in a linear fashion as in this example. It might, therefore, be necessary to fit a quadratic equation or some other functional form to these estimates in order to predict the seasonal factors for future time periods.

PROBLEMS

1. Verify that the fourth, fifth, and sixth moving averages in Table 6.1 are 26.25, 26.50, and 27.00.

2. Verify that the centered moving average corresponding to the second quarter in year 3 is 28.25 as given in Table 6.1.

3. Verify that the centered moving average corresponding to the third quarter in year 3 is 28.75 as given in Table 6.1.

4. Verify that the estimate of $CL_t \times IR_t$ for the first quarter of year 2 is 0.94 as given in Table 6.3.

5. Verify that the estimate of $CL_t \times IR_t$ for the third quarter of year 3 is 0.98 as given in Table 6.3.

6. Verify that cl_t, the estimate of CL_t, for the fourth quarter of year 2 is 0.99 as given in Table 6.4.

7. Verify that ir_t, the estimate of IR_t, for the fourth quarter of year 2 is 0.98 as given in Table 6.4.

8. Consider Table 6.6 of Example 6.2, in which we forecasted the seminar demand time series using the multiplicative decomposition method. Verify that forecasts for periods 18, 19, and 20 are 40.54, 56.82, and 22.02.

9. Consider the following data concerning quarterly sales of the popular game Oligopoly at a variety store.

Year	Quarter	Oligopoly Sales
1	1	20
	2	25
	3	35
	4	44
2	1	28
	2	29
	3	43
	4	48
3	1	24
	2	37
	3	39
	4	56

a. Calculate the moving average and centered moving averages needed to decompose this time series.
b. Calculate the seasonal factors for quarters 1, 2, 3, and 4.
c. Find the trend line to be used in forecasting this time series.
d. Calculate forecasts for the four quarters of year 4 (using trend and seasonal factors only).

10. Consider Table 6.9 of Example 6.3, in which we forecasted the Rola-Cola sales time series using the multiplicative decomposition method. Verify that a forecast for period 38 is 441.48.

*11. Consider Table 6.9 of Example 6.3. Verify that an approximate 95 percent confidence interval for Rola-Cola sales in period 38 is [414.56, 468.40].

12. Consider Table 6.9 of Example 6.3. Find a forecast for cumulative Rola-Cola sales in periods 40 through 45.

* Starred problem is only for students who have read Section *6–3.

CHAPTER 7

Winters' Method

7-1 INTRODUCTION

In this chapter we again discuss analyzing and forecasting a time series that has multiplicative seasonal variation. The forecasting technique we will present is an exponential smoothing procedure that is best used to forecast a time series with a linear trend and multiplicative seasonal variation. The method was developed by P. R. Winters [12] and is known as Winters' Method.

In Section 7–2 we will describe how Winters' Method can be used to estimate the trend and seasonal factors, TR_t and SN_t, of a multiplicative time series. We will also show how these estimates are used to obtain point forecasts for future values of a time series. In Section 7–3 we will describe an intuitive method that can be used to calculate "fairly accurate" confidence interval forecasts when using Winters' Method. In Section 7–4 we will discuss the reasoning behind this intuitive method for calculating approximate confidence intervals.

7-2 POINT FORECASTS

Winters' Method is an exponential smoothing procedure that is best used to forecast a time series with a linear trend and multiplicative seasonal variation. The method assumes that the time series to be forecasted can be adequately described using the model

$$y_t = (\beta_0 + \beta_1 t) \times SN_t + \epsilon_t$$

7–2.1   UPDATING THE ESTIMATES

Let us suppose that at the end of period $T - 1$ we have estimates of the model parameters β_0, β_1, and SN_t. In period T we obtain a new observation y_T, and we would like to update the estimates in light of this new observation. We shall denote these updated estimates as $b_0(T)$, $b_1(T)$, and $sn_T(T)$. Here $b_0(T)$ is the estimate of the intercept of the trend line, where we define the intercept to be the intercept at the original origin of time. The estimate of the intercept using the current time period as the origin shall be denoted $a_0(T)$.

Before we proceed further, it should be noted that we have several seasonal factors. Let us denote the number of seasons that must occur before the first season repeats itself as L. Then, for monthly time series, $L = 12$; and for quarterly time series, $L = 4$. We shall define the seasonal factors and the original estimates of the seasonal factors so that they sum to L. That is, we shall have

$$\sum_{t=1}^{L} SN_t = L \quad \text{and} \quad \sum_{t=1}^{L} sn_t(0) = L$$

Now, given the new observation y_T, we wish to obtain the estimates $a_0(T)$, $b_1(T)$, and $sn_T(T)$ by updating $a_0(T - 1)$ and $b_1(T - 1)$, the estimates obtained in the previous period, and by updating $sn_T(T - L)$, the last estimate of the current seasonal factor obtained L periods ago. In order to update the estimate of β_0, often called the *permanent component*, we use the following equation.

$$a_0(T) = \alpha \frac{y_T}{sn_T(T - L)} + (1 - \alpha)[a_0(T - 1) + b_1(T - 1)]$$

where $a_0(T)$ is the new estimate of β_0 and α is a smoothing constant between 0 and 1. Here $[a_0(T - 1) + b_1(T - 1)]$ is simply the estimate of the average level of the time series at time T, as calculated using the estimates computed in the last period. We divide y_T by $sn_T(T - L)$, the estimate of the seasonal factor for the current season computed L periods ago, to deseasonalize y_T and hence prevent the seasonal factor from influencing the estimate of the permanent component. See Figure 7.1.

We obtain the updated estimate of β_1, often called the *trend*, by using the equation

$$b_1(T) = \beta[a_0(T) - a_0(T - 1)] + (1 - \beta)b_1(T - 1)$$

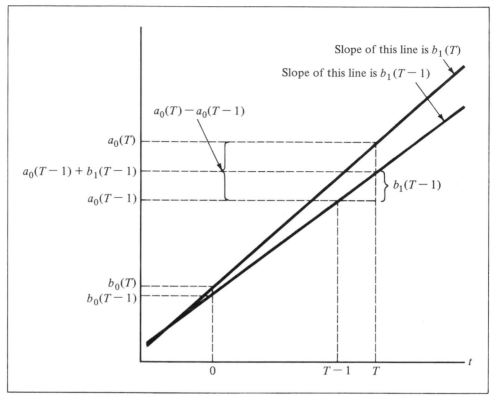

FIGURE 7.1   *Updating the Estimates in Winters' Method*

where β is a smoothing constant between 0 and 1. Here $[a_0(T) - a_0(T - 1)]$ is simply the difference between the estimate of the permanent component made in the current period and the estimate of the permanent component made in the last period. This difference and the estimate of the trend made last period yield the estimate of the trend in the current period. Again, see Figure 7.1.

The updated estimate of the *seasonal factor* is obtained using the equation

$$sn_T(T) = \gamma \frac{y_T}{a_0(T)} + (1 - \gamma)sn_T(T - L)$$

where γ is a smoothing constant between 0 and 1. Here we obtain the updated estimate by smoothing the estimate of the current season's factor computed L periods ago with the current observed seasonal variation. The estimate made L periods ago is used since this was the last time this particular season was observed. The current observed seasonal variation is obtained by dividing the observation y_T by the current estimate of the average level of the time series, $a_0(T)$. Note that we will force the initial values of the seasonal factors to add to L; but, after the smoothing procedure has begun, the seasonal factors do not necessarily add to L.

Once the updated estimates $a_0(T)$, $b_1(T)$, and $sn_T(T)$ are obtained, forecasts are generated using the following equation, which yields a τ period ahead forecast of $y_{T+\tau}$.

$$\hat{y}_{T+\tau}(T) = [a_0(T) + b_1(T)\tau]sn_{T+\tau}(T + \tau - L)$$

Note that in order to forecast for period $T + \tau$, we use the estimate of the seasonal factor computed in period $T + \tau - L$, which is denoted $sn_{T+\tau}(T + \tau - L)$. However, if a forecast is to be made for a period more than L periods in the future, the index $T + \tau - L$ refers to a seasonal factor that has not yet been computed. In such a case the latest estimate of the appropriate seasonal factor should be used in calculating the forecast.

7-2.2   THE INITIAL ESTIMATES

In order to begin the smoothing procedure, initial values for the estimates must be supplied. Thus we must determine $a_0(0)$, $b_1(0)$, and $sn_t(0)$ for $t = 1, 2, \ldots, L$, where L is the number of different seasons. In the discussion that follows we will assume that seasons repeat annually so that L is the number of seasons in a year. One way to determine the initial estimates is to use the multiplicative decomposition method. It should be noted that if b_0, b_1, and sn_t, for $t = 1, 2, \ldots, L$, represent the estimates obtained using the multiplicative decomposition method, then $a_0(0) = b_0$, $b_1(0) = b_1$, and $sn_t(0) = sn_t$, for $t = 1, 2, \ldots, L$.

Although the multiplicative decomposition method can be used to obtain the initial estimates, the method described below is more frequently used. This method is presented in Johnson and Montgomery [8] and is similar to a method presented by Winters [12]. Suppose that historical data for the last m years is available. We shall define $\bar{y}_i$ to be the average of the observations in the ith year, where i takes on values $1, 2, \ldots, m$. The initial estimate of the trend component, β_1, is given by the equation

$$b_1(0) = \frac{\bar{y}_m - \bar{y}_1}{(m - 1)L}$$

Here $\bar{y}_m$, the average of the observations in year m, measures the average level of the time series in the middle of year m. Similarly, $\bar{y}_1$ measures the average level of the time series in the middle of year 1. Thus $\bar{y}_m - \bar{y}_1$ measures the difference in these average levels. The total number of seasons elapsed between the middle of year 1 and the middle of year m is $(m - 1)L$. Thus the initial estimate $b_1(0)$ is simply the change in average level per season from the middle of year 1 to the middle of year m. The initial estimate of the permanent component, β_0, is given by

$$a_0(0) = \bar{y}_1 - \frac{L}{2} b_1(0)$$

Again, $\bar{y}_1$, the average of the observations in year 1, measures the average level of the time series in the middle of year 1. The number of seasons that have elapsed from the start of year 1 to the middle of year 1 is given by $L/2$. The initial estimate $a_0(0)$ is, therefore, the average level of the time series at the middle of year 1 less the amount this average level has changed from the start of year 1 to the middle of year 1.

Obtaining initial estimates for the L seasonal factors is done as follows. The expression

$$S_t = \frac{y_t}{\bar{y}_i - [(L + 1)/2 - j]b_1(0)}$$

must be computed for each season (month, quarter, etc.) t occurring in years 1 through m. Here $\bar{y}_i$ is the average of the observations for the year in which season t occurs. So, if $1 \leq t \leq L$, then $i = 1$; if $L + 1 \leq t \leq 2L$, then $i = 2$; etc. Thus $\bar{y}_i$ measures the average level of the time series in the middle of the year in which season t occurs. The letter j denotes the position of season t within the year. If the time series consists of monthly data, then, for January $j = 1$, for February $j = 2$, for March $j = 3$, and so on. Thus

$$- [(L + 1)/2 - j]$$

measures the number of seasons that season t is from the middle of the year. A positive value for this expression indicates that season t occurs after the middle of the year, while a negative value indicates that season t occurs before the middle of the year. Hence, the expression

$$\bar{y}_i - [(L + 1)/2 - j]b_1(0)$$

measures the average level of the time series in season t. If season t occurs before the middle of the year, we subtract the appropriate trend from the average level at midyear in order to obtain the average level in season t. If season t occurs after the middle of the year, we add the appropriate trend to the average level in the middle of the year to obtain the average level in season t.

The expression S_t is then the ratio of the observation in season t to the average level of the time series in season t. It is the factor by which we must multiply the average level in order to obtain the observation, and hence S_t represents factors not accounted for in the average level of the time series. Since the average level is determined by the permanent component and the trend, S_t represents seasonal factors and the error term ϵ_t.

The equation for S_t yields m estimates of each distinct seasonal factor, one for each year. If the time series is monthly, we obtain m estimates of the seasonal factor for each month. These m estimates are then averaged to yield one estimate for each different season. Thus we obtain

$$\overline{sn}_t = \frac{1}{m} \sum_{K=0}^{m-1} S_{t+KL} \qquad \text{for } t = 1, 2, \ldots, L$$

which is the average seasonal index for each different season. So, if the time series is monthly, we obtain 12 seasonal factors, one for each month. This averaging procedure does more than yield one estimate for each distinct season. Hopefully, it also removes the influence of the error term ϵ_t. This is reasonable since we assume the expected value of ϵ_t to be zero. Thus the factors obtained should represent seasonal effects only.

Finally, the seasonal factors are normalized so that they add to L. This is done by multiplying each average seasonal index by a quantity equal to L divided by the sum of the average seasonal indexes. Thus the initial estimate $sn_t(0)$ is given by the formula

$$sn_t(0) = \overline{sn}_t \left[\frac{L}{\displaystyle\sum_{t=1}^{L} \overline{sn}_t} \right] \qquad \text{for } t = 1, \ldots, L$$

We thus obtain the initial estimates $a_0(0)$, $b_1(0)$, and $sn_t(0)$, for $t = 1, 2, \ldots,$ L. It should be noted that these estimates are based on a time origin that immediately precedes the first period accounted for in the m years of historical data used in obtaining them. For forecasting purposes, however, a time origin at the end of the m years of historical data is desired. Estimates at this point in time can be generated by repeatedly smoothing the initial estimates using the updating equations given previously. Upon completion of this smoothing procedure, the origin of time can be redefined to be the end of the m years of historical data. At this point, the forecasting procedure may begin.

Simulation of an historical time series can be used to determine the optimal combination of smoothing constants α, β, and γ to be used in Winters' Method. A set of forecasts is generated for each combination in a set of combinations of α, β, and γ. The set of combinations of α, β, and γ can be formed by sequentially ranging each of α, β, and γ from .01 to .30 in steps of a given size. The forecasts generated by each combination are compared to the actual observations in the historical time series. The particular combination minimizing the sum of squared forecast errors is the combination to be used in the actual forecasting of future values of the time series. The simulations determining the optimal combination of α, β, and γ are initialized with starting values for $a_0(0)$, $b_1(0)$, and $sn_t(0)$, for $t = 1, \ldots, L$. These starting values are obtained using the first m_1 years out of the m years of historical data. It was found in Section 5–3 that while the accuracy of the forecasts generated by triple exponential smoothing can be quite sensitive to such a choice (that is, the choice of n_1, which is analogous to m_1), the accuracy of the forecasts generated by single and double exponential smoothing does not seem to be so sensitive. Winters' Method updates the permanent component and the trend component using the equations

$$a_0(T) = \alpha \frac{y_T}{sn_T(T - L)} + (1 - \alpha)[a_0(T - 1) + b_1(T - 1)]$$

and

$$b_1(T) = \beta[a_0(T) - a_0(T - 1)] + (1 - \beta)b_1(T - 1)$$

It can be shown that if $\alpha = \beta$, then the above updating procedure is equivalent to double exponential smoothing, which uses only one smoothing constant. If $\alpha \neq \beta$, we may consider the above procedure a modified double exponential smoothing. Hence, because of the results of Section 5–3, we would hope that the choice of m_1 will not drastically affect the accuracy of the resulting forecasts. Furthermore, since we must obtain initial estimates of the seasonal factors, since for such purposes each complete year of history is much like a single observation, and since we may not have many years of history, it may not be possible to let m_1 be much, if at all, less than m. In our example of Winters' Method, Example 7.1, we will in fact let $m_1 = m$.

7–2.3 THE NO TREND CASE

Before proceeding to Example 7.1 it should be mentioned that Winters' Method can also be used to forecast a time series having no trend. In this case, the updating equations are

$$a_0(T) = \alpha \frac{y_T}{sn_T(T - L)} + (1 - \alpha)a_0(T - 1)$$

and

$$sn_T(T) = \gamma \frac{y_T}{a_0(T)} + (1 - \gamma)sn_T(T - L)$$

Here the initial estimate of the permanent component, $a_0(0)$, is just the average of the values of the observations in the m years of historical data. The initial estimate of SN_t, $sn_t(0)$, is determined using the method previously discussed for the situation in which there is a linear trend, with the exception that we compute S_t using the formula

$$S_t = \frac{y_t}{a_0(0)}$$

Winters' Method, then, is an intuitive modification of either single or double exponential smoothing. It is used to forecast a time series with seasonal variation and with either no trend or a linear trend.

7–2.4   USING WINTERS' METHOD TO FORECAST ROLA-COLA SALES

Example 7.1
We now consider Winters' Method as a forecasting tool to be used in forecasting the time series concerning Rola-Cola sales given in Table III.2 and illustrated in Figure III.5.

We first obtain the initial estimates $a_0(0)$, $b_1(0)$, and $sn_t(0)$, for $t = 1, 2, \ldots, L$, by using the entire three years of historical data. If these estimates are found using the multiplicative decomposition method, we obtain the following estimates from Example 6.3, Table 6.7.

$$a_0(0) = 380.163 \qquad b_1(0) = 9.489$$

$$sn_1(0) = .493 \qquad sn_7(0) = 1.467$$

$$sn_2(0) = .596 \qquad sn_8(0) = 1.693$$

$$sn_3(0) = .595 \qquad sn_9(0) = 1.990$$

$$sn_4(0) = .680 \qquad sn_{10}(0) = 1.307$$

$$sn_5(0) = .564 \qquad sn_{11}(0) = 1.029$$

$$sn_6(0) = .986 \qquad sn_{12}(0) = .600$$

We next illustrate how to obtain these estimates using the procedure discussed in this section. We shall first obtain the initial estimate $b_1(0)$. Since the historical data consists of 3 years of monthly data, we have $m = 3$ and $L = 12$. The average sales for year 1 is $\bar{y}_1 = 447.82$, while the average sales for year 3 is $\bar{y}_3 = 677.58$. Thus, the initial estimate of the trend component is

$$b_1(0) = \frac{\bar{y}_m - \bar{y}_1}{(m-1)L} = \frac{677.58 - 447.82}{(3-1)12} = 9.57$$

The initial estimate of the permanent component is

$$a_0(0) = \bar{y}_1 - \frac{L}{2}(b_1(0)) = 447.82 - \frac{12}{2}(9.57) = 390.40$$

We now derive initial estimates of the seasonal factors. In this case a seasonal factor for each month must be derived. In order to illustrate the calculations we shall obtain the seasonal factor for the month of April. First, S_t must be calculated for each April in the three years of the sales history. For April of year 1 we have

$$S_4 = \frac{y_4}{\bar{y}_1 - [(L+1)/2 - j]b_1(0)}$$

$$= \frac{289}{447.82 - [(12+1)/2 - 4](9.57)}$$

$$= .6818$$

For April of year 2 we have

$$S_{16} = \frac{y_{16}}{\bar{y}_2 - [(L+1)/2 - j]b_1(0)}$$

$$= \frac{370}{564.83 - [(12+1)/2 - 4](9.57)}$$

$$= .6840$$

For April of year 3 we have

$$S_{28} = \frac{y_{28}}{\bar{y}_3 - [(L+1)/2 - j]b_1(0)}$$

$$= \frac{443}{677.58 - [(12+1)/2 - 4](9.57)}$$

$$= .6777$$

We now average these three values to obtain

$$\overline{sn}_4 = \frac{S_4 + S_{16} + S_{28}}{3} - \frac{.6818 + .6840 + .6777}{3} = .6812$$

Similarly, values of $\overline{sn}_t$ may be obtained for the other eleven months. If this is done, we find that

$$\sum_{t=1}^{12} \overline{sn}_t = 11.8285$$

Thus, the initial estimate $sn_t(0)$ is given by

$$sn_t(0) = \overline{sn}_t \left[\frac{L}{\sum_{t=1}^{L} \overline{sn}_t} \right] = \overline{sn}_t \left(\frac{12}{11.8285} \right) = 1.0145 \overline{sn}_t$$

For example,

$$sn_4(0) = 1.0145 \overline{sn}_4 = 1.0145(.6812) = .6911$$

The initial estimates obtained using the above method are

$$a_0(0) = 390.40 \qquad b_1(0) = 9.57$$

$$sn_1(0) = .4841 \qquad sn_7(0) = 1.4842$$

$$sn_2(0) = .5847 \qquad sn_8(0) = 1.6927$$

$$sn_3(0) = .6022 \qquad sn_9(0) = 1.9869$$

$$sn_4(0) = .6911 \qquad sn_{10}(0) = 1.2897$$

$$sn_5(0) = .5859 \qquad sn_{11}(0) = 1.0074$$

$$sn_6(0) = .9965 \qquad sn_{12}(0) = .5946$$

Notice that these initial estimates are quite close to the estimates obtained using the multiplicative decomposition method.

We will now use the above initial estimates to (1) initialize the simulations determining the optimal combination of α, β, and γ and (2) begin the actual forecasting procedure.

We first determine the optimum combination of α, β, and γ to be used in the actual forecasting of future values of the time series. We will form the set of combinations of α, β, and γ from which the optimum combination will be chosen by arbitrarily ranging α, β, and γ from .05 to .25 in increments of .05. This gives 125 combinations of α, β, and γ. Starting with the initial values just obtained we then, using each combination of α, β, and γ, generate a sequence of one-period-ahead forecasts for the 36 months of historical data. The combination of α, β, and γ that minimizes the sum of squared forecast errors is the combination to be used in the actual forecasting of future values of the time series. To demonstrate the procedure involved, let us consider the combination $\alpha = .2$, $\beta = .15$, and $\gamma = .05$. Using the initial estimates just obtained, that is,

$$a_0(0) = 390.40, \quad b_1(0) = 9.57, \quad \text{and} \quad sn_1(0) = .4841$$

a forecast for the value of the time series in January of year 1 is

$$\hat{y}_1(0) = [a_0(0) + b_1(0) \cdot 1]sn_1(0)$$
$$= [390.40 + 9.57][.4841]$$
$$= 193.63$$

for which, since the observation in January of year 1 is $y_1 = 189$, we have a forecast error of $y_1 - \hat{y}_1(0) = 189 - 193.63 = -4.63$.

We can now obtain the updated estimates, $a_0(1)$, $b_1(1)$, and $sn_1(1)$, as shown below.

$$a_0(1) = \alpha \frac{y_1}{sn_1(0)} + (1 - \alpha)[a_0(0) + b_1(0)]$$
$$= .2\left[\frac{189}{.4841}\right] + .8[390.40 + 9.57]$$
$$= 398.06$$

$$b_1(1) = \beta[a_0(1) - a_0(0)] + (1 - \beta)b_1(0)$$
$$= .15[398.06 - 390.40] + .85(9.57)$$
$$= 9.28$$

$$sn_1(1) = \gamma \frac{y_1}{a_0(1)} + (1 - \gamma)sn_1(0)$$
$$= .05\left[\frac{189}{398.06}\right] + .95(.4841)$$
$$= .4836$$

Using these updated initial estimates, a forecast made in period 1, January of year 1, for period 2, February of year 1, is

$$\hat{y}_2(1) = [a_0(1) + b_1(1) \cdot 1]sn_2(0)$$
$$= [398.06 + 9.28][.5847]$$
$$= 238.17$$

for which, since the observation in February of year 1 is $y_2 = 229$, we have a forecast error of $y_2 - \hat{y}_2(1) = 229 - 238.17 = -9.17$.

Using the estimates just obtained, we can go on to compute the updated estimates, $a_0(2)$, $b_1(2)$, and $sn_2(2)$, as shown below.

$$a_0(2) = \alpha \frac{y_2}{sn_2(0)} + (1 - \alpha)[a_0(1) + b_1(1)]$$

$$= .2 \left[\frac{229}{.5847} \right] + .8[398.06 + 9.28]$$

$$= 404.20$$

$$b_1(2) = \beta[a_0(2) - a_0(1)] + (1 - \beta)b_1(1)$$

$$= .15[404.20 - 398.06] + .85(9.28)$$

$$= 8.81$$

$$sn_2(2) = \gamma \frac{y_2}{a_0(2)} + (1 - \gamma)sn_2(0)$$

$$= .05 \left[\frac{229}{404.20} \right] + (.95)(.5847)$$

$$= .5838$$

This procedure is continued through the entire three years (36 periods) of historical data. The results are summarized in Table 7.1. In this table, $sn_T(T - L)$ is considered to be obtained before y_T is observed and is used to find the forecast $\hat{y}_T(T - 1)$ of y_T. For $T = 1, 2, \ldots, 12$, we have

$$sn_T(T - L) = sn_T(0)$$

When y_T is observed, $sn_T(T - L)$ is updated to $sn_T(T)$.

Using the results in Table 7.1, we find that, for $\alpha = .2$, $\beta = .15$, and $\gamma = .05$, the sum of the squared forecast errors is

$$\sum_{T=1}^{36} (y_T - \hat{y}_T(T - 1))^2 = 2254.3228$$

To find the optimal combination of α, β, and γ, we evaluate the sum of the squared forecast errors for the other 124 combinations of α, β, and γ. It is found that the combination of α, β, and γ that minimizes the sum of the squared forecast errors is $\alpha = .2$, $\beta = .15$, and $\gamma = .05$. This is the combination for which we have demonstrated the relevant calculations and which we will use to forecast future values of the time series.

Referring to Table 7.1, we have

TABLE 7.1   *Simulating One-Period-Ahead Forecasting of the Historical Rola-Cola Sales*
Time Series Using Winters' Method with $\alpha = .2$, $\beta = .15$, *and* $\gamma = .05$

T	y_T	$a_0(T)$	$b_1(T)$	$sn_T(T-12)$	$sn_T(T)$	$\hat{y}_T(T-1)$	$y_T - \hat{y}_T(T-1)$
		$a_0(0) = 390.40$	$b_1(0) = 9.57$				
1	189	398.06	9.28	.4841	.4836	193.63	−4.63
2	229	404.20	8.81	.5847	.5838	238.17	−9.17
3	249	413.11	8.83	.6022	.6022	248.72	.28
4	289	421.18	8.71	.6911	.6909	291.60	−2.60
5	260	432.67	9.13	.5859	.5867	251.88	8.12
6	431	439.94	8.85	.9965	.9957	440.25	−9.25
7	660	447.97	8.73	1.4842	1.4837	666.10	−6.10
8	777	457.16	8.80	1.6927	1.6930	773.05	3.95
9	915	464.87	8.63	1.9869	1.9860	925.82	−10.82
10	613	473.87	8.69	1.2897	1.2899	610.68	2.32
11	485	482.33	8.65	1.0074	1.0073	486.12	−1.12
12	277	485.96	7.90	.5946	.5934	291.94	−14.94
13	244	495.99	8.22	.4836	.4841	238.85	5.15
14	296	504.77	8.30	.5838	.5839	294.35	1.65
15	319	516.40	8.80	.6022	.6030	308.99	10.01
16	370	527.28	9.11	.6909	.6914	362.84	7.16
17	313	535.82	9.03	.5867	.5865	314.68	−1.68
18	556	547.56	9.44	.9957	.9966	542.48	13.52
19	831	557.62	9.53	1.4837	1.4840	826.39	4.61
20	960	567.12	9.52	1.6930	1.6930	960.21	−.21
21	1152	577.33	9.63	1.9860	1.9864	1145.21	6.79
22	759	587.25	9.67	1.2899	1.2900	757.12	1.88
23	607	598.06	9.84	1.0073	1.0077	601.28	5.72
24	371	611.37	10.36	.5934	.5940	360.71	10.29
25	298	620.51	10.18	.4841	.4839	300.95	−2.95
26	378	634.02	10.68	.5839	.5845	368.27	9.73
27	373	639.47	9.89	.6030	.6020	388.75	−15.75
28	443	647.64	9.64	.6914	.6910	448.97	−5.97
29	374	653.35	9.05	.5865	.5858	385.51	−11.51
30	660	662.36	9.04	.9966	.9966	660.18	−.18
31	1004	672.43	9.20	1.4840	1.4844	996.35	7.65
32	1153	681.51	9.18	1.6930	1.6930	1154.02	−1.02
33	1388	692.30	9.42	1.9864	1.9874	1372.01	15.99
34	904	701.53	9.39	1.2900	1.2900	905.23	−1.23
35	715	710.64	9.35	1.0077	1.0076	716.38	−1.38
36	441	724.47	10.02	.5940	.5948	427.71	13.29

$$\Sigma_{T=1}^{36}(y_T - \hat{y}_T(T-1))^2 = 2254.3228$$

$$a_0(36) = 724.47, \quad b_1(36) = 10.02, \quad \text{and} \quad sn_{37}(25) = .4839$$

Hence, a forecast made in period 36, December of year 3, for period 37, January of year 4, is

$$\hat{y}_{37}(36) = [a_0(36) + b_1(36) \cdot 1] sn_{37}(25)$$
$$= [724.47 + 10.02][.4839]$$
$$= 355.39$$

Since the observation in January of year 4 is $y_{37} = 352$ (Table 7.2), we have a forecast error of $y_{37} - \hat{y}_{37}(36) = 352 - 355.39 = -3.39$.

TABLE 7.2   *One-Period-Ahead Forecasts of Future Values of Rola-Cola Sales Calculated Using Winters' Method with $\alpha = .2$, $\beta = .15$, and $\gamma = .05$*

T	y_T	$a_0(T)$	$b_1(T)$	$sn_T(T - 12)$	$sn_T(T)$	$\hat{y}_T(T - 1)$	$y_T - \hat{y}_T(T - 1)$
		$a_0(36) = 724.47$		$b_1(36) = 10.02$			
37	352	733.09	9.81	.4839	.4837	355.39	−3.39
38	445	746.57	10.36	.5845	.5851	434.25	10.75
39	453	756.04	10.23	.6020	.6019	455.69	−2.69
40	541	769.60	10.73	.6910	.6916	529.51	11.49
41	457	780.28	10.72	.5858	.5858	457.13	−.13
42	762	785.71	9.93	.9966	.9953	788.34	−26.34
43	1194	797.38	10.19	1.4844	1.4851	1181.08	12.92
44	1361	806.84	10.08	1.6930	1.6927	1367.19	−6.19
45	1615	816.06	9.95	1.9874	1.9869	1623.51	−8.51
46	1059	825.00	9.80	1.2900	1.2896	1065.51	−6.51
47	824	831.39	9.29	1.0076	1.0068	841.15	−17.15
48	495	838.99	9.03	.5948	.5945	500.02	−5.02

$$\Sigma_{T=37}^{48} (y_T - \hat{y}_T(T - 1))^2 = 1599.4889$$

We next obtain the updated estimates, $a_0(37)$, $b_1(37)$, and $sn_{37}(37)$, as shown below.

$$a_0(37) = \alpha \frac{y_{37}}{sn_{37}(25)} + (1 - \alpha)[a_0(36) + b_1(36)]$$
$$= .2 \left[\frac{352}{.4839} \right] + .8 [724.47 + 10.02]$$
$$= 733.09$$

$$b_1(37) = \beta[a_0(37) - a_0(36)] + (1 - \beta)b_1(36)$$

$$= .15[733.09 - 724.47] + (.85)(10.02)$$

$$= 9.81$$

$$sn_{37}(37) = \gamma\frac{y_{37}}{a_0(37)} + (1 - \gamma)sn_{37}(25)$$

$$= .05\left[\frac{352}{733.09}\right] + (.95)(.4839)$$

$$= .4837$$

This procedure may be continued for as many future periods as desired. Table 7.2 summarizes the one-period-ahead forecasts made for year 4. Note that the sum of the squared forecast errors is

$$\sum_{T=37}^{48} (y_T - \hat{y}_T(T - 1))^2 = 1599.4889$$

In Figure 7.2 the observed values of the time series and the one-period-ahead forecasts are plotted.

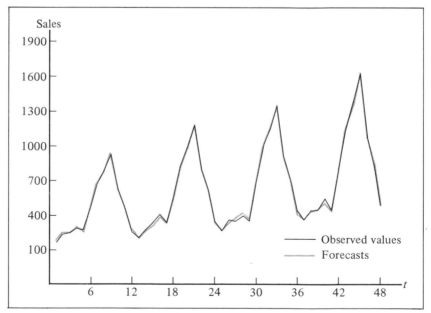

FIGURE 7.2   *One-Period-Ahead Forecasts of Future Values of Rola-Cola Sales Calculated Using Winters' Method with $\alpha = .2$, $\beta = .15$, and $\gamma = .05$*

7–3 APPROXIMATE CONFIDENCE INTERVALS

Although it can be shown (see Section 4–3) that there is a theoretical, statistical basis behind single, double, and triple exponential smoothing, there is no theoretical, statistical basis behind Winters' Method, which is an intuitive modification of double exponential smoothing. Hence, there is no theoretically correct confidence interval for $y_{T+\tau}$. However, an intuitive method yields "fairly accurate" confidence intervals. An approximate $(100 - \alpha)$ percent confidence interval made at time T for $y_{T+\tau}$ is

$$[\hat{y}_{T+\tau}(T) - E_{T+\tau}^{(100-\alpha)}(T),\ \hat{y}_{T+\tau}(T) + E_{T+\tau}^{(100-\alpha)}(T)]$$

where

$$\hat{y}_{T+\tau}(T) = [a_0(T) + b_1(T)\tau]sn_{T+\tau}(T + \tau - L)$$

$$E_{T+\tau}^{(100-\alpha)}(T) = z_{\alpha/2}d_\tau\,\Delta_W(T)$$

Here $z_{\alpha/2}$ is the point on the scale of the normal curve having mean 0 and variance 1 such that there is an area of $(100 - \alpha)/100$ under this normal curve between $-z_{\alpha/2}$ and $z_{\alpha/2}$; d_τ is a constant, depending only on τ and not upon T, which is given by the formula

$$d_\tau = 1.25 \left[\frac{1 + \dfrac{\theta}{(1 + v)^3}[(1 + 4v + 5v^2) + 2\theta(1 + 3v)\tau + 2\theta^2\tau^2]}{1 + \dfrac{\theta}{(1 + v)^3}[(1 + 4v + 5v^2) + 2\theta(1 + 3v) + 2\theta^2]} \right]^{1/2}$$

where θ equals the maximum of α, β, and γ, the smoothing constants employed by Winters' Method, and where $v = 1 - \theta$; and

$$\Delta_W(T) = \frac{\displaystyle\sum_{t=1}^{T} \left| \frac{y_t}{sn_t(t - L)} - (a_0(t - 1) + b_1(t - 1)) \right|}{T}$$

where $sn_t(t - L)$ is the latest estimate of the appropriate seasonal factor for period

t, an estimate obtained before y_t is observed. Since $\Delta_W(T)$ can be easily updated to $\Delta_W(T + 1)$ using the formula

$$\Delta_W(T + 1) = \frac{T \, \Delta_W(T) + [y_{T+1}/(sn_{T+1}(T + 1 - L)) - (a_0(T) + b_1(T))]}{T + 1}$$

the error $E_{T+\tau}^{(100-\alpha)}(T)$ can be easily updated to $E_{T+1+\tau}^{(100-\alpha)}(T + 1)$ using the formula

$$E_{T+1+\tau}^{(100-\alpha)}(T + 1) = z_{\alpha/2} d_\tau \, \Delta_W(T + 1)$$

Hence, we can update the previous approximate $(100 - \alpha)$ percent confidence interval, which is made at time T for $y_{T+\tau}$, to the following approximate $(100 - \alpha)$ percent confidence interval, which is made at time $T + 1$ for $y_{T+1+\tau}$.

$$[\hat{y}_{T+1+\tau}(T + 1) - E_{T+1+\tau}^{(100-\alpha)}(T + 1), \, \hat{y}_{T+1+\tau}(T + 1) + E_{T+1+\tau}^{(100-\alpha)}(T + 1)]$$

Here

$$\hat{y}_{T+1+\tau}(T + 1) = [a_0(T + 1) + b_1(T + 1)\tau] sn_{T+1+\tau}(T + 1 + \tau - L)$$

The intuitive reasoning behind this method will be discussed in Section 7–4.

Example 7.2   We will now continue Example 7.1 and find approximate 95 percent one-period-ahead and three-period-ahead confidence intervals for Rola-Cola sales in months 37 through 48. Since 1.96 is the point on the scale of the normal curve having mean 0 and variance 1 such that there is an area of .95 under this normal curve between -1.96 and 1.96, we have $z_{5/2} = 1.96$. Since $\theta = .2$ is the maximum of the smoothing constants used in the final forecasting of the Rola-Cola sales, we have $v = 1 - \theta = .8$. Hence, for generating one-period-ahead forecasts ($\tau = 1$) we have

$$d_1 = 1.25$$

and

$$E_{T+1}^{(95)}(T) = z_{5/2} d_1 \, \Delta_W(T) = 1.96(1.25) \, \Delta_W(T) = 2.45 \, \Delta_W(T)$$

For generating three-period-ahead forecasts ($\tau = 3$) we have

$$d_3 = 1.25 \left[\frac{1 + \dfrac{.2}{(1.8)^3} [(1 + 4(.8) + 5(.8)^2) + 2(.2)(1 + 3(.8))(3) + 2(.2)^2(3)^2]}{1 + \dfrac{.2}{(1.8)^3} [(1 + 4(.8) + 5(.8)^2) + 2(.2)(1 + 3(.8)) + 2(.2)^2]} \right]^{1/2}$$

$$= 1.3041$$

and

$$E_{T+3}^{(95)}(T) = z_{5/2} d_3 \, \Delta_W(T) = 1.96(1.3041) \, \Delta_W(T) = 2.556036 \, \Delta_W(T)$$

Since we will begin the generation of confidence intervals from a time origin at $T = 36$, we now refer to Table 7.1 and compute

$$\Delta_W(36) = \frac{\displaystyle\sum_{t=1}^{36} \left| \frac{y_t}{sn_t(t-12)} - (a_0(t-1) + b_1(t-1)) \right|}{36}$$

$$= \frac{1}{36} \left\{ \left| \frac{189}{.4841} - (390.40 + 9.57) \right| \right.$$

$$+ \left| \frac{229}{.5847} - (398.06 + 9.28) \right| + \cdots$$

$$\left. + \left| \frac{441}{.5940} - (710.64 + 9.35) \right| \right\}$$

$$= 8.2354$$

Hence,

$$E_{37}^{(95)}(36) = 2.45 \, \Delta_W(36) = 20.1767$$

and

$$E_{39}^{(95)}(36) = 2.556036 \, \Delta_W(36) = 21.05$$

Since

$$\hat{y}_{37}(36) = [a_0(36) + b_1(36) \cdot 1] sn_{37}(25)$$
$$= [724.47 + 10.02(1)][.4839]$$
$$= 355.39$$

an approximate 95 percent confidence interval made at time 36 for y_{37} is

$$[355.39 - 20.1767, \ 355.39 + 20.1767] \quad \text{or} \quad [335.21, 375.57]$$

And since

$$\hat{y}_{39}(36) = [a_0(36) + b_1(36) \cdot 3] sn_{39}(27)$$
$$= [724.47 + (10.02)3] (.6020)$$
$$= 454.2271$$

an approximate 95 percent confidence interval made at time 36 for y_{39} is

$$[454.2271 - 21.05, 454.2271 + 21.05] \qquad \text{or} \qquad [433.18, 475.28]$$

Assuming we now observe $y_{37} = 352$, we can update $\Delta_W(36) = 8.2354$ to $\Delta_W(37)$ using the formula

$$\Delta_W(37) = \frac{36 \, \Delta_W(36) + |352/.4839 - (724.47 + 10.02)|}{37} = 8.2038$$

Hence,

$$E_{38}^{(95)}(37) = 2.45 \, \Delta_W(37) = 20.0993$$

and

$$E_{40}^{(95)}(37) = 2.556036 \, \Delta_W(37) = 20.9692$$

Referring to Tables 7.1 and 7.2,

$$\hat{y}_{38}(37) = [a_0(37) + b_1(37) \cdot 1] sn_{38}(26)$$
$$= [733.09 + 9.81][.5845]$$
$$= 434.25$$

So an approximate 95 percent confidence interval made at time 37 for y_{38} is

$$[434.25 - 20.0993, 434.25 + 20.0993] \qquad \text{or} \qquad [414.15, 454.35]$$

And since

$$\hat{y}_{40}(37) = [a_0(37) + b_1(37) \cdot 3] sn_{40}(28)$$
$$= [733.09 + 9.81(3)][.6910]$$
$$= 526.9013$$

an approximate 95 percent confidence interval made at time 37 for y_{40} is

$$[526.9013 - 20.9692, 526.9013 + 20.9692] \qquad \text{or} \qquad [505.93, 547.87]$$

Table 7.3 summarizes the approximate 95 percent one-period-ahead and three-period-ahead confidence intervals for Rola-Cola sales in months 37 through 48. It should be noted that 11 out of 12 of the one-period-ahead confidence intervals and 10 out of 11 of the three-period-ahead confidence intervals contain the actual observed values of the time series.

TABLE 7.3 One-Period-Ahead and Three-Period-Ahead Forecasts of Future Values of Rola-Cola Sales Calculated Using Winters' Method with $\alpha = .2$, $\beta = .15$, and $\gamma = .05$

T	y_T	$a_0(T)$	$b_1(T)$	$sn_T(T-12)$	$\Delta_W(T)$	$E_T^{(95)}(T-1)$	$\hat{y}_T(T-1)$	$[\hat{y}_T(T-1) - E_T^{(95)}(T-1),\ \hat{y}_T(T-1) + E_T^{(95)}(T-1)]$	$E_T^{(95)}(T-3)$	$\hat{y}_T(T-3)$	$[\hat{y}_T(T-3) - E_T^{(95)}(T-3),\ \hat{y}_T(T-3) + E_T^{(95)}(T-3)]$
36		724.47	10.02		8.2354						
37	352	733.09	9.81	.4839	8.2038	20.1767	355.39	[335.21, 375.57]			
38	445	746.57	10.36	.5845	8.4730	20.0993	434.25	[414.15, 454.35]			
39	453	756.04	10.23	.6020	8.3696	20.7589	455.67	[434.91, 476.43]	21.05	454.2271	[433.18, 475.28]
40	541	769.60	10.73	.6910	8.5767	20.5055	529.51	[509.00, 550.02]	20.9692	526.9013	[505.93, 547.87]
41	457	780.28	10.72	.5858	8.3724	21.0130	457.13	[436.12, 478.14]	21.6573	455.5474	[433.89, 477.20]
42	762	785.71	9.93	.9966	8.8016	20.5124	788.34	[767.83, 808.85]	21.3930	784.0551	[762.66, 805.45]
43	1194	797.38	10.19	1.4844	8.7998	21.5639	1181.08	[1159.52, 1202.64]	21.9224	1190.1770	[1168.25, 1212.10]
44	1361	806.84	10.08	1.6930	8.6833	21.5596	1367.19	[1345.63, 1388.75]	21.4002	1375.4609	[1354.06, 1396.86]
45	1615	816.06	9.95	1.9874	8.5859	21.2740	1623.51	[1602.24, 1644.78]	22.4972	1620.7247	[1598.23, 1643.22]
46	1059	825.00	9.80	1.2900	8.5097	21.0355	1065.51	[1044.47, 1086.54]	22.4926	1068.0555	[1045.56, 1090.55]
47	824	831.39	9.29	1.0076	8.6906	20.8487	841.15	[820.30, 862.00]	22.1948	843.4418	[821.25, 865.64]
48	495	838.99	9.03	.5948	8.6860	21.2921	500.02	[478.73, 521.31]	21.9459	503.1473	[481.20, 525.09]

A forecast made at time T for $\Sigma_{t=1}^{\tau} y_{T+t}$, the sum of the values of the time series from period $T + 1$ to $T + \tau$, is

$$\sum_{t=1}^{\tau} \hat{y}_{T+t}(T) = \sum_{t=1}^{\tau} (a_0(T) + b_1(T)t)sn_{T+t}(T + t - L)$$

An intuitive method yields "fairly accurate" confidence intervals for this sum. An approximate $(100 - \alpha)$ percent confidence interval made at time T for $\Sigma_{t=1}^{\tau} y_{T+t}$, is

$$\left[\sum_{t=1}^{\tau} \hat{y}_{T+t}(T) - E_{[T+1,T+\tau]}^{(100-\alpha)}(T), \sum_{t=1}^{\tau} \hat{y}_{T+t}(T) + E_{[T+1,T+\tau]}^{(100-\alpha)}(T)\right]$$

where

$$E_{[T+1,T+\tau]}^{(100-\alpha)}(T) = z_{\alpha/2} q_\tau \Delta_W(T)$$

Here $z_{\alpha/2}$ and $\Delta_W(T)$ are as previously defined, and

$$q_\tau = 1.25 \left\{ \frac{\tau + \dfrac{\theta\tau^2}{2(1 + v)^3}[5(1 + 2v + v^2) + 4(1 - v^2)\tau + \theta^2\tau^2]}{1 + \dfrac{\theta}{(1 + v)^3}[(1 + 4v + 5v^2) + 2\theta(1 + 3v) + 2\theta^2]} \right\}^{1/2}$$

where θ equals the maximum of α, β, and γ, the smoothing constants employed by Winters' Method, and where $v = 1 - \theta$. Since, as previously discussed, $\Delta_W(T)$ can be easily updated to $\Delta_W(T + 1)$, the error $E_{[T+1,T+\tau]}^{(100-\alpha)}(T)$ can be easily updated to $E_{[T+2,T+1+\tau]}^{(100-\alpha)}(T + 1)$ using the formula

$$E_{[T+2,T+1+\tau]}^{(100-\alpha)}(T + 1) = z_{\alpha/2} q_\tau \Delta_W(T + 1)$$

Hence, we can update the above approximate $(100 - \alpha)$ percent confidence interval made at time T for $\Sigma_{t=1}^{\tau} y_{T+t}$ to the following approximate $(100 - \alpha)$ percent confidence interval made at time $T + 1$ for $\Sigma_{t=1}^{\tau} y_{T+1+t}$.

$$\left[\sum_{t=1}^{\tau} \hat{y}_{T+1+t}(T + 1) - E_{[T+2,T+1+\tau]}^{(100-\alpha)}(T + 1), \sum_{t=1}^{\tau} \hat{y}_{T+1+t}(T + 1) + E_{[T+2,T+1+\tau]}^{(100-\alpha)}(T + 1)\right]$$

Here

$$\hat{y}_{T+1+t}(T + 1) = [a_0(T + 1) + b_1(T + 1)t]sn_{T+1+t}(T + 1 + t - L)$$

The intuitive reasoning behind this method will be discussed in Section 7–4.

Example 7.3   We continue Example 7.2 and find approximate 95 percent confidence intervals for three-period cumulative sales of Rola-Cola in months 37 through 48. As in Example 7.2, $z_{\alpha/2} = z_{5/2} = 1.96$ and $\theta = .2$ and $v = .8$, which implies (since $\tau = 3$) that

$$q_3 = 1.25 \left\{ \frac{3 + \dfrac{.2(3)^2}{2(1.8)^3}[5(1 + 2(.8) + (.8)^2) + 4(1 - (.8)^2)3 + (.2)^2(3)^2]}{1 + \dfrac{.2}{(1.8)^3}[(1 + 4(.8) + 5(.8)^2) + 2(.2)(1 + 3(.8)) + 2(.2)^2]} \right\}^{1/2}$$

$$= 2.7314,$$

and

$$E^{(95)}_{[T+1,T+7]}(T) = z_{5/2}q_3\, \Delta_W(T) = 1.96(2.7314)\, \Delta_W(T) = 5.353544\, \Delta_W(T)$$

We will begin the generation of confidence intervals from a time origin of $T = 36$. Since, as shown in Example 7.2, $\Delta_W(36) = 8.2354$, we have

$$E^{(95)}_{[37,39]}(36) = 5.353544\, \Delta_W(36) = 44.0886$$

Since a forecast made at time 36 for $(y_{37} + y_{38} + y_{39})$ is

$$\begin{aligned}
\hat{y}_{37}(36) + \hat{y}_{38}(36) + \hat{y}_{39}(36) &= [a_0(36) + b_1(36) \cdot 1]sn_{37}(25) \\
&\quad + [a_0(36) + b_1(36) \cdot 2]sn_{38}(26) \\
&\quad + [a_0(36) + b_1(36) \cdot 3]sn_{39}(27) \\
&= [724.47 + 10.02(1)][.4839] \\
&\quad + [724.47 + 10.02(2)][.5845] \\
&\quad + [724.47 + 10.02(3)][.6020] \\
&= 1244.8128
\end{aligned}$$

an approximate 95 percent confidence interval, which is made at time 36 for $(y_{37} + y_{38} + y_{39})$, is

$$[1244.8128 - 44.0886,\ 1244.8128 + 44.0886] \quad \text{or} \quad [1200.7242,\ 1288.9014]$$

Assuming we now observe $y_{37} = 352$, we can update $\Delta_W(36) = 8.2354$ to $\Delta_W(37) = 8.2038$, as shown in Example 7.2. Hence

$$E^{(95)}_{[38,40]}(37) = 5.353544\, \Delta_W(37) = 43.9194$$

Since a forecast made at time 37 for $(y_{38} + y_{39} + y_{40})$ is

$$\hat{y}_{38}(37) + \hat{y}_{39}(37) + \hat{y}_{40}(37) = [a_0(37) + b_1(37) \cdot 1]sn_{38}(26)$$
$$+ [a_0(37) + b_1(37) \cdot 2]sn_{39}(27)$$
$$+ [a_0(37) + b_1(37) \cdot 3]sn_{40}(28)$$
$$= [733.09 + 9.81][.5845] + [733.09 + 9.81(2)][.6020]$$
$$+ [733.09 + 9.81(3)][.6910]$$
$$= 1414.2577$$

an approximate 95 percent confidence interval made at time 37 for $(y_{38} + y_{39} + y_{40})$ is

$$[1414.2577 - 43.9194, \ 1414.2577 + 43.9194] \quad \text{or} \quad [1370.3383, \ 1458.1771]$$

Table 7.4 summarizes the approximate 95 percent confidence intervals for three-period cumulative sales in months 37 through 48. It should be noted that all 10 of the confidence intervals contain the actual cumulative sales figures.

7–4 THE REASONING BEHIND CONFIDENCE INTERVALS FOR WINTERS' METHOD

In this section we wish to discuss the reasoning behind the procedure given in Section 7–3, which yields approximate confidence intervals for values of a time series forecasted by Winters' Method. Winters' Method yields a forecast made at time T for $y_{T+\tau}$ of

$$\hat{y}_{T+\tau}(T) = [a_0(T) + b_1(T)\tau]sn_{T+\tau}(T + \tau - L)$$

The estimate of the permanent component, $a_0(T)$, and the estimate of the trend component, $b_1(T)$, are updated using the equations

$$a_0(T + 1) = \alpha \left[\frac{y_{T+1}}{sn_{T+1}(T + 1 - L)} \right] + (1 - \alpha)[a_0(T) + b_1(T)]$$

$$b_1(T + 1) = \beta[a_0(T + 1) - a_0(T)] + (1 - \beta)b_1(T)$$

If we assume that $\alpha = \beta$, it can be shown that the above smoothing operations are equivalent to double exponential smoothing. Examining the Winters' Method smoothing equations, we note that the equivalent double exponential smoothing procedure is applied to the deseasonalized observations. The deseasonalized observation at time $T + 1$ is

TABLE 7.4   *Forecasts of Cumulative Rola-Cola Sales in the Next Three Periods Calculated Using Winters' Method with* $\alpha = .2$, $\beta = .15$, *and* $\gamma = .05$

T	y_T	$\sum\limits_{t=1}^{3} y_{T-3+t}$	$\Delta_W(T)$	$E^{(95)}_{[T-2,T]}(T-3)$	$\sum\limits_{t=1}^{3} \hat{y}_{T-3+t}(T-3)$	$[\sum_{t=1}^{3}\hat{y}_{T-3+t}(T-3) - E^{(95)}_{[T-2,T]}(T-3),$ $\sum_{t=1}^{3}\hat{y}_{T-3+t}(T-3) + E^{(95)}_{[T-2,T]}(T-3)]$
36			8.2354			
37	352		8.2038			
38	445		8.4730			
39	453	1250	8.3696	44.0886	1244.8128	[1200.7242, 1288.9014]
40	541	1439	8.5767	43.9194	1414.2577	[1370.3383, 1458.1771]
41	457	1451	8.3724	45.3606	1441.4168	[1396.0562, 1486.7774]
42	762	1760	8.8016	44.8070	1768.4213	[1723.6143, 1813.2283]
43	1194	2413	8.7998	45.9157	2435.6647	[2389.7490, 2481.5804]
44	1361	3317	8.6833	44.8220	3353.8446	[3309.0226, 3398.6666]
45	1615	4170	8.5859	47.1198	4165.6027	[4118.4829, 4212.7225]
46	1059	4035	8.5097	47.1101	4060.4877	[4013.3776, 4107.5978]
47	824	3498	8.6906	46.4864	3533.8186	[3487.3322, 3580.3050]
48	495	2378	8.6860	45.9650	2411.0133	[2365.0483, 2456.9783]

$$\frac{y_{T+1}}{sn_{T+1}(T+1-L)}$$

Now in Section 4–3 it was stated that, using double exponential smoothing, the errors in approximate $(100 - \alpha)$ percent confidence intervals for $y_{T+\tau}$ and $\sum_{t=1}^{\tau} y_{T+t}$ made at time T are, respectively,

$$E^{(100-\alpha)}_{T+\tau}(T) = z_{\alpha/2} d_\tau \, \Delta(T)$$

and

$$E^{(100-\alpha)}_{[T+1,T+\tau]}(T) = z_{\alpha/2} q_\tau \, \Delta(T)$$

Here, d_τ and q_τ are as given in Section 4–3 and are functions of τ and the smoothing constant α used in double exponential smoothing. Furthermore,

$$\Delta(T) = \frac{\sum\limits_{t=1}^{T} |y_t - \hat{y}_t(t-1)|}{T}$$

where y_t is the observed value of the time series at time t and $\hat{y}_t(t-1)$ is the forecast made at time origin $t - 1$ by double exponential smoothing for y_t. Since, if $\alpha = \beta$, the permanent component and trend component are updated in Winters' Method by applying double exponential smoothing to the deseasonalized data, it intuitively follows that the errors in $(100 - \alpha)$ percent confidence intervals made at time T for the deseasonalized values of $y_{T+\tau}$ and $\Sigma_{t=1}^{\tau} y_{T+t}$ are, respectively,

$$E_{T+\tau}^{(100-\alpha)}(T) = z_{\alpha/2} d_\tau \, \Delta_W(T)$$

and

$$E_{[T+1, T+\tau]}^{(100-\alpha)}(T) = z_{\alpha/2} q_\tau \, \Delta_W(T)$$

Here $\Delta_W(T)$ is the "analogue" of $\Delta(T)$, with the deseasonalized observation replacing y_t and the one-period-ahead forecast of the deseasonalized observation replacing $\hat{y}_t(t - 1)$. Hence,

$$\Delta_W(T) = \frac{\sum\limits_{t=1}^{T} \left| \dfrac{y_t}{sn_t(t - L)} - (a_0(t - 1) + b_1(t - 1)) \right|}{T}$$

Although the above errors are used to construct $(100 - \alpha)$ percent confidence intervals for the deseasonalized observations, when Winters' Method is being used it seems reasonable to use these errors to construct $(100 - \alpha)$ percent confidence intervals for the actual time series values, $y_{T+\tau}$ and $\Sigma_{t=1}^{\tau} y_{T+t}$. Indeed, we do much the same thing in finding confidence intervals when the time series decomposition method is used as a forecasting technique. If one returns to the sections describing Winters' Method and the intuitive methods for obtaining confidence intervals when using Winters' Method, it is apparent that these intuitive methods do indeed employ the errors

$$E_{T+\tau}^{(100-\alpha)}(T) = z_{\alpha/2} d_\tau \, \Delta_W(T)$$

and

$$E_{[T+1, T+\tau]}^{(100-\alpha)}(T) = z_{\alpha/2} q_\tau \, \Delta_W(T)$$

to construct $(100 - \alpha)$ percent confidence intervals made at time T for, respectively, $y_{T+\tau}$ and $\Sigma_{t=1}^{\tau} y_{T+t}$. Moreover, in these sections, d_τ and q_τ are defined to be functions of τ and θ, where θ is the maximum of the smoothing constants α, β, and γ employed in Winters' Method. The constant θ is defined in this manner because in Winters' Method the smoothing constants α, β, and γ are not always equal. Since it can be shown that d_τ and q_τ are increasing functions of the parameter θ, it seems reasonable to be conservative and define the parameter θ to be the maximum of the smoothing constants α, β, and γ employed by Winters' Method. Thus d_τ and q_τ are as large as might reasonably be expected.

Since the confidence intervals for Winters' Method and the decomposition technique are intuitive, one might ask whether they are "accurate." It should be noted that, while these intuitive confidence intervals do take into account the uncertainty with respect to the permanent and trend components, they do not take into account the uncertainty with respect to the seasonal component. Hence, it follows that we would expect the intuitive methods for obtaining $(100 - \alpha)$ percent confidence intervals to produce intervals that contain somewhat fewer than $(100 - \alpha)$ percent of future observed values of the time series. Although we have not yet carried out extensive computer simulations, the results we do have to date indicate that the intuitive methods for obtaining $(100 - \alpha)$ percent confidence intervals produce confidence intervals that contain close to $(100 - \alpha)$ percent of "future observed values" of time series that have been artificially generated according to the "Winters' models" and then analyzed using Winters' Method and the intuitive methods for constructing confidence intervals. It should be noted that the error terms generated for the "Winters' models" in our simulations are independent and identically distributed according to a normal distribution. In fact, the 20 observations of seminar demand and the 48 observations of Rola-Cola sales were generated in such a manner.

PROBLEMS

1. Consider Example 7.1 in which we forecasted the Rola-Cola sales time series using Winters' Method. Verify that the initial estimate of the seasonal factor for the month of June is $sn_6(0) = .9965$.

2. Consider Table 7.2 in Example 7.1. Given the observation of Rola-Cola sales in February of year 4, which is $y_{38} = 445$, verify that the updated estimates $a_0(38)$, $b_1(38)$, and $sn_{38}(38)$ are

    $$a_0(38) = 746.57, \quad b_1(38) = 10.36, \quad \text{and} \quad sn_{38}(38) = .5851$$

3. Consider Table 7.2 in Example 7.1. Verify that the forecast of Rola-Cola sales which is made in period 38 for y_{39} is $\hat{y}_{39}(38) = 455.69$.

4. Consider Table 7.2 in Example 7.1. Given the observation of Rola-Cola sales in March of year 4, which is $y_{39} = 453$, verify that the updated estimates $a_0(39)$, $b_1(39)$, and $sn_{39}(39)$ are

    $$a_0(39) = 756.04, \quad b_1(39) = 10.23, \quad \text{and} \quad sn_{39}(39) = .6019$$

5. Consider Table 7.2 in Example 7.1. Verify that the forecast of Rola-Cola sales which is made in period 39 for y_{40} is $\hat{y}_{40}(39) = 529.51$.

6. Calculate a forecast of Rola-Cola sales which is made in period 39 for y_{44}.

7. Suppose that Winters' Method is being used to forecast a quarterly time series having no trend. Given the estimates $a_0(8) = 15$, $sn_5(5) = .75$, $sn_6(6) = 1.25$, $sn_7(7) = 1.50$, and $sn_8(8) = .50$, calculate forecasts made in period 8 for y_9, y_{10}, y_{11}, and y_{12}.

8. Consider Example 7.2 in which confidence intervals for future values of the Rola-Cola sales time series presented in Table III.2 were calculated.

    a. Suppose that we observe a sales figure for period 38 of $y_{38} = 445$. Verify that when $\Delta_W(37)$ is updated, we obtain $\Delta_W(38) = 8.4730$ as presented in Table 7.3.

    b. Verify that the updated error in an approximate 95 percent confidence interval for y_{39} is $E_{39}^{(95)}(38) = 20.7589$.

    c. Verify that the updated error in an approximate 95 percent confidence interval for y_{41} is $E_{41}^{(95)}(38) = 21.6573$.

    d. Verify that an updated approximate 95 percent confidence interval for y_{39} is $[434.91, 476.43]$.

    e. Verify that an updated approximate 95 percent confidence interval for y_{41} is $[433.89, 477.20]$.

9. Consider Example 7.3 in which confidence intervals for cumulative values of the Rola-Cola sales time series presented in Table III.2 were calculated.

    a. Suppose that we observe a sales figure for period 38 of $y_{38} = 445$. Verify that the updated error in an approximate 95 percent confidence interval for

$$\sum_{t=1}^{3} y_{38+t} = y_{39} + y_{40} + y_{41}$$

    is $E_{[39,41]}^{(95)}(38) = 45.3606$.

    b. Verify that an updated approximate 95 percent confidence interval for $y_{39} + y_{40} + y_{41}$ is $[1396.0562, 1486.774]$.

CHAPTER 8

Forecasting Time Series with Additive Seasonal Variation

8–1 INTRODUCTION

In this chapter we discuss analyzing and forecasting a time series that has additive seasonal variation. If a time series displays additive seasonal variation, the magnitude of the "seasonal swing" of the time series is independent of the average level as determined by the trend. Thus it follows that a reasonable model to use when analyzing a time series with additive seasonal variation is

$$y_t = TR_t + SN_t + \epsilon_t$$

where the meaning of the notation above has been previously discussed.

In Section 8–2 we present a regression approach to forecasting time series with additive seasonal variation. This approach employs "dummy variables" that account for the seasonality in a time series. In Section 8–3 we discuss Additive Winters' Method. This method is an exponential smoothing approach that can be used to forecast time series with additive seasonal variation. We also present an intuitive method that can be used to calculate "fairly accurate" confidence interval forecasts when using Additive Winters' Method.

8–2 REGRESSION USING DUMMY VARIABLES

In this section we discuss the use of regression analysis in the estimation of TR_t and SN_t in the model

$$y_t = TR_t + SN_t + \epsilon_t$$

We will now assume that TR_t is given by one of the following equations.

$$TR_t = \beta_0 \qquad TR_t = \beta_0 + \beta_1 t \qquad TR_t = \beta_0 + \beta_1 t + \beta_2 t^2$$

Moreover, assuming there are L seasons per year, we will assume that SN_t is given by the equation

$$SN_t = \beta_{S2} x_{S2,t} + \beta_{S3} x_{S3,t} + \cdots + \beta_{SL} x_{SL,t}$$

where $x_{S2,t}, x_{S3,t}, \ldots, x_{SL,t}$ are *dummy variables* given by the equations

$$x_{S2,t} = \begin{cases} 1 & \text{if period } t \text{ is season 2} \\ 0 & \text{otherwise} \end{cases}$$

$$x_{S3,t} = \begin{cases} 1 & \text{if period } t \text{ is season 3} \\ 0 & \text{otherwise} \end{cases}$$

$$\vdots$$

$$x_{SL,t} = \begin{cases} 1 & \text{if period } t \text{ is season } L \\ 0 & \text{otherwise} \end{cases}$$

For example, then, if $L = 12$ and period t is season 3, we have

$$y_t = TR_t + SN_t + \epsilon_t$$
$$= TR_t + \beta_{S2} x_{S2,t} + \beta_{S3} x_{S3,t} + \beta_{S4} x_{S4,t} + \cdots + \beta_{S12} x_{S12,t} + \epsilon_t$$
$$= TR_t + \beta_{S2}(0) + \beta_{S3}(1) + \beta_{S4}(0) + \cdots + \beta_{S12}(0) + \epsilon_t$$
$$= TR_t + \beta_{S3} + \epsilon_t$$

So the use of the dummy variables insures that a seasonal factor (or parameter) for season 3 is added to the trend level of the time series for each time period that is season 3. This seasonal factor, β_{S3}, accounts for the seasonality of the time series in season 3.

In general, the purpose of the dummy variables is to insure that the appropriate seasonal parameter is included in the regression model in each time period. It should be noticed that use of the dummy variables also assures us that $(TR_t + SN_t)$ is a linear function of the parameters $\beta_0, \beta_1, \beta_2, \beta_{S2}, \beta_{S3}, \ldots, \beta_{SL}$, and hence that the model

$$y_t = TR_t + SN_t + \epsilon_t$$

is a regression model for which we can find the least squares estimates $b_0, b_1, b_2, b_{S2}, b_{S3}, \ldots, b_{SL}$ of the above respective parameters. Then tr_t, the estimate of TR_t, is given by one of the following equations.

$$tr_t = b_0 \qquad tr_t = b_0 + b_1 t \qquad tr_t = b_0 + b_1 t + b_2 t^2$$

Moreover, sn_t, the estimate of SN_t, is given by the equation

$$sn_t = b_{S2}x_{S2,t} + b_{S3}x_{S3,t} + \cdots + b_{SL}x_{SL,t}$$

Hence, the forecast of a value y_t of the time series is

$$\hat{y}_t = tr_t + sn_t$$

For example, then, if period t is

(1) season 1, then

$$y_t = TR_t + \epsilon_t$$

and a forecast of y_t is

$$\hat{y}_t = tr_t$$

(2) season 2, then

$$y_t = TR_t + \beta_{S2} + \epsilon_t$$

and a forecast of y_t is

$$\hat{y}_t = tr_t + b_{S2}$$

(3)   season 3, then

$$y_t = TR_t + \beta_{S3} + \epsilon_t$$

and a forecast of y_t is

$$\hat{y}_t = tr_t + b_{S3}$$

.

.

.

(L)   season L, then

$$y_t = TR_t + \beta_{SL} + \epsilon_t$$

and a forecast of y_t is

$$\hat{y}_t = tr_t + b_{SL}$$

We see that we have, quite arbitrarily, set the seasonal factor for season 1 equal to 0. Thus, the other seasonal factors—β_{S2}, β_{S3}, . . . , β_{SL}—are defined with respect to season 1. Intuitively, β_{Sj} is the difference, excluding trend, between the "expected value" of the time series in season j and the "expected value" of the time series in season 1. If β_{Sj} is positive, this implies that, excluding trend, the value of the time series in season j can be expected to be greater than the value of the time series in season 1. If β_{Sj} is negative, this implies that, excluding trend, the value of the time series in season j can be expected to be smaller than the value of the time series in season 1. We do not have to set the seasonal factor for season 1 equal to 0. We can set the seasonal factor for any particular season equal to 0 and thus define the other seasonal factors with respect to that particular season. However, we must arbitrarily set one of the seasonal factors equal to 0, for if we do not, it can be shown that we have specified a regression model the parameters of which cannot be estimated using the methods presented in this book. In this book we will always set the seasonal factor for season 1 equal to 0.

Since the additive time series model is a regression model based on statistical theory, we can construct "statistically correct" confidence intervals for y_t and $\Sigma_{t=l}^{l+q} y_t$. That is, there exist errors $E_t(100 - \alpha)$ and $E_{(l, l+q)}(100 - \alpha)$ such that

$$[\hat{y}_t - E_t(100 - \alpha), \hat{y}_t + E_t(100 - \alpha)]$$

and

$$\left[\sum_{t=l}^{l+q} \hat{y}_t - E_{(l, l+q)}(100 - \alpha), \sum_{t=l}^{l+q} \hat{y}_t + E_{(l, l+q)}(100 - \alpha) \right]$$

are $(100 - \alpha)$ percent confidence intervals for, respectively, y_t and $\Sigma_{t=l}^{l+q} y_t$.

Example   Let us first repeat the time series concerning seminar demand previously given
8.1   in Table III.1.

Quarter	Year 1	Year 2	Year 3	Year 4
1	10	11	13	15
2	31	33	34	37
3	43	45	48	51
4	16	17	19	21

The plot of this data, given in Figure III.4, indicates that the data can be fairly well described by a model that employs a linear trend, $TR_t = \beta_0 + \beta_1 t$, and additive seasonal variation given by the equation

$$SN_t = \beta_{S2} x_{S2,t} + \beta_{S3} x_{S3,t} + \beta_{S4} x_{S4,t}$$

where $x_{S2,t}$, $x_{S3,t}$, and $x_{S4,t}$ are dummy variables given by the equations

$$x_{S2,t} = \begin{cases} 1 & \text{if period } t \text{ is quarter 2} \\ 0 & \text{otherwise} \end{cases}$$

$$x_{S3,t} = \begin{cases} 1 & \text{if period } t \text{ is quarter 3} \\ 0 & \text{otherwise} \end{cases}$$

$$x_{S4,t} = \begin{cases} 1 & \text{if period } t \text{ is quarter 4} \\ 0 & \text{otherwise} \end{cases}$$

The following regression model describes the above data.

$$\begin{aligned} y_t &= TR_t + SN_t + \epsilon_t \\ &= \beta_0 + \beta_1 t + \beta_{S2} x_{S2,t} + \beta_{S3} x_{S3,t} + \beta_{S4} x_{S4,t} + \epsilon_t \end{aligned}$$

Using the observed data, the least squares estimates of β_0, β_1, β_{S2}, β_{S3}, and β_{S4} are found to be, respectively, $b_0 = 8.75$, $b_1 = .5$, $b_{S2} = 21$, $b_{S3} = 33.5$, and $b_{S4} = 4.5$. The interpretation of the last three of these estimates is as follows. We estimate that, excluding trend, expected seminar demand in quarter 2 (spring), quarter 3 (summer), and quarter 4 (fall) are, respectively, 21, 33.5, and 4.5 greater than expected seminar demand in quarter 1 (winter). Thus the estimate of TR_t is

$$tr_t = b_0 + b_1 t = 8.75 + .5t$$

and the estimate of SN_t is

$$sn_t = b_{S2} x_{S2,t} + b_{S3} x_{S3,t} + b_{S4} x_{S4,t}$$

$$= \begin{cases} 0 & \text{if period } t \text{ is quarter 1} \\ b_{S2} = 21 & \text{if period } t \text{ is quarter 2} \\ b_{S3} = 33.5 & \text{if period } t \text{ is quarter 3} \\ b_{S4} = 4.5 & \text{if period } t \text{ is quarter 4} \end{cases}$$

Hence, a forecast of y_t, seminar demand in period t, is given by

$$\hat{y}_t = tr_t + sn_t$$

Table 8.1 summarizes the observed value y_t and the predicted value $\hat{y}_t$ for periods 1 through 16. Note that, utilizing the additive regression model, we have obtained

$$\sum_{t=1}^{16} (y_t - \hat{y}_t)^2 = 5$$

which is smaller than the sum of squared forecast errors

$$\sum_{t=1}^{16} (y_t - \hat{y}_t)^2 = 17.234$$

obtained by utilizing the multiplicative decomposition method as we did in Example 6.2. It can also be shown that $R^2 = .9983$, which indicates that the regression model fits the observed data quite well.

Let us next consider making forecasts of future values of seminar demand. Forecasts and 95 percent confidence intervals for seminar demand in periods 17, 18, 19, and 20 are summarized in Table 8.2. It should be emphasized that these

TABLE 8.1   *Forecasts of Historical Seminar Demand Calculated Using a Regression Model with Dummy Variables*

t	y_t	$tr_t = 8.75 + .5t$	sn_t	$\hat{y}_t = tr_t + sn_t$	$(y_t - \hat{y}_t)^2$
1	10	9.25	0	9.25	$(.75)^2$
2	31	9.75	21	30.75	$(.25)^2$
3	43	10.25	33.5	43.75	$(-.75)^2$
4	16	10.75	4.5	15.25	$(.75)^2$
5	11	11.25	0	11.25	$(-.25)^2$
6	33	11.75	21	32.75	$(.25)^2$
7	45	12.25	33.5	45.75	$(-.75)^2$
8	17	12.75	4.5	17.25	$(-.25)^2$
9	13	13.25	0	13.25	$(-.25)^2$
10	34	13.75	21	34.75	$(-.75)^2$
11	48	14.25	33.5	47.75	$(.25)^2$
12	19	14.75	4.5	19.25	$(-.25)^2$
13	15	15.25	0	15.25	$(-.25)^2$
14	37	15.75	21	36.75	$(.25)^2$
15	51	16.25	33.5	49.75	$(1.25)^2$
16	21	16.75	4.5	21.25	$(-.25)^2$

$$\sum_{t=1}^{16} (y_t - \hat{y}_t)^2 = 5.00$$

TABLE 8.2   *Forecasts of Future Values of Seminar Demand Calculated Using a Regression Model with Dummy Variables*

t	$tr_t = 8.75$ $+ .5t$	sn_t	$\hat{y}_t = tr_t$ $+ sn_t$	$E_t(95)$	$[\hat{y}_t - E_t(95),$ $\hat{y}_t + E_t(95)]$	y_t	$(y_t - \hat{y}_t)^2$
17	17.25	0	17.25	1.85	[15.40, 19.10]	16	$(1.25)^2$
18	17.75	21	38.75	1.85	[36.90, 40.60]	39	$(-.25)^2$
19	18.25	33.5	51.75	1.85	[49.90, 53.60]	53	$(-1.25)^2$
20	18.75	4.5	23.25	1.85	[21.40, 25.10]	22	$(1.25)^2$

$$\sum_{t=17}^{20} (y_t - \hat{y}_t)^2 = 4.75$$

forecasts and confidence intervals are made using the first 16 periods of data. Assume now that the actual values of seminar demand in periods 17, 18, 19, and 20 are observed to be $y_{17} = 16$, $y_{18} = 39$, $y_{19} = 53$, and $y_{20} = 22$. Note from Table 8.2 that

$$\sum_{t=17}^{20} (y_t - \hat{y}_t)^2 = 4.75$$

which is smaller than the sum of squared forecast errors

$$\sum_{t=17}^{20} (y_t - \hat{y}_t)^2 = 17.945$$

obtained by utilizing the multiplicative decomposition method. In Figure 8.1 the observed value y_t and the (rounded) predicted value $\hat{y}_t$ for periods $t = 1, 2, \ldots, 20$ are plotted.

Next, a forecast of $(y_{17} + y_{18} + y_{19})$, the total seminar demand in periods 17 through 19, is

$$\hat{y}_{17} + \hat{y}_{18} + \hat{y}_{19} = 17.25 + 38.75 + 51.75 = 107.75$$

Since the error in a 95 percent confidence interval for $(y_{17} + y_{18} + y_{19})$ is shown in Section *8–2 to be $E_{(17,19)}(95) = 3.79$, it follows that a 95 percent confidence interval for $(y_{17} + y_{18} + y_{19})$ is

$$[107.75 - 3.79, 107.75 + 3.79] \quad \text{or} \quad [103.96, 111.54]$$

Note that the error $E_{(17,19)}(95) = 3.79$ is not equal to, but rather is smaller than, the sum of the individual errors

$$E_{17}(95) + E_{18}(95) + E_{19}(95) = 1.85 + 1.85 + 1.85 = 5.55$$

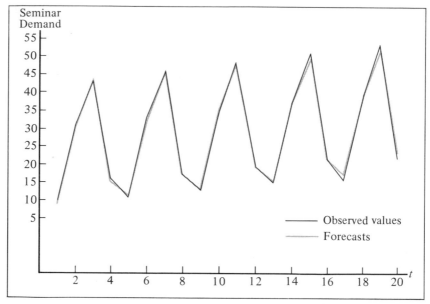

FIGURE 8.1   *Forecasts of Historical and Future Values of Seminar Demand*
Calculated Using a Regression Model with Dummy Variables

*8–2 THE REGRESSION CALCULATIONS

Using the data of Table III.1, we see that in order to find the least squares estimates of β_0, β_1, β_{S2}, β_{S3}, and β_{S4} in the regression model

$$y_t = \beta_0 + \beta_1 t + \beta_{S2} x_{S2,t} + \beta_{S3} x_{S3,t} + \beta_{S4} x_{S4,t} + \epsilon_t$$

we must use

$$\mathbf{y} = \begin{bmatrix} 10 \\ 31 \\ 43 \\ 16 \\ 11 \\ 33 \\ 45 \\ 17 \\ 13 \\ 34 \\ 48 \\ 19 \\ 15 \\ 37 \\ 51 \\ 21 \end{bmatrix} \qquad \mathbf{X} = \begin{bmatrix} 1 & 1 & 0 & 0 & 0 \\ 1 & 2 & 1 & 0 & 0 \\ 1 & 3 & 0 & 1 & 0 \\ 1 & 4 & 0 & 0 & 1 \\ 1 & 5 & 0 & 0 & 0 \\ 1 & 6 & 1 & 0 & 0 \\ 1 & 7 & 0 & 1 & 0 \\ 1 & 8 & 0 & 0 & 1 \\ 1 & 9 & 0 & 0 & 0 \\ 1 & 10 & 1 & 0 & 0 \\ 1 & 11 & 0 & 1 & 0 \\ 1 & 12 & 0 & 0 & 1 \\ 1 & 13 & 0 & 0 & 0 \\ 1 & 14 & 1 & 0 & 0 \\ 1 & 15 & 0 & 1 & 0 \\ 1 & 16 & 0 & 0 & 1 \end{bmatrix}$$

Thus the least squares estimates of β_0, β_1, β_{S2}, β_{S3}, and β_{S4} are

$$
\begin{bmatrix} b_0 \\ b_1 \\ b_{S2} \\ b_{S3} \\ b_{S4} \end{bmatrix} = (\mathbf{X'X})^{-1}\mathbf{X'y} = \begin{bmatrix} 8.75 \\ .5 \\ 21 \\ 33.5 \\ 4.5 \end{bmatrix}
$$

Moreover, a 95 percent confidence interval for y_t is

$$[\hat{y}_t - E_t(95),\ \hat{y}_t + E_t(95)]$$

where

$$\hat{y}_t = 8.75 + .5t + 21x_{S2,t} + 33.5x_{S3,t} + 4.5x_{S4,t}$$

and

$$E_t(95) = t_{5/2}(16 - 5) \cdot s \cdot f_t$$

Since $\mathbf{x'}_{17} = [1\ 17\ 0\ 0\ 0]$, $\mathbf{x'}_{18} = [1\ 18\ 1\ 0\ 0]$, $\mathbf{x'}_{19} = [1\ 19\ 0\ 1\ 0]$, and $\mathbf{x'}_{20} = [1\ 20\ 0\ 0\ 1]$, and

$$t_{5/2}(16 - 5) = t_{5/2}(11) = 2.201$$

and

$$
s = \left[\frac{\sum\limits_{t=1}^{16} (y_t - \hat{y}_t)^2}{16 - 5} \right]^{1/2} = \left(\frac{5}{11}\right)^{1/2} = .6742
$$

we have

$$
\hat{y}_t = \begin{cases} 17.25 & \text{if } t = 17 \\ 38.75 & \text{if } t = 18 \\ 51.75 & \text{if } t = 19 \\ 23.25 & \text{if } t = 20 \end{cases}
$$

$$f_t = \sqrt{1 + \mathbf{x'}_t(\mathbf{X'X})^{-1}\mathbf{x}_t} = 1.2467 \qquad \text{if } t = 17, 18, 19, 20$$

$$E_t(95) = t_{5/2}(11) \cdot s \cdot f_t = 1.85 \qquad \text{if } t = 17, 18, 19, 20$$

A 95 percent confidence interval for

$$\sum_{t=l=17}^{t=l+q=17+2=19} y_t$$

is

$$\left[\sum_{t=17}^{19} \hat{y}_t - E_{(17,17+2)}(95), \sum_{t=17}^{19} \hat{y}_t + E_{(17,17+2)}(95) \right]$$

where

$$E_{(17,17+2)}(95) = t_{5/2}(11) \cdot s \cdot \left[(2 + 1) + \left(\sum_{t=17}^{19} \mathbf{x}_t' \right) \left(\mathbf{X}'\mathbf{X} \right)^{-1} \left(\sum_{t=17}^{19} \mathbf{x}_t \right) \right]^{1/2}$$

$$= 3.79$$

since

$$\sum_{t=17}^{19} \mathbf{x}_t' = [1\ 17\ 0\ 0\ 0] + [1\ 18\ 1\ 0\ 0] + [1\ 19\ 0\ 1\ 0]$$

$$= [3\ 54\ 1\ 1\ 0]$$

8-3 ADDITIVE WINTERS' METHOD

In this section we discuss an exponential smoothing procedure that is best used to forecast a time series with a linear trend and additive seasonal variation. This method is a modification of the Winters' Method presented in Section 7–2, and assumes that the time series to be forecasted is adequately represented by the model

$$y_t = \beta_0 + \beta_1 t + SN_t + \epsilon_t$$

Unless indicated otherwise, we will use the same notation and definitions used in Sections 7–2 and 7–3. In these sections the seasonal variation is assumed to be multiplicative, whereas in this section the seasonal variation is assumed to be additive. It follows that we can obtain the updating and forecasting equations for this section from the equations in Sections 7–2 and 7–3 by replacing division operations with subtraction operations and multiplication operations with addition operations, as indicated in the following discussion.

Assume that at time T we have estimates $a_0(T)$, $b_1(T)$, and $sn_{T+\tau}(T + \tau - L)$. Then, a forecast of $y_{T+\tau}$ made at time T is

$$\hat{y}_{T+\tau}(T) = a_0(T) + b_1(T)\tau + sn_{T|\tau}(T + \tau - L)$$

Moreover, an intuitive method, which yields "fairly accurate" confidence intervals, gives an approximate $(100 - \alpha)$ percent confidence interval made at time T for $y_{T+\tau}$. This approximate confidence interval is

$$[\hat{y}_{T+\tau}(T) - E_{T+\tau}^{(100-\alpha)}(T), \hat{y}_{T+\tau}(T) + E_{T+\tau}^{(100-\alpha)}(T)]$$

where

$$E_{T+\tau}^{(100-\alpha)}(T) = z_{\alpha/2}d_\tau \, \Delta_W(T)$$

and

$$\Delta_W(T) = \frac{\displaystyle\sum_{t=1}^{T} |y_t - sn_t(t - L) - (a_0(t - 1) + b_1(t - 1))|}{T}$$

A forecast made at time T for the sum of the values of the time series from period $T + 1$ to $T + \tau$, $\sum_{t=1}^{\tau} y_{T+t}$, is

$$\sum_{t=1}^{\tau} \hat{y}_{T+t}(T) = \sum_{t=1}^{\tau} (a_0(T) + b_1(T)t + sn_{T+t}(T + t - L))$$

There is also an intuitive method that yields "fairly accurate" confidence intervals and gives an approximate $(100 - \alpha)$ percent confidence interval for this sum. This approximate confidence interval is

$$\left[\sum_{t=1}^{\tau} \hat{y}_{T+t}(T) - E_{[T+1,T+\tau]}^{(100-\alpha)}(T), \sum_{t=1}^{\tau} \hat{y}_{T+t}(T) + E_{[T+1,T+\tau]}^{(100-\alpha)}(T)\right]$$

where

$$E_{[T+1,T+\tau]}^{(100-\alpha)}(T) = z_{\alpha/2}q_\tau \, \Delta_W(T)$$

If we observe y_{T+1} at time $T + 1$, we can update $a_0(T)$, $b_1(T)$, $sn_{T+1}(T + 1 - L)$, and $\Delta_W(T)$ using the equations

$$a_0(T + 1) = \alpha[y_{T+1} - sn_{T+1}(T + 1 - L)] + (1 - \alpha)[a_0(T) + b_1(T)]$$

$$b_1(T + 1) = \beta[a_0(T + 1) - a_0(T)] + (1 - \beta)b_1(T)$$

$$sn_{T+1}(T + 1) = \gamma[y_{T+1} - a_0(T + 1)] + (1 - \gamma)sn_{T+1}(T + 1 - L)$$

$$\Delta_W(T + 1) = \frac{T\,\Delta_W(T) + [y_{T+1} - sn_{T+1}(T + 1 - L) - (a_0(T) + b_1(T))]}{T + 1}$$

Then, we can update the previous approximate $(100 - \alpha)$ percent confidence intervals that are made at time T for $y_{T+\tau}$ and $\Sigma_{t=1}^{\tau} y_{T+t}$ to approximate $(100 - \alpha)$ percent confidence intervals that are made at time $T + 1$ for, respectively, $y_{T+1+\tau}$ and $\Sigma_{t=1}^{\tau} y_{T+1+t}$. These intervals are

$$[\hat{y}_{T+1+\tau}(T + 1) - E_{T+1+\tau}^{(100-\alpha)}(T + 1), \hat{y}_{T+1+\tau}(T + 1) + E_{T+1+\tau}^{(100-\alpha)}(T + 1)]$$

and

$$\left[\sum_{t=1}^{\tau} \hat{y}_{T+1+t}(T + 1) - E_{[T+2,T+1+\tau]}^{(100-\alpha)}(T + 1), \sum_{t=1}^{\tau} \hat{y}_{T+1+t}(T + 1) + E_{[T+2,T+1+\tau]}^{(100-\alpha)}(T + 1)\right]$$

where

$$\hat{y}_{T+1+t}(T + 1) = a_0(T + 1) + b_1(T + 1)t + sn_{T+1+t}(T + 1 + t - L)$$

$$E_{T+1+\tau}^{(100-\alpha)}(T + 1) = z_{\alpha/2}d_\tau\,\Delta_W(T + 1)$$

$$E_{[T+2,T+1+\tau]}^{(100-\alpha)}(T + 1) = z_{\alpha/2}q_\tau\,\Delta_W(T + 1)$$

We obtain initial estimates $a_0(0)$, $b_1(0)$, and $sn_t(0)$ for $t = 1, 2, \ldots, L$ by using the regression approach discussed in Section 8–2 on the m years of historical data. Thus, if b_0, b_1, b_{S2}, b_{S3}, $\ldots$, b_{SL} are the least squares estimates of β_0, β_1, β_{S2}, β_{S3}, $\ldots$, β_{SL} in the regression model

$$y_t = \beta_0 + \beta_1 t + \beta_{S2}x_{S2,t} + \beta_{S3}x_{S3,t} + \cdots + \beta_{SL}x_{SL,t} + \epsilon_t$$

where $x_{S2,t}$, $x_{S3,t}$, $\ldots$, $x_{SL,t}$ are the previously discussed dummy variables, then the initial estimates for Winters' Method are given by

$$a_0(0) = b_0$$

$$b_1(0) = b_1$$

$$sn_1(0) = 0$$

$$sn_2(0) = b_{S2}$$

$$sn_3(0) = b_{S3}$$

.

.

.

$$sn_L(0) = b_{SL}$$

It should be noted that the comments made in Section 7–2 concerning smoothing these initial estimates to the current origin of time and concerning the choice of the optimal combination of smoothing constants α, β, and γ also apply to this section.

Example 8.2   We now consider Winters' Method as a forecasting tool to be used in forecasting the time series concerning seminar demand given in Table III.1 and illustrated in Figure III.5. In Example 8.1 we found the least squares estimates of β_0, β_1, β_{S2}, β_{S3}, and β_{S4} in the regression model

$$y_t = \beta_0 + \beta_1 t + \beta_{S2} x_{S2,t} + \beta_{S3} x_{S3,t} + \beta_{S4} x_{S4,t} + \epsilon_t$$

where $x_{S2,t}$, $x_{S3,t}$, and $x_{S4,t}$ are the previously discussed dummy variables. These estimates serve as the initial estimates in Winters' Method and are

$$a_0(0) = b_0 = 8.75$$

$$b_1(0) = b_1 = .5$$

$$sn_1(0) = 0$$

$$sn_2(0) = b_{S2} = 21$$

$$sn_3(0) = b_{S3} = 33.5$$

$$sn_4(0) = b_{S4} = 4.5$$

Although a simulation of the historical data should be carried out to determine the optimum combination of the smoothing constants α, β, and γ, in this example we will arbitrarily set $\alpha = .2$, $\beta = .1$, and $\gamma = .1$.

Having obtained the initial estimates and the smoothing constants, we now smooth the initial estimates to the end of the historical data, period 16, so that we may begin forecasting future values of the time series. Since the observation in period 1 is $y_1 = 10$, we obtain the updated estimates, $a_0(1)$, $b_1(1)$, and $sn_1(1)$, as follows.

$$a_0(1) = \alpha[y_1 - sn_1(0)] + (1 - \alpha)[a_0(0) + b_1(0)]$$

$$= .2[10 - 0] + .8[8.75 + .5]$$

$$= 9.4$$

$$b_1(1) = \beta[a_0(1) - a_0(0)] + (1 - \beta)b_1(0)$$

$$= .1[9.4 - 8.75] + .9(.5)$$

$$= .515$$

$$sn_1(1) = \gamma[y_1 - a_0(1)] + (1 - \gamma)sn_1(0)$$

$$= .1[10 - 9.4] + .9(0)$$

$$= .06$$

Since the observation in period 2 is $y_2 = 31$, we next obtain the updated estimates, $a_0(2)$, $b_1(2)$, and $sn_2(2)$, as follows.

$$a_0(2) = \alpha[y_2 - sn_2(0)] + (1 - \alpha)[a_0(1) + b_1(1)]$$

$$= .2[31 - 21] + .8[9.4 + .515]$$

$$= 9.932$$

$$b_1(2) = \beta[a_0(2) - a_0(1)] + (1 - \beta)b_1(1)$$

$$= .1[9.932 - 9.4] + .9(.515)$$

$$= .5167$$

$$sn_2(2) = \gamma[y_2 - a_0(2)] + (1 - \gamma)sn_2(0)$$

$$= .1[31 - 9.932] + .9(21)$$

$$= 21.0068$$

This process is continued up to the end of the historical data, period 16. The results obtained are summarized in Table 8.3.

Assume we now wish to find approximate 95 percent one-period-ahead, three-period-ahead, and cumulative three-period-ahead forecasts for seminar demand in periods 17, 18, 19, and 20. Since the maximum of the smoothing constants $\alpha = .2$, $\beta = .1$, and $\gamma = .1$ is $\theta = .2$, and since $\theta = .2$ is equal to the value of θ used in Examples 7.2 and 7.3, the calculation of d_1, d_3, q_3 and therefore of $E_{T+1}^{(95)}(T)$, $E_{T+3}^{(95)}(T)$, and $E_{[T+1,T+3]}^{(95)}(T)$ done in these examples applies to this example. So we have

$$E_{T+1}^{(95)}(T) = 2.45\,\Delta_W(T)$$

$$E_{T+3}^{(95)}(T) = 2.556036\,\Delta_W(T)$$

$$E_{[T+1,T+3]}^{(95)}(T) = 5.353544\,\Delta_W(T)$$

TABLE 8.3   *Updating the Estimates Using Winters' Method with $\alpha = .2$, $\beta =$*
.1, and $\gamma - .1$ on the Historical Seminar Demand Time Series

T	y_T	$a_0(T)$	$b_1(T)$	$sn_T(T-4)$	$sn_T(T)$
		$a_0(0) = 8.75$	$b_1(0) = .5$		
1	10	9.4	.515	0	.06
2	31	9.932	.5167	21	21.0068
3	43	10.259	.4977	33.5	33.4241
4	16	10.9054	.5126	4.5	4.5595
5	11	11.3224	.503	.06	.0218
6	33	11.859	.5064	21.0068	21.0202
7	45	12.2075	.4906	33.4241	33.3609
8	17	12.6466	.4855	4.5595	4.5389
9	13	13.1013	.4824	.0218	.0095
10	34	13.4629	.4703	21.0202	20.9719
11	48	14.0744	.4844	33.3609	33.4174
12	19	14.5393	.4825	4.5389	4.5311
13	15	15.0155	.4819	.0095	.0070
14	37	15.6035	.4925	20.9719	21.0144
15	51	16.3933	.5222	33.4174	33.5363
16	21	16.8262	.5133	4.5311	4.4954

Since we will begin the generation of confidence intervals from a time origin of $T = 16$, we now refer to Table 8.3 and compute

$$\Delta_W(16) = \frac{\sum_{t=1}^{16} |y_t - sn_t(t-4) - (a_0(t-1) + b_1(t-1))|}{16}$$

$$= \frac{1}{16} \{ |10 - 0 - (8.75 + .5)| + |31 - 21 - (9.4 + .515)|$$

$$+ \cdots + |21 - 4.5311 - (16.3933 + .5222)|\}$$

$$= .5173$$

Hence,

$$E_{17}^{(95)}(16) = 2.45 \, \Delta_W(16) = 1.2674$$

$$E_{19}^{(95)}(16) = 2.556\,036 \, \Delta_W(16) = 1.3222$$

$$E_{[17,19]}^{(95)}(16) = 5.353544 \, \Delta_W(16) = 2.7694$$

Moreover, from Table 8.3 we have

$$\hat{y}_{17}(16) = a_0(16) + b_1(16) + sn_{17}(13)$$
$$= 16.8262 + .5133 + .0070$$
$$= 17.3465$$

$$\hat{y}_{18}(16) = a_0(16) + b_1(16) \cdot 2 + sn_{18}(14)$$
$$= 16.8262 + (.5133)2 + 21.0144$$
$$= 38.8672$$

$$\hat{y}_{19}(16) = a_0(16) + b_1(16) \cdot 3 + sn_{19}(15)$$
$$= 16.8262 + (.5133)3 + 33.5363$$
$$= 51.9024$$

and

$$\hat{y}_{17}(16) + \hat{y}_{18}(16) + \hat{y}_{19}(16) = 108.1161$$

Hence, approximate 95 percent confidence intervals made at time 16 for y_{17}, y_{19}, and ($y_{17} + y_{18} + y_{19}$) are, respectively,

$$[17.3465 - 1.2674, 17.3465 + 1.2674] \quad \text{or} \quad [16.0791, 18.6139]$$

$$[51.9024 - 1.3222, 51.9024 + 1.3222] \quad \text{or} \quad [50.5802, 53.2246]$$

$$[108.1161 - 2.7694, 108.1161 + 2.7694] \quad \text{or} \quad [105.3467, 110.8855]$$

Assuming we now observe $y_{17} = 16$, we obtain updated estimates as follows.

$$a_0(17) = \alpha[y_{17} - sn_{17}(13)] + (1 - \alpha)[a_0(16) + b_1(16)]$$
$$= .2[16 - .0070] + .8[16.8262 + .5133]$$
$$= 17.0702$$

$$b_1(17) = \beta[a_0(17) - a_0(16)] + (1 - \beta)b_1(16)$$
$$= .1[17.0702 - 16.8262] + .9(.5133)$$
$$= .4864$$

$$sn_{17}(17) = \gamma[y_{17} - a_0(17)] + (1 - \gamma)sn_{17}(13)$$
$$= .1[16 - 17.0702] + .9(.0070)$$
$$= -.1007$$

$$\Delta_W(17) = \frac{16\,\Delta_W(16) + |y_{17} - sn_{17}(13) - (a_0(16) + b_1(16))|}{17}$$

$$= \frac{16(.5173) + |16 - .0070 - (16.8262 + .5133)|}{17}$$

$$= .5661$$

and

$$E_{18}^{(95)}(17) = 2.45\,\Delta_W(17) = 1.3869$$

$$E_{20}^{(95)}(17) = 2.556036\,\Delta_W(17) = 1.4470$$

$$E_{[18,20]}^{(95)}(17) = 5.353544\,\Delta_W(17) = 3.0306$$

So,

$$\hat{y}_{18}(17) = a_0(17) + b_1(17) + sn_{18}(14)$$

$$= 17.0702 + .4864 + 21.0144$$

$$= 38.571$$

$$\hat{y}_{19}(17) = a_0(17) + b_1(17) \cdot 2 + sn_{19}(15)$$

$$= 17.0702 + .4864(2) + 33.5363$$

$$= 51.5793$$

$$\hat{y}_{20}(17) = a_0(17) + b_1(17) \cdot 3 + sn_{20}(16)$$

$$= 17.0702 + .4864(3) + 4.4954$$

$$= 23.0248$$

and

$$\hat{y}_{18}(17) + \hat{y}_{19}(17) + \hat{y}_{20}(17) = 113.1751$$

Hence, approximate 95 percent confidence intervals made at time 17 for y_{18}, y_{20}, and $(y_{18} + y_{19} + y_{20})$ are, respectively,

$$[38.571 - 1.3869, 38.571 + 1.3869] \qquad \text{or} \qquad [37.1841, 39.9579]$$

$$[23.0248 - 1.4470, 23.0248 + 1.4470] \qquad \text{or} \qquad [21.5778, 24.4718]$$

$$[113.1751 - 3.0306, 113.1751 + 3.0306] \qquad \text{or} \qquad [110.1445, 116.2057]$$

Forecasts made for periods 17, 18, 19, and 20 are summarized in Table 8.4 and Table 8.5. Note that of the 8 confidence intervals constructed, 7 contain the

TABLE 8.4   *One-Period-Ahead and Three-Period-Ahead Forecasts of Future Values of Seminar Demand Calculated Using Winters' Method with $\alpha = .2$, $\beta = .1$, and $\gamma = .1$*

T	y_T	$a_0(T)$	$b_1(T)$	$sn_T(T-4)$	$sn_T(T)$	$\Delta_W(T)$
16	21	16.8262	.5133	4.5311	4.4954	.5173
17	16	17.0702	.4864	.0070	−.1007	.5661
18	39	17.6424	.4950	21.0144	21.0487	.5585
19	53	18.4027	.5215	33.5363	33.6424	.5989
20	22	18.6403	.4931	4.4954	4.3818	.6399

T	y_T	$E_T^{(95)}(T-1)$	$\hat{y}_T(T-1)$	$[\hat{y}_T(T-1) - E_T^{(95)}(T-1),$ $\hat{y}_T(T-1) + E_T^{(95)}(T-1)]$
16	21	—	21.4466	—
17	16	1.2674	17.3465	[16.0791, 18.6139]
18	39	1.3869	38.5710	[37.1841, 39.9579]
19	53	1.3683	51.6737	[50.3054, 53.0420]
20	22	1.4673	23.4196	[21.9523, 24.8869]

T	y_T	$E_T^{(95)}(T-3)$	$\hat{y}_T(T-3)$	$[\hat{y}_T(T-3) - E_T^{(95)}(T-3),$ $\hat{y}_T(T-3) + E_T^{(95)}(T-3)]$
16	21	—	—	—
17	16	—	—	—
18	39	—	—	—
19	53	1.3222	51.9024	[50.5802, 53.2246]
20	22	1.4470	23.0248	[21.5778, 24.4718]

TABLE 8.5   *Forecasts of Cumulative Seminar Demand in the Next Three Periods Calculated Using Winters' Method with $\alpha = .2$, $\beta = .1$, and $\gamma = .1$*

T	y_T	$\sum_{t=1}^{3} y_{T-3+t}$	$\Delta_W(T)$	$E_{[T-2,T]}^{(95)}(T-3)$	$\sum_{t=1}^{3} \hat{y}_{T-3+t}(T-3)$	$\left[\sum_{t=1}^{3} \hat{y}_{T-3+t}(T-3) - E_{[T-2,T]}^{(95)}(T-3) \right.$ $\left. \sum_{t=1}^{3} \hat{y}_{T-3+t}(T-3) + E_{[T-2,T]}^{(95)}(T-3) \right]$
16	—	—	.5173	—	—	—
17	16	—	.5661	—	—	—
18	39	—	.5585	—	—	—
19	53	108	.5989	2.7694	108.1161	[105.3467, 110.8855]
20	22	114	.6399	3.0306	113.1751	[110.1445, 116.2057]

observed value of the time series. Furthermore, if the lower end of the confidence interval for y_{17}, 16.0791, is rounded down to 16, then all 8 of the confidence intervals contain the observed values of the times series.

PROBLEMS

1. Consider Example 8.1 in which the regression approach is used to forecast the seminar demand time series of Table III.1. Using this approach, determine forecasts of seminar demand for periods 21, 22, 23, and 24.

*2. Suppose that the regression model

$$y_t = \beta_0 + \beta_1 t + \beta_{S2}x_{S2,t} + \beta_{S3}x_{S3,t} + \beta_{S4}x_{S4,t} + \epsilon_t$$

where $x_{S2,t}$, $x_{S3,t}$, and $x_{S4,t}$ are dummy variables, describes the following data concerning sales of lawn mowers in a department store.

Quarter	Year 1	Year 2
1	15	19
2	63	66
3	31	32
4	27	30

Determine the vector **y** and matrix **X** which should be used to calculate the least squares estimates of β_0, β_1, β_{S2}, β_{S3}, and β_{S4}.

*3. Consider the regression calculations in Section *8–2 that have been carried out for Example 8.1.

 a. Determine the vectors x'_{21}, x'_{22}, x'_{23}, and x'_{24} to be used in calculating 95 percent confidence intervals for y_{21}, y_{22}, y_{23}, and y_{24}.

 b. Determine the vector

$$\sum_{t=21}^{24} x'_t$$

 to be used in calculating a 95 percent confidence interval for cumulative seminar demand in periods 21 through 24.

---

* Starred problems are only for students who have read Section *8–2.

4. Consider Table 8.4 of Example 8.2 in which we forecasted the seminar demand time series using Winters' Method. Given the seminar demand observation for period 18, which is $y_{18} = 39$, verify that the updated estimates $a_0(18)$, $b_1(18)$, and $sn_{18}(18)$ are

$$a_0(18) = 17.6424, \quad b_1(18) = .4950, \quad \text{and} \quad sn_{18}(18) = 21.0487$$

5. Consider Table 8.4 of Example 8.2. Given the seminar demand observation for period 18, which is $y_{18} = 39$, verify that when $\Delta_W(17)$ is updated, we obtain $\Delta_W(18) = .5585$.

6. Consider Table 8.4 in Example 8.2. Verify that the seminar demand forecast for y_{19} made in period 18 is $\hat{y}_{19}(18) = 51.6737$.

7. Consider Table 8.4 in Example 8.2. Verify that an approximate 95 percent confidence interval for y_{19} is [50.5802, 53.2246].

8. Consider Example 8.2. Find the seminar demand forecast for y_{21} that is made in period 18.

9. Consider Example 8.2. Find an approximate 95 percent confidence interval for y_{21} calculated in period 18.

10. Consider Example 8.2. Find an approximate 95 percent confidence interval, calculated in period 18, for

$$\sum_{t=1}^{3} y_{18+t} = y_{19} + y_{20} + y_{21}$$

CHAPTER 9

Further Topics Concerning Seasonal Variation

9–1 INTRODUCTION

In this chapter we discuss some additional topics concerning seasonal variation. In Chapters 6, 7, and 8 we have forecasted the Rola-Cola sales time series and the seminar demand time series using several forecasting methods. In Section 9–2 we will summarize our modeling and forecasting of these time series.

Sections 9–3 and 9–4 discuss two types of regression models that can be useful in forecasting time series having seasonal variation. In some situations, regression models that involve trigonometric terms can be used to accurately forecast time series with additive or multiplicative seasonal variation. We will discuss regression models of this type in Section 9–3. Section 9–4 presents another useful regression model, which employs both causal variables and mathematical functions of time in forecasting a time series of interest.

9–2 A FURTHER DISCUSSION OF MODELING ADDITIVE AND MULTIPLICATIVE SEASONAL VARIATION

We now review the analyses that have been carried out for the seminar demand time series given in Table III.1 and the Rola-Cola sales time series given in Table III.2. We do this in order to emphasize the importance of the choice of the model to be used in analyzing and forecasting a time series.

In Sections 6–2 and 6–3 we have analyzed and forecasted the seminar demand time series of Table III.1 using the multiplicative decomposition method. For the 16 periods of historical data, the sum of the squared differences between the observed and predicted values is

$$\sum_{t=1}^{16} (y_t - \hat{y}_t)^2 = 17.234$$

For the 4 future periods of seminar demand, the sum of the squared differences between the observed and predicted values is

$$\sum_{t=17}^{20} (y_t - \hat{y}_t)^2 = 17.945$$

In Section 8–2 we have analyzed and forecasted the seminar demand time series of Table III.1 using an additive regression model. For the 16 periods of historical data, the sum of the squared differences between the observed and predicted values is

$$\sum_{t=1}^{16} (y_t - \hat{y}_t)^2 = 5$$

For the 4 future periods of seminar demand, the sum of the squared differences between the observed and predicted values is

$$\sum_{t=17}^{20} (y_t - \hat{y}_t)^2 = 4.75$$

Note that the additive regression model both fits the observed historical data and predicts future values of the time series better than the multiplicative decomposition model. This is as one would expect, since the seminar demand time series approximately possesses additive seasonal variation, and the additive regression model is best used to model additive seasonal variation; whereas the multiplicative decomposition model is best used to model multiplicative seasonal variation.

In Section 6–3 we have analyzed and forecasted the Rola-Cola sales time series of Table III.2 using the multiplicative decomposition method. For the 36 periods of historical data, the sum of the squared differences between the observed and predicted values is

$$\sum_{t=1}^{36} (y_t - \hat{y}_t)^2 = 3154.7919$$

For the 12 future periods of sales, the sum of the squared differences between the observed and predicted values is

$$\sum_{t=37}^{48} (y_t - \hat{y}_t)^2 = 3693.5263$$

We will subsequently analyze and forecast the Rola-Cola sales time series of Table III.2 using an additive regression model. We will find that for the 36 periods of historical data, the sum of the squared differences between the observed and predicted values is

$$\sum_{t=1}^{36} (y_t - \hat{y}_t)^2 = 76874.144$$

For the 12 future periods of sales, we will find that the sum of the squared differences between the observed and predicted values is

$$\sum_{t=37}^{48} (y_t - \hat{y}_t)^2 = 156{,}451.33$$

Note that the multiplicative decomposition model both fits the observed historical data and predicts future values of the time series better than the additive regression model. This is as one would expect, since the Rola-Cola sales time series approximately possesses multiplicative seasonal variation, and the multiplicative decomposition model is best used to model multiplicative seasonal variation, whereas the additive regression model is best used to model additive seasonal variation.

We now consider using the additive regression model

$$y_t = TR_t + SN_t + \epsilon_t$$

to analyze and forecast the Rola-Cola sales time series of Table III.2. Since this time series exhibits a linear trend and consists of monthly observations, we have

$$TR_t = \beta_0 + \beta_1 t$$

and

$$SN_t = \beta_{S2} x_{S2,t} + \beta_{S3} x_{S3,t} + \beta_{S4} x_{S4,t} + \beta_{S5} x_{S5,t} + \beta_{S6} x_{S6,t} + \beta_{S7} x_{S7,t}$$
$$+ \beta_{S8} x_{S8,t} + \beta_{S9} x_{S9,t} + \beta_{S10} x_{S10,t} + \beta_{S11} x_{S11,t} + \beta_{S12} x_{S12,t}$$

where for $j = 2, \ldots, 12$

$$x_{Sj} = \begin{cases} 1 & \text{if period } t \text{ is month } j \\ 0 & \text{otherwise} \end{cases}$$

Using regression analysis, the least squares estimates of β_0, β_1, β_{S2}, β_{S3}, β_{S4}, β_{S5}, β_{S6}, β_{S7}, β_{S8}, β_{S9}, β_{S10}, β_{S11}, β_{S12} can be found. These estimates yield the following forecasting equation for a future value of the time series.

$$\hat{y}_t = tr_t + sn_t$$

where

$$tr_t = 119.22 + 9.57t$$

$$sn_t = 47.76x_{S2,t} + 50.85x_{S3,t} + 94.95x_{S4,t} + 33.71x_{S5,t}$$
$$+ 257.47x_{S6,t} + 530.56x_{S7,t} + 652.66x_{S8,t} + 831.42x_{S9,t}$$
$$+ 428.84x_{S10,t} + 262.94x_{S11,t} + 14.03x_{S12,t}$$

For example, a forecast of Rola-Cola sales in March of year 4 (period 39) is

$$\hat{y}_{39} = 119.22 + 9.57(39) + 50.85(1) = 543.42$$

The observed value y_t and the predicted value $\hat{y}_t$ are given in Table 9.1 for the historical data and are given in Table 9.2 for the 12 future values of the time series. These values are plotted in Figure 9.1. Note that fitting a model that is best used to model additive seasonal variation to a time series exhibiting multiplicative seasonal variation yields forecasts of future values of the time series that "fall behind" the "swings" of the multiplicative seasonal variation. This is because the seasonal swings in the multiplicative time series increase in magnitude as the average level of the time series increases, while the additive model does not reflect this increasing seasonal variation.

It is also possible to analyze and forecast a time series with multiplicative seasonal variation by fitting an additive regression model to the natural logarithms

TABLE 9.1   *Forecasts of Historical Rola-Cola Sales Calculated Using a Regression Model with Dummy Variables*

t	y_t	$\hat{y}_t$	t	y_t	$\hat{y}_t$
1	189	128.79	19	831	831.67
2	229	186.13	20	960	963.33
3	249	198.79	21	1152	1151.67
4	289	252.46	22	759	758.67
5	260	200.79	23	607	602.33
6	431	434.13	24	371	363.00
7	660	716.79	25	298	358.54
8	777	848.46	26	378	415.87
9	915	1036.79	27	373	428.54
10	613	643.79	28	443	482.21
11	485	487.46	29	374	430.54
12	277	248.13	30	660	663.87
13	244	243.67	31	1004	946.54
14	296	301.00	32	1153	1078.21
15	319	313.67	33	1388	1266.54
16	370	367.33	34	904	873.54
17	313	315.67	35	715	717.21
18	556	549.00	36	441	477.87

$$\sum_{t=1}^{36} (y_t - \hat{y}_t)^2 = 76874.144$$

TABLE 9.2	*Forecasts of Future Values of Rola-Cola Sales Calculated Using a Regression Model with Dummy Variables*	
t	y_t	$\hat{y}_t$
37	352	473.42
38	445	530.75
39	453	543.42
40	541	597.08
41	457	545.42
42	762	778.75
43	1194	1061.42
44	1361	1193.08
45	1615	1381.42
46	1059	988.42
47	824	832.08
48	495	592.75

$$\sum_{t=37}^{48} (y_t - \hat{y}_t)^2 = 156{,}451.33$$

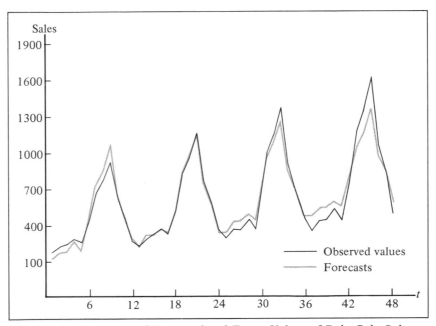

FIGURE 9.1   *Forecasts of Historical and Future Values of Rola-Cola Sales Calculated Using a Regression Model with Dummy Variables*

of the values of the time series. In order to use this approach to forecast future values of the Rola-Cola sales time series, we would employ the model

$$y^*_t = TR_t + SN_t + \epsilon_t$$

where TR_t and SN_t are previously defined and where $y^*_t = \ln y_t$. Using regression analysis, the least squares estimates of the parameters in the above model can be found. These estimates yield a forecasting equation for $y^*_t = \ln y_t$. This forecasting equation is

$$\hat{y}^*_t = tr_t + sn_t$$

where

$$tr_t = 5.2507 + .0175t$$

$$\begin{aligned} sn_t = {} & .1901x_{S2,t} + .2210x_{S3,t} + .3599x_{S4,t} + .1949x_{S5,t} \\ & + .7267x_{S6,t} + 1.1250x_{S7,t} + 1.2561x_{S8,t} + 1.4156x_{S9,t} \\ & + .9826x_{S10,t} + .7343x_{S11,t} + .2049x_{S12,t} \end{aligned}$$

For example, a forecast of $y^*_{39} = \ln y_{39}$, the natural logarithm of Rola-Cola sales in March of year 4 (period 39), is

$$\hat{y}^*_{39} = 5.2507 + .0175(39) + .2210(1) = 6.15557$$

Hence, a forecast of y_{39}, the actual Rola-Cola sales in March of year 4, is

$$\hat{y}_{39} = \exp(\hat{y}^*_{39}) = e^{6.15557} = 471.34$$

The observed value y_t and the predicted value $\hat{y}_t$ are given in Table 9.3 for the historical data and are given in Table 9.4 for the 12 future values of the time series. Then, these values are plotted in Figure 9.2. Note that, while the fit to the historical data is reasonably good, the forecasts of the future values of the time series are all greater than the actual observed future values. This is because we are fitting the following model to the data.

$$\ln y_t = y^*_t = TR_t + SN_t + \epsilon_t$$

which is equivalent to

$$y_t = (e^{TR_t})(e^{SN_t})(e^{\epsilon_t})$$

We see, then, that we are fitting a model that is appropriate for analyzing and forecasting multiplicative seasonal variation. However, the trend we are using is an exponential trend. When fitting this model to historical data, it is possible to "force" the exponential trend to approximate a linear trend, and hence we get a reasonably good fit to the historical data. However, when using the fitted model to forecast future values of the time series, this exponential trend "takes off" and causes the forecasts to be greater than the actual observed values of the time series. This phenomenon is illustrated in Figure 9.2.

TABLE 9.3   *Forecasts of Historical Rola-Cola Sales Calculated Using a "Logged" Regression Model with Dummy Variables*

t	y_t	$\hat{y}_t$	t	y_t	$\hat{y}_t$
1	189	194.08	19	831	819.65
2	229	238.87	20	960	950.99
3	249	250.72	21	1152	1135.24
4	289	293.17	22	759	749.24
5	260	252.98	23	607	594.86
6	431	438.18	24	371	356.53
7	660	664.12	25	298	295.62
8	777	770.53	26	378	363.85
9	915	919.82	27	373	381.90
10	613	607.07	28	443	446.56
11	485	481.98	29	374	385.34
12	277	288.88	30	660	667.44
13	244	239.53	31	1004	1011.60
14	296	294.81	32	1153	1173.7
15	319	309.43	33	1388	1401.1
16	370	361.83	34	904	924.71
17	313	312.22	35	715	734.17
18	556	540.79	36	441	440.03

$$\sum_{t=1}^{36}(y_t - \hat{y}_t)^2 = 3803.4060$$

TABLE 9.4   *Forecasts of Future Values of Rola-Cola Sales Calculated Using a "Logged" Regression Model with Dummy Variables*

t	y_t	$\hat{y}_t$
37	352	364.86
38	445	449.06
39	453	471.34
40	541	551.15
41	457	475.59
42	762	823.75
43	1194	1248.51
44	1361	1448.57
45	1615	1729.23
46	1059	1141.27
47	824	906.10
48	495	543.08

$$\sum_{t=37}^{48}(y_t - \hat{y}_t)^2 = 44288.674$$

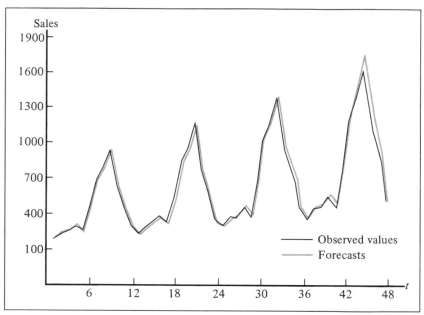

FIGURE 9.2   *Forecasts of Historical and Future Values of Rola-Cola Sales Calculated Using a "Logged" Regression Model with Dummy Variables*

*9–2   THE REGRESSION CALCULATIONS

The least squares estimates of the parameters in the model

$$y_t = TR_t + SN_t + \epsilon_t$$

are given by the matrix algebra equation

$$\mathbf{b} = (\mathbf{X'X})^{-1}\mathbf{X'y}$$

where **y** is a column vector consisting of the 36 observed values of the Rola-Cola sales time series. The least squares estimates of the parameters in the model

$$\ln y_t = y^*{}_t = TR_t + SN_t + \epsilon_t$$

are given by the matrix algebra equation

$$\mathbf{b^*} = (\mathbf{X'X})^{-1}\mathbf{X'y^*}$$

where **y*** is a column vector consisting of the natural logarithms of the 36 observed values of the Rola-Cola sales time series. In both cases **X** equals the matrix in Box A.

BOX A

$$
\mathbf{X} =
\begin{bmatrix}
1 & 1 & 0 & 0 & 0 & 0 & 0 & 0 & 0 & 0 & 0 & 0 & 0 \\
1 & 2 & 1 & 0 & 0 & 0 & 0 & 0 & 0 & 0 & 0 & 0 & 0 \\
1 & 3 & 0 & 1 & 0 & 0 & 0 & 0 & 0 & 0 & 0 & 0 & 0 \\
1 & 4 & 0 & 0 & 1 & 0 & 0 & 0 & 0 & 0 & 0 & 0 & 0 \\
1 & 5 & 0 & 0 & 0 & 1 & 0 & 0 & 0 & 0 & 0 & 0 & 0 \\
1 & 6 & 0 & 0 & 0 & 0 & 1 & 0 & 0 & 0 & 0 & 0 & 0 \\
1 & 7 & 0 & 0 & 0 & 0 & 0 & 1 & 0 & 0 & 0 & 0 & 0 \\
1 & 8 & 0 & 0 & 0 & 0 & 0 & 0 & 1 & 0 & 0 & 0 & 0 \\
1 & 9 & 0 & 0 & 0 & 0 & 0 & 0 & 0 & 1 & 0 & 0 & 0 \\
1 & 10 & 0 & 0 & 0 & 0 & 0 & 0 & 0 & 0 & 1 & 0 & 0 \\
1 & 11 & 0 & 0 & 0 & 0 & 0 & 0 & 0 & 0 & 0 & 1 & 0 \\
1 & 12 & 0 & 0 & 0 & 0 & 0 & 0 & 0 & 0 & 0 & 0 & 1 \\
1 & 13 & 0 & 0 & 0 & 0 & 0 & 0 & 0 & 0 & 0 & 0 & 0 \\
1 & 14 & 1 & 0 & 0 & 0 & 0 & 0 & 0 & 0 & 0 & 0 & 0 \\
1 & 15 & 0 & 1 & 0 & 0 & 0 & 0 & 0 & 0 & 0 & 0 & 0 \\
1 & 16 & 0 & 0 & 1 & 0 & 0 & 0 & 0 & 0 & 0 & 0 & 0 \\
1 & 17 & 0 & 0 & 0 & 1 & 0 & 0 & 0 & 0 & 0 & 0 & 0 \\
1 & 18 & 0 & 0 & 0 & 0 & 1 & 0 & 0 & 0 & 0 & 0 & 0 \\
1 & 19 & 0 & 0 & 0 & 0 & 0 & 1 & 0 & 0 & 0 & 0 & 0 \\
1 & 20 & 0 & 0 & 0 & 0 & 0 & 0 & 1 & 0 & 0 & 0 & 0 \\
1 & 21 & 0 & 0 & 0 & 0 & 0 & 0 & 0 & 1 & 0 & 0 & 0 \\
1 & 22 & 0 & 0 & 0 & 0 & 0 & 0 & 0 & 0 & 1 & 0 & 0 \\
1 & 23 & 0 & 0 & 0 & 0 & 0 & 0 & 0 & 0 & 0 & 1 & 0 \\
1 & 24 & 0 & 0 & 0 & 0 & 0 & 0 & 0 & 0 & 0 & 0 & 1 \\
1 & 25 & 0 & 0 & 0 & 0 & 0 & 0 & 0 & 0 & 0 & 0 & 0 \\
1 & 26 & 1 & 0 & 0 & 0 & 0 & 0 & 0 & 0 & 0 & 0 & 0 \\
1 & 27 & 0 & 1 & 0 & 0 & 0 & 0 & 0 & 0 & 0 & 0 & 0 \\
1 & 28 & 0 & 0 & 1 & 0 & 0 & 0 & 0 & 0 & 0 & 0 & 0 \\
1 & 29 & 0 & 0 & 0 & 1 & 0 & 0 & 0 & 0 & 0 & 0 & 0 \\
1 & 30 & 0 & 0 & 0 & 0 & 1 & 0 & 0 & 0 & 0 & 0 & 0 \\
1 & 31 & 0 & 0 & 0 & 0 & 0 & 1 & 0 & 0 & 0 & 0 & 0 \\
1 & 32 & 0 & 0 & 0 & 0 & 0 & 0 & 1 & 0 & 0 & 0 & 0 \\
1 & 33 & 0 & 0 & 0 & 0 & 0 & 0 & 0 & 1 & 0 & 0 & 0 \\
1 & 34 & 0 & 0 & 0 & 0 & 0 & 0 & 0 & 0 & 1 & 0 & 0 \\
1 & 35 & 0 & 0 & 0 & 0 & 0 & 0 & 0 & 0 & 0 & 1 & 0 \\
1 & 36 & 0 & 0 & 0 & 0 & 0 & 0 & 0 & 0 & 0 & 0 & 1 \\
\end{bmatrix}
$$

9–3 THE USE OF MODELS HAVING TRIGONOMETRIC TERMS

Time series with additive or multiplicative seasonal variation can sometimes be analyzed and accurately forecasted by using a combination of polynomial functions and trigonometric functions as the functions $f_1(t), f_2(t), \ldots, f_p(t)$ in the general time series regression model

$$y_t = \beta_0 + \beta_1 f_1(t) + \beta_2 f_2(t) + \cdots + \beta_p f_p(t) + \epsilon_t$$

At the end of this section we will discuss some of the more useful models of this type. First, however, it should be noted that each of the models to be discussed is a special case of the general time series regression model. Thus the estimates of the parameters $\beta_0, \beta_1, \ldots, \beta_p$ can be found using least squares regression analysis. Alternatively, these estimates can be found and updated using general direct smoothing. In Section *9–3 we will discuss calculation of the least squares estimates of $\beta_0, \beta_1, \ldots, \beta_p$. The reader interested in the application of direct smoothing to updating the estimates of $\beta_0, \beta_1, \ldots, \beta_p$ is referred to Brown [3] or Johnson and Montgomery [8]. One advantage of these models is that, since they are a special case of the general time series regression model, they yield statistically correct confidence intervals for future values of the time series. It should be remembered that we have presented methods for obtaining "intuitive confidence intervals" for values of the time series when the time series decomposition model and the models assumed in Winters' Method are being used. The reason that we have concentrated on the time series decomposition model and Winters' models in this book is that, whereas the models to be presented containing polynomial and trigonometric functions are fairly effective in modeling additive or multiplicative seasonal variation that is fairly smooth and regular, we feel that the time series decomposition model and the models assumed in Winters' Method are more effective in modeling all types of additive or multiplicative seasonal variation, whether regular or irregular. We feel that this is true because the time series decomposition model and the Winters' models use a different parameter to model the effect of each different season in a year.

In the following discussion of time series regression models containing polynomial and trigonometric terms, we will assume that the trend is given by the equation

$$TR_t = \beta_0 + \beta_1 t$$

Other types of trend can, of course, be used.

Let L be the number of seasons in a year. Thus $L = 4$ for quarterly data and $L = 12$ for monthly data. Then, two models that are useful in modeling additive seasonal variation are

1. $y_t = \beta_0 + \beta_1 t + \beta_2 \sin \dfrac{2\pi t}{L} + \beta_3 \cos \dfrac{2\pi t}{L} + \epsilon_t$

2. $y_t = \beta_0 + \beta_1 t + \beta_2 \sin \dfrac{2\pi t}{L} + \beta_3 \cos \dfrac{2\pi t}{L} + \beta_4 \sin \dfrac{4\pi t}{L}$

$\quad + \beta_5 \cos \dfrac{4\pi t}{L} + \epsilon_t$

The first of these models is useful in modeling a very regular additive seasonal pattern, while the second model possesses terms allowing reasonably accurate modeling of a more complicated additive seasonal pattern.

Two models that are useful in modeling multiplicative seasonal variation are

3. $y_t = \beta_0 + \beta_1 t + \beta_2 \dfrac{\sin 2\pi t}{L} + \beta_3 t \dfrac{\sin 2\pi t}{L} + \beta_4 \dfrac{\cos 2\pi t}{L}$

$\quad + \beta_5 t \dfrac{\cos 2\pi t}{L} + \epsilon_t$

4. $y_t = \beta_0 + \beta_1 t + \beta_2 \dfrac{\sin 2\pi t}{L} + \beta_3 t \dfrac{\sin 2\pi t}{L} + \beta_4 \dfrac{\cos 2\pi t}{L}$

$\quad + \beta_5 t \dfrac{\cos 2\pi t}{L} + \beta_6 \dfrac{\sin 4\pi t}{L} + \beta_7 t \dfrac{\sin 4\pi t}{L}$

$\quad + \beta_8 \dfrac{\cos 4\pi t}{L} + \beta_9 t \dfrac{\cos 4\pi t}{L} + \epsilon_t$

The first of these models is useful in modeling a very regular multiplicative seasonal pattern, while the second model possesses terms allowing reasonably accurate modeling of a more complicated multiplicative seasonal pattern. For convenience in the examples to follow we will refer to these four models as trigonometric regression models 1, 2, 3, and 4.

Example 9.1   In this example we consider fitting trigonometric regression models 1, 2, 3, and 4 to the Rola-Cola sales data of Table III.2. When these models are fit to the Rola-Cola sales data, the following values of R^2 and s are obtained.

Model 1: $R^2 = .8363$ $s = 131.23$

Model 2: $R^2 = .9508$ $s = 74.31$

Model 3: $R^2 = .8737$ $s = 119.06$

Model 4: $R^2 = .9718$ $s = 60.44$

Inspecting these R^2 and s values, we see that we obtain the best overall fit to the historical Rola-Cola sales data using model 4. This model gives the highest R^2 value and the lowest s value. Because the Rola-Cola sales data approximately possess multiplicative seasonal variation, this result should be expected. As noted previously, models 3 and 4 possess terms that allow the modeling of multiplicative seasonal variation, while models 1 and 2 are more appropriate for modeling additive seasonal variation. Thus, we might expect the overall fit of models 3 and 4 to be superior to models 1 and 2. Since model 4 has more terms than model 3, the value of R^2 must be larger for model 4 than for model 3. Since the value of s also is smaller for model 4 than for model 3 (a result that is not necessary but that perhaps could be expected), we conclude that model 4 yields the best overall fit.

Thus, we will use trigonometric regression model 4 to forecast future values of Rola-Cola sales. The least squares estimates of β_0, β_1, β_2, β_3, β_4, β_5, β_6, β_7, β_8, and β_9 in model 4 are as follows.

$$b_0 = 376.6856 \qquad b_5 = -1.3005$$

$$b_1 = 9.6565 \qquad b_6 = 31.9609$$

$$b_2 = -223.3434 \qquad b_7 = 1.3512$$

$$b_3 = -5.7743 \qquad b_8 = -91.4370$$

$$b_4 = -82.0982 \qquad b_9 = -1.7742$$

These estimates give the following prediction equation.

$$\hat{y}_t = 376.6856 + 9.6565t - 223.3434 \sin \frac{2\pi t}{12} - 5.7743t \sin \frac{2\pi t}{12}$$

$$- 82.0982 \cos \frac{2\pi t}{12} - 1.3005t \cos \frac{2\pi t}{12} + 31.9609 \sin \frac{4\pi t}{12}$$

$$+ 1.3512t \sin \frac{4\pi t}{12} - 91.4370 \cos \frac{4\pi t}{12} - 1.7742t \cos \frac{4\pi t}{12}$$

Note here that $L = 12$ since the Rola-Cola sales data are monthly data.

The observed value y_t of Rola-Cola sales and the predicted value $\hat{y}_t$ are given in Table 9.5 for the historical data and are given in Table 9.6 for the 12 future values of the time series. Table 9.6 also gives 95 percent confidence interval forecasts for future Rola-Cola sales. The observed and predicted Rola-Cola sales are plotted in Figure 9.3.

It is interesting to compare the results obtained using this trigonometric model with the results obtained when the Rola-Cola sales time series was forecasted by the multiplicative decomposition method. For the trigonometric regression model just discussed, the sum of the squared differences between the observed and predicted values for the 36 periods of historical data is

$$\sum_{t=1}^{36} (y_t - \hat{y}_t)^2 = 94,976$$

For the 12 future periods of sales, the sum of squared differences between the observed and predicted values is

TABLE 9.5 *Forecasts of Historical Rola-Cola Sales Calculated Using Trigonometric Regression Model 4*

t	y_t	$\hat{y}_t$	t	y_t	$\hat{y}_t$
1	189	181.78	19	831	806.02
2	229	227.71	20	960	1031.54
3	249	261.77	21	1152	1053.02
4	289	262.53	22	759	849.64
5	260	291.98	23	607	559.51
6	431	422.40	24	371	360.80
7	660	638.94	25	298	323.28
8	777	823.36	26	378	372.80
9	915	846.39	27	373	397.87
10	613	684.57	28	443	383.88
11	485	446.98	29	374	432.55
12	277	282.01	30	660	642.42
13	244	252.60	31	1004	972.97
14	296	300.28	32	1153	1239.66
15	319	329.78	33	1388	1259.73
16	370	323.14	34	904	1014.84
17	313	362.24	35	715	672.09
18	556	532.45	36	441	439.50

$$\sum_{t=1}^{36} (y_t - \hat{y}_t)^2 = 94{,}976$$

TABLE 9.6 *Forecasts of Future Values of Rola-Cola Sales Calculated Using Trigonometric Regression Model 4*

t	y_t	$\hat{y}_t$	95 Percent Confidence Interval
37	352	395.14	(234.07, 556.22)
38	445	446.01	(275.08, 616.93)
39	453	465.38	(292.55, 638.21)
40	541	443.29	(270.03, 616.23)
41	457	502.09	(328.17, 676.00)
42	762	753.03	(579.13, 926.93)
43	1194	1141.40	(967.04, 1315.75)
44	1361	1448.60	(1273.92, 1623.27)
45	1615	1465.70	(1291.00, 1640.40)
46	1059	1178.39	(1002.69, 1354.09)
47	824	783.79	(608.60, 958.98)
48	495	519.07	(339.71, 698.44)

$$\sum_{t=37}^{48} (y_t - \hat{y}_t)^2 = 62{,}857.35$$

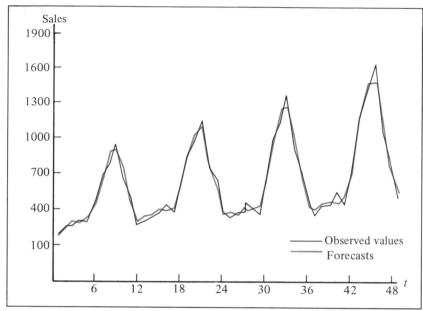

FIGURE 9.3   *Forecasts of Historical and Future Values of Rola-Cola Sales Calculated Using Trigonometric Regression Model 4*

$$\sum_{t=37}^{48} (y_t - \hat{y}_t)^2 = 62{,}857.35$$

When using the multiplicative decomposition method, the sum of the squared differences between the observed and predicted values for the 36 periods of historical data is

$$\sum_{t=1}^{36} (y_t - \hat{y}_t)^2 = 3154.7919$$

For the 12 future periods of sales, the sum of squared differences between the observed and predicted values is

$$\sum_{t=37}^{48} (y_t - \hat{y}_t)^2 = 3693.5263$$

Thus, the multiplicative decomposition method has forecasted Rola-Cola sales more accurately than the trigonometric regression model. It is also interesting to note that the confidence interval forecasts obtained using the trigonometric regression model, although theoretically correct, are quite a bit wider than the intuitive confidence intervals obtained using the multiplicative decomposition method (refer to Table 6.9).

Example
9.2   We now consider fitting trigonometric regression models 1 and 2 to the seminar demand data of Table III.1. When these models are fit to the seminar demand data, the following values of R^2 and s are obtained.

$$\text{Model 1:}\qquad R^2 = 0.98 \qquad s = 2.38$$

$$\text{Model 2:}\qquad R^2 = 0.99 \qquad s = 0.65$$

It should be noticed that we are restricting our attention to trigonometric regression models 1 and 2 in this example because the seminar demand time series approximately possesses additive seasonal variation and because trigonometric regression models 1 and 2 are most appropriate when modeling and forecasting additive seasonal variation. Examining the above R^2 and s values, we see that we obtain the best overall fit to the historical seminar demand data using model 2, which has a higher value of R^2 and a lower value of s.

Thus we will use trigonometric regression model 2 to forecast future values of seminar demand. The least squares estimates of β_0, β_1, β_2, β_3, β_4, and β_5 in model 2 are as follows.

$$b_0 = 23.53 \qquad b_3 = -8.17$$

$$b_1 = 0.50 \qquad b_4 = -1.23$$

$$b_2 = -16.71 \qquad b_5 = -1.54$$

These estimates give the following forecasting equation.

$$\hat{y}_t = 23.53 + 0.50t - 16.71 \sin \frac{2\pi t}{4} - 8.17 \cos \frac{2\pi t}{4}$$

$$- 1.23 \sin \frac{4\pi t}{4} - 1.54 \cos \frac{4\pi t}{4}$$

Note here that $L = 4$ since the seminar demand data are quarterly data.

The observed value y_t of seminar demand and the predicted value $\hat{y}_t$ are given in Table 9.7 for the historical data and are given in Table 9.8 for the 4 future values of the time series. Table 9.8 also gives 95 percent confidence interval forecasts for future seminar demand. The observed and predicted seminar demands are plotted in Figure 9.4.

It is interesting to compare the results obtained using this trigonometric model with the results obtained when the seminar demand time series was forecasted using the additive dummy-variable regression model of Section 8–2. For the trigonometric regression model just discussed, the sum of the squared differences between the observed and predicted values for the 16 periods of historical data is

$$\sum_{t=1}^{16} (y_t - \hat{y}_t)^2 = 4.24$$

TABLE 9.7 *Forecasts of Historical Seminar Demand Calculated Using Trigonometric Regression Model 2*

t	y_t	$\hat{y}_t$
1	10	8.94
2	31	31.03
3	43	43.46
4	16	15.57
5	11	11.15
6	33	32.84
7	45	45.65
8	17	17.36
9	13	13.35
10	34	34.66
11	48	47.85
12	19	19.14
13	15	15.56
14	37	36.47
15	51	50.04
16	21	20.93

$$\sum_{t=1}^{16} (y_t - \hat{y}_t)^2 = 4.24$$

TABLE 9.8 *Forecasts of Future Values of Seminar Demand Calculated Using Trigonometric Regression Model 2*

t	y_t	$\hat{y}_t$	95 Percent Confidence Interval
17	16	16.89	(14.99, 18.79)
18	39	39.12	(37.20, 41.05)
19	53	51.31	(49.36, 53.27)
20	22	23.78	(21.80, 25.77)

$$\sum_{t=17}^{20} (y_t - \hat{y}_t)^2 = 6.83$$

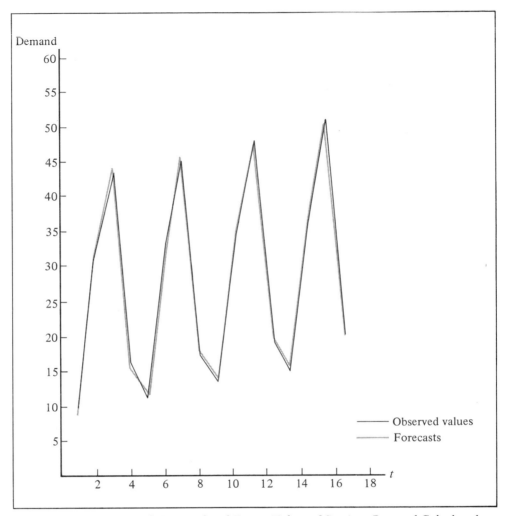

FIGURE 9.4   *Forecasts of Historical and Future Values of Seminar Demand Calculated Using Trigonometric Regression Model 2*

For the 4 future periods of seminar demand, the sum of squared differences between the observed and predicted values is

$$\sum_{t=17}^{20} (y_t - \hat{y}_t)^2 = 6.83$$

When using the additive dummy-variable regression model, the sum of the squared differences between the observed and predicted values for the 16 periods of historical data is

$$\sum_{t=1}^{16} (y_t - \hat{y}_t)^2 = 5$$

For the 4 future periods of seminar demand, the sum of squared differences between the observed and predicted values is

$$\sum_{t=17}^{20} (y_t - \hat{y}_t)^2 = 4.75$$

So, in this situation the trigonometric regression model forecasted the historical data more accurately than did the dummy-variable regression model. However, the dummy-variable regression model forecasted future values of seminar demand more accurately than did the trigonometric regression model.

*9–3   THE REGRESSION CALCULATIONS

As an example of the regression calculations, assume that we have three years of quarterly seasonal time series data, and it is believed that the model

$$y_t = \beta_0 + \beta_1 t + \beta_2 \sin\frac{2\pi t}{L} + \beta_3 t\, \frac{\sin 2\pi t}{L} + \beta_4 \cos\frac{2\pi t}{L} + \beta_5 t\, \frac{\cos 2\pi t}{L} + \epsilon_t$$

describes this data fairly well. Then, since $L = 4$, the least squares estimates b_0, b_1, b_2, b_3, b_4, and b_5 of β_0, β_1, β_2, β_3, β_4, and β_5 are the elements contained in the column vector **b**, which is calculated as

$$\mathbf{b} = (\mathbf{X'X})^{-1}\mathbf{X'y}$$

where **y** is a column vector containing the 12 observed values of the time series, and **X** equals the matrix in Box B.

9–4   COMBINING CAUSAL MODELS AND TIME SERIES MODELS

The following example illustrates the use of regression models that employ both causal variables and mathematical functions of time to forecast a time series.

Example 9.3   Alluring Tackle, Inc., a manufacturer of fishing equipment, makes the Bass Grabber, a type of fishing lure. The company would like to develop a prediction

BOX B

$$\mathbf{X} = \begin{bmatrix} 1 & 1 & \sin\dfrac{2\pi \cdot 1}{4} & \sin\dfrac{2\pi \cdot 1}{4} & \cos\dfrac{2\pi \cdot 1}{4} & \cos\dfrac{2\pi \cdot 1}{4} \\[2ex] 1 & 2 & \sin\dfrac{2\pi \cdot 2}{4} & 2\sin\dfrac{2\pi \cdot 2}{4} & \cos\dfrac{2\pi \cdot 2}{4} & 2\cos\dfrac{2\pi \cdot 2}{4} \\[2ex] 1 & 3 & \sin\dfrac{2\pi \cdot 3}{4} & 3\sin\dfrac{2\pi \cdot 3}{4} & \cos\dfrac{2\pi \cdot 3}{4} & 3\cos\dfrac{2\pi \cdot 3}{4} \\[2ex] 1 & 4 & \sin\dfrac{2\pi \cdot 4}{4} & 4\sin\dfrac{2\pi \cdot 4}{4} & \cos\dfrac{2\pi \cdot 4}{4} & 4\cos\dfrac{2\pi \cdot 4}{4} \\[2ex] 1 & 5 & \sin\dfrac{2\pi \cdot 5}{4} & 5\sin\dfrac{2\pi \cdot 5}{4} & \cos\dfrac{2\pi \cdot 5}{4} & 5\cos\dfrac{2\pi \cdot 5}{4} \\[2ex] 1 & 6 & \sin\dfrac{2\pi \cdot 6}{4} & 6\sin\dfrac{2\pi \cdot 6}{4} & \cos\dfrac{2\pi \cdot 6}{4} & 6\cos\dfrac{2\pi \cdot 6}{4} \\[2ex] 1 & 7 & \sin\dfrac{2\pi \cdot 7}{4} & 7\sin\dfrac{2\pi \cdot 7}{4} & \cos\dfrac{2\pi \cdot 7}{4} & 7\cos\dfrac{2\pi \cdot 7}{4} \\[2ex] 1 & 8 & \sin\dfrac{2\pi \cdot 8}{4} & 8\sin\dfrac{2\pi \cdot 8}{4} & \cos\dfrac{2\pi \cdot 8}{4} & 8\cos\dfrac{2\pi \cdot 8}{4} \\[2ex] 1 & 9 & \sin\dfrac{2\pi \cdot 9}{4} & 9\sin\dfrac{2\pi \cdot 9}{4} & \cos\dfrac{2\pi \cdot 9}{4} & 9\cos\dfrac{2\pi \cdot 9}{4} \\[2ex] 1 & 10 & \sin\dfrac{2\pi \cdot 10}{4} & 10\sin\dfrac{2\pi \cdot 10}{4} & \cos\dfrac{2\pi \cdot 10}{4} & 10\cos\dfrac{2\pi \cdot 10}{4} \\[2ex] 1 & 11 & \sin\dfrac{2\pi \cdot 11}{4} & 11\sin\dfrac{2\pi \cdot 11}{4} & \cos\dfrac{2\pi \cdot 11}{4} & 11\cos\dfrac{2\pi \cdot 11}{4} \\[2ex] 1 & 12 & \sin\dfrac{2\pi \cdot 12}{4} & 12\sin\dfrac{2\pi \cdot 12}{4} & \cos\dfrac{2\pi \cdot 12}{4} & 12\cos\dfrac{2\pi \cdot 12}{4} \end{bmatrix}$$

model that can be used to obtain point forecasts and confidence interval forecasts of the sales (in tens of thousands of lures) of the Bass Grabber. With a reliable model, Alluring Tackle, Inc. can more effectively plan its production schedule, plan its budget, and estimate requirements for producing and storing this product. The sales (in tens of thousands of lures) of the Bass Grabber in "sales period t," where each sales period is defined to last 4 weeks, is denoted by the symbol y_t and is believed to be partially determined by one or more of the "independent variables" x_{t1} = the price in period t of the Bass Grabber as offered by Alluring Tackle, Inc. (in dollars); x_{t2} = the average industry price in period t of competitors' similar lures (in dollars); and x_{t3} = the advertising expenditure in period t of Alluring Tackle, Inc. to promote the Bass Grabber (in thousands of dollars). Assume the data in Table 9.9 has been observed over the past 30 sales periods.

The data in Table 9.9 indicates that sales of the Bass Grabber have been increasing in a linear fashion over time and have been of a seasonal nature, with sales of the lure being largest in the spring and summer, when most recreational fishing takes place. Alluring Tackle, Inc. believes that this pattern will continue in the future. Hence, remembering that each year consists of thirteen, four-week seasons, a reasonable model forecasting the time series is the following multiple regression model.

$$y_t = \beta_0 + \beta_1 x_{t1} + \beta_2 x_{t2} + \beta_3 x_{t3} + \beta_4 t + \beta_{S2} x_{S2,t} + \beta_{S3} x_{S3,t}$$
$$+ \beta_{S4} x_{S4,t} + \beta_{S5} x_{S5,t} + \beta_{S6} x_{S6,t} + \beta_{S7} x_{S7,t} + \beta_{S8} x_{S8,t}$$
$$+ \beta_{S9} x_{S9,t} + \beta_{S10} x_{S10,t} + \beta_{S11} x_{S11,t} + \beta_{S12} x_{S12,t}$$
$$+ \beta_{S13} x_{S13,t} + \epsilon_t$$

where for $j = 2, \ldots, 13$

$$x_{Sj,t} = \begin{cases} 1 & \text{if sales period } t \text{ is season } j \\ 0 & \text{otherwise} \end{cases}$$

When the least squares estimates of the parameters in this model are computed, they yield the following forecasting equation for y_t. Note that the t-statistic for each estimate is given in parentheses below the estimates. The forecasting equation is

$$\hat{y}_t = .1776 + .4071 x_{t1} - .7837 x_{t2} + .9934 x_{t3} + .0435 t$$
$$\quad (0.05) \quad (0.42) \qquad (-1.51) \qquad (4.89) \qquad (6.49)$$

$$+ .7800 x_{S2,t} + 2.373 x_{S3,t} + 3.488 x_{S4,t} + 3.805 x_{S5,t}$$
$$\quad (3.16) \qquad (9.28) \qquad (12.88) \qquad (13.01)$$

$$+ 5.673 x_{S6,t} + 6.738 x_{S7,t} + 6.097 x_{S8,t} + 4.301 x_{S9,t}$$
$$\quad (19.41) \qquad (23.23) \qquad (21.47) \qquad (14.80)$$

$$+ 3.856 x_{S10,t} + 2.621 x_{S11,t} + .9969 x_{S12,t} - 1.467 x_{S13,t}$$
$$\quad (13.89) \qquad (9.24) \qquad (3.50) \qquad (-4.70)$$

Inspection of the t values shows that all coefficients for the seasonal dummy

TABLE 9.9 *Sales of the Bass Grabber (in tens of thousands of lures)*

Period t	Sales y_t	Price x_{t1}	Average Industry Price x_{t2}	Advertising Expenditure x_{t3}
1	4.797	3.85	3.80	5.50
2	6.297	3.75	4.00	6.75
3	8.010	3.70	4.30	7.25
4	7.800	3.70	3.70	5.50
5	9.690	3.60	3.85	7.00
6	10.871	3.60	3.80	6.50
7	12.425	3.60	3.75	6.75
8	10.310	3.80	3.85	5.25
9	8.307	3.80	3.65	5.25
10	8.960	3.85	4.00	6.00
11	7.969	3.90	4.10	6.50
12	6.276	3.90	4.00	6.25
13	4.580	3.70	4.10	7.00
14	5.759	3.75	4.20	6.90
15	6.586	3.75	4.10	6.80
16	8.199	3.80	4.10	6.80
17	9.630	3.70	4.20	7.10
18	9.810	3.80	4.30	7.00
19	11.913	3.70	4.10	6.80
20	12.879	3.80	3.75	6.50
21	12.065	3.80	3.75	6.25
22	10.530	3.75	3.65	6.00
23	9.845	3.70	3.90	6.50
24	9.524	3.55	3.65	7.00
25	7.354	3.60	4.10	6.80
26	4.697	3.65	4.25	6.80
27	6.052	3.70	3.65	6.50
28	6.416	3.75	3.75	5.75
29	8.253	3.80	3.85	5.80
30	10.057	3.70	4.25	6.80

variables $x_{S2,t}, \ldots , x_{S13,t}$ are significantly different from zero. Of the other variables, the coefficient for x_{t3} (advertising expenditure) is significantly different from zero, while the coefficients for x_{t1} (price) and perhaps x_{t2} (average industry price) are probably not significantly different from zero. Moreover, the multiple coefficient of determination can be computed to be

$$R^2 = \frac{\sum\limits_{t=1}^{30} (\hat{y}_t - \bar{y}_t)^2}{\sum\limits_{t=1}^{30} (y_t - \bar{y}_t)^2} = .9926$$

Observe that if either x_{t1} or x_{t2} or both x_{t1} and x_{t2} are removed from the regression model, R^2 is decreased by a very small amount. We do not claim that the forecasting equation given above optimally balances the fit of the regression model to the observed data and the cost of using the model. For example, it might be very difficult to predict values of the independent variables needed to predict the dependent variable sales. In such a case, the use of these independent variables may not be worth the cost and effort. In order to build an optimal model, various models should be fitted to the data, and appropriate comparisons should be made as discussed in Chapter 2. Our only objective in this section is to give an example in which both causal variables and mathematical functions of time are used to forecast a time series. As an interesting exercise, the reader might wish to use the principles of Chapter 2 in an attempt to find a more appropriate regression model for forecasting sales of the Bass Grabber.

As examples of the use of the above forecasting equation, forecasts and 95 percent confidence intervals for y_{31}, y_{32}, y_{33}, and $(y_{31} + y_{32} + y_{33})$ are given below.

For sales period 31, which is in the fifth season of the third year, assume $x_{t1} = 3.80$, $x_{t2} = 3.90$, and $x_{t3} = 6.80$. Then, a forecast and 95 percent confidence interval for y_{31} are, respectively,

$$\hat{y}_{31} = .1776 + .4071(3.80) - .7837(3.90)$$
$$+ .9934(6.80) + .0435(31) + 3.805$$
$$= 10.578$$

and

$$[10.578 - .895, 10.578 + .895] \quad \text{or} \quad [9.683, 11.473]$$

For sales period 32, which is in the sixth season of the third year, assume $x_{t1} = 3.75$, $x_{t2} = 3.85$, and $x_{t3} = 6.85$. Then, a forecast and 95 percent confidence interval for y_{32} are, respectively,

$$\hat{y}_{32} = .1776 + .4071(3.75) - .7837(3.85)$$
$$+ .9934(6.85) + .0435(32) + 5.673$$
$$= 12.559$$

and

$$[12.559 - .911, 12.559 + .911] \quad \text{or} \quad [11.648, 13.470]$$

For sales period 33, which is in the seventh season of the third year, assume $x_{t1} = 3.85$, $x_{t2} = 3.90$, and $x_{t3} = 6.90$. Then, a forecast and 95 percent confidence interval for y_{33} are, respectively,

$$\hat{y}_{33} = .1776 + .4071(3.85) - .7837(3.90)$$
$$+ .9934(6.90) + .0435(33) + 6.738$$
$$= 13.718$$

and

$$[13.718 - .9, 13.718 + .9] \quad \text{or} \quad [12.818, 14.618]$$

A forecast and 95 percent confidence interval for $(y_{31} + y_{32} + y_{33})$ are, respectively,

$$\hat{y}_{31} + \hat{y}_{32} + \hat{y}_{33} = 36.855$$

and

$$[36.855 - 1.848, 36.855 + 1.848] \quad \text{or} \quad [35.007, 38.703]$$

*9–4 THE REGRESSION CALCULATIONS

The least squares estimates of the parameters of the regression model in Example 9.3 are the elements in the column vector **b**, where

$$\mathbf{b} = (\mathbf{X'X})^{-1}\mathbf{X'y}$$

Here **y** is a column vector containing the 30 observed values of the time series, and **X** equals the matrix in Box C. Furthermore, a 95 percent confidence interval for y_t is

$$[\hat{y}_t - E_t(95), \hat{y}_t + E_t(95)]$$

where

$$E_t(95) = t_{5/2}(30 - 17) \cdot s \cdot f_t$$

Assume

$$\mathbf{x}'_{31} = [1 \; 3.80 \; 3.90 \; 6.80 \; 31 \; 0 \; 0 \; 0 \; 1 \; 0 \; 0 \; 0 \; 0 \; 0 \; 0 \; 0 \; 0]$$

$$\mathbf{x}'_{32} = [1 \; 3.75 \; 3.85 \; 6.85 \; 32 \; 0 \; 0 \; 0 \; 0 \; 1 \; 0 \; 0 \; 0 \; 0 \; 0 \; 0 \; 0]$$

$$\mathbf{x}'_{33} = [1 \; 3.85 \; 3.90 \; 6.90 \; 33 \; 0 \; 0 \; 0 \; 0 \; 0 \; 1 \; 0 \; 0 \; 0 \; 0 \; 0 \; 0]$$

Determining that

$$t_{5/2}(30 - 17) = t_{5/2}(13) = 2.160$$

and calculating

$$s = \left(\frac{\sum_{t=1}^{30} (y_t - \hat{y}_t)^2}{30 - 17} \right)^{1/2} = \left(\frac{1.172}{13} \right)^{1/2} = .3003$$

BOX C

$\mathbf{X} =$

1	3.85	3.80	5.50	1	0	0	0	0	0	0	0	0	0	0	0	0
1	3.75	4.00	6.75	2	1	0	0	0	0	0	0	0	0	0	0	0
1	3.70	4.30	7.25	3	0	1	0	0	0	0	0	0	0	0	0	0
1	3.70	3.70	5.50	4	0	0	1	0	0	0	0	0	0	0	0	0
1	3.60	3.85	7.00	5	0	0	0	1	0	0	0	0	0	0	0	0
1	3.60	3.80	6.50	6	0	0	0	0	1	0	0	0	0	0	0	0
1	3.60	3.75	6.75	7	0	0	0	0	0	1	0	0	0	0	0	0
1	3.80	3.85	5.25	8	0	0	0	0	0	0	1	0	0	0	0	0
1	3.80	3.65	5.25	9	0	0	0	0	0	0	0	1	0	0	0	0
1	3.85	4.00	6.00	10	0	0	0	0	0	0	0	0	1	0	0	0
1	3.90	4.10	6.50	11	0	0	0	0	0	0	0	0	0	1	0	0
1	3.90	4.00	6.25	12	0	0	0	0	0	0	0	0	0	0	1	0
1	3.70	4.10	7.00	13	0	0	0	0	0	0	0	0	0	0	0	1
1	3.75	4.20	6.90	14	0	0	0	0	0	0	0	0	0	0	0	0
1	3.75	4.10	6.80	15	1	0	0	0	0	0	0	0	0	0	0	0
1	3.80	4.10	6.80	16	0	1	0	0	0	0	0	0	0	0	0	0
1	3.70	4.20	7.10	17	0	0	1	0	0	0	0	0	0	0	0	0
1	3.80	4.30	7.00	18	0	0	0	1	0	0	0	0	0	0	0	0
1	3.70	4.10	6.80	19	0	0	0	0	1	0	0	0	0	0	0	0
1	3.80	3.75	6.50	20	0	0	0	0	0	1	0	0	0	0	0	0
1	3.80	3.75	6.25	21	0	0	0	0	0	0	1	0	0	0	0	0
1	3.75	3.65	6.00	22	0	0	0	0	0	0	0	1	0	0	0	0
1	3.70	3.90	6.50	23	0	0	0	0	0	0	0	0	1	0	0	0
1	3.55	3.65	7.00	24	0	0	0	0	0	0	0	0	0	1	0	0
1	3.60	4.10	6.80	25	0	0	0	0	0	0	0	0	0	0	1	0
1	3.65	4.25	6.80	26	0	0	0	0	0	0	0	0	0	0	0	1
1	3.70	3.65	6.50	27	0	0	0	0	0	0	0	0	0	0	0	0
1	3.75	3.75	5.75	28	1	0	0	0	0	0	0	0	0	0	0	0
1	3.80	3.85	5.80	29	0	1	0	0	0	0	0	0	0	0	0	0
1	3.70	4.25	6.80	30	0	0	1	0	0	0	0	0	0	0	0	0

we have

$$
\hat{y}_t = \begin{cases} 10.578 & \text{if } t = 31 \\ 12.559 & \text{if } t = 32 \\ 13.718 & \text{if } t = 33 \end{cases}
$$

$$
f_t = (1 + \mathbf{x}_t' (\mathbf{X}'\mathbf{X})^{-1} \mathbf{x}_t)^{1/2} = \begin{cases} 1.380 & \text{if } t = 31 \\ 1.404 & \text{if } t = 32 \\ 1.388 & \text{if } t = 33 \end{cases}
$$

$$
E_t(95) = t_{5/2}(13) \cdot s \cdot f_t = \begin{cases} .895 & \text{if } t = 31 \\ .911 & \text{if } t = 32 \\ .9 & \text{if } t = 33 \end{cases}
$$

A 95 percent confidence interval for

$$
\overset{t=l+q=31+2=33}{\underset{t=l=31}{\sum}} y_t
$$

is

$$
\left[\sum_{t=31}^{33} \hat{y}_t - E_{(31,31+2)}(95), \; \sum_{t=31}^{33} \hat{y}_t + E_{(31,31+2)}(95) \right]
$$

where

$$
E_{(31,31+2)}(95) = t_{5/2}(13) \cdot s \cdot \left\{ (2 + 1) + \left(\sum_{t=31}^{33} \mathbf{x}_t' \right) (\mathbf{X}'\mathbf{X})^{-1} \left(\sum_{t=31}^{33} \mathbf{x}_t \right) \right\}^{1/2}
$$

since

$$
\sum_{t=31}^{33} \mathbf{x}_t' = [1\ 3.80\ 3.90\ 6.80\ 31\ 0\ 0\ 0\ 1\ 0\ 0\ 0\ 0\ 0\ 0\ 0]
$$
$$
+ [1\ 3.75\ 3.85\ 6.85\ 32\ 0\ 0\ 0\ 0\ 1\ 0\ 0\ 0\ 0\ 0\ 0]
$$
$$
+ [1\ 3.85\ 3.90\ 6.90\ 33\ 0\ 0\ 0\ 0\ 0\ 1\ 0\ 0\ 0\ 0\ 0]
$$
$$
= [3\ 11.40\ 11.65\ 20.55\ 96\ 0\ 0\ 0\ 1\ 1\ 1\ 0\ 0\ 0\ 0\ 0]
$$

PROBLEMS

1. Suppose that it is believed that a time series possesses a quadratic trend and a very regular additive seasonal pattern. A regression model having trigonometric terms is to be used to forecast future values of this time series. Determine a regression model that might be used to forecast this time series.

*2. Consider the regression model determined in Problem 1. If the parameters of this model are to be estimated using four years of quarterly data, find the matrix **X** to be used in calculating the least squares estimates.

3.   Consider the regression model of Example 9.3 in which we forecasted sales of the Bass Grabber.

  a. Assume that in period 34 the price of the Bass Grabber as offered by Alluring Tackle, Inc., will be \$3.90; the average industry price of competitors' similar lures will be \$3.80; and the advertising expenditure of Alluring Tackle, Inc., will be \$7,000. Find a forecast for Bass Grabber sales in period 34 using this regression model.

  b. Assume that in period 36 the price of the Bass Grabber as offered by Alluring Tackle, Inc., will be \$3.90; the average industry price of competitors' similar lures will be \$3.80; and the advertising expenditure of Alluring Tackle, Inc., will be \$5,000. Find a forecast for Bass Grabber sales in period 36 using this regression model.

*4. Suppose that a firm believes that demand for a product is partially determined by one or both of the independent variables x_{t1} = the price in period t of the product as offered by the company (in dollars) and x_{t2} = the advertising expenditure in period t of the company to promote the product (in hundreds of thousands of dollars). Suppose also that the firm believes that demand for the product is increasing in a quadratic fashion over time and is seasonal.

  a. Formulate a multiple regression model that might be used to forecast demand for this product. Assume that quarterly data will be used to estimate the parameters in this model.

  b. Using the following observed values of the independent variables for the past three years, determine the **X** matrix to be used in calculating the least squares estimates of the parameters in this model.

Year	Quarter	Price x_{t1}	Advertising x_{t2}
1	1	3.50	8.50
	2	3.50	8.25
	3	3.60	8.00
	4	3.45	8.40
2	1	3.65	8.70
	2	3.60	8.60
	3	3.70	8.90
	4	3.55	8.65

---

* Starred problems are only for students who have read Sections *9–2, *9–3, and *9–4.

Year	Quarter	Price x_{t1}	Advertising x_{t2}
3	1	3.75	8.75
	2	3.65	8.00
	3	3.55	8.15
	4	3.65	9.00

*5. Suppose that predicted values for the independent variables x_{t1} and x_{t2} in the regression model of Problem *4 are as follows.

Year	Quarter	Predicted Price	Predicted Advertising
4	1	3.70	9.00
	2	3.70	8.80
	3	3.75	9.00
	4	3.80	9.10

a. Determine the vectors $\mathbf{x}'_{13}$, $\mathbf{x}'_{14}$, $\mathbf{x}'_{15}$, $\mathbf{x}'_{16}$ to be used in calculating 95 percent confidence intervals for y_{13}, y_{14}, y_{15}, and y_{16}.

b. Determine the vector $\Sigma^{16}_{t=13}\mathbf{x}'_t$ to be used in calculating a 95 percent confidence interval for cumulative demand in periods 13 through 16.

PART III SUMMARY

In Part III you learned several approaches that can be used to forecast time series having seasonal variation. First, you learned that it is important to recognize the difference between additive seasonal variation and multiplicative seasonal variation. If a time series displays additive seasonal variation, the magnitude of the seasonal swing of the time series is independent of the average level as determined by the trend. If a time series displays multiplicative seasonal variation, the magnitude of the seasonal swing of the time series is proportional to the average level as determined by the trend. You learned that in order to choose an appropriate model to be used in forecasting a time series, it is often very helpful to attempt to classify the time series as approximately possessing additive or multiplicative seasonal variation. You also learned that numbers called "seasonal factors" are used to model seasonal time series. These factors, which are correction factors that are used to account for the seasonality of a time series, are not known by the forecaster, and, hence, must be estimated along with the trend in order to forecast future values of a time series. In Part III you studied several techniques that can be used to estimate the trend and seasonal factors of time series possessing seasonal variation.

In Chapter 6 you studied the multiplicative decomposition method. This technique decomposes a time series so that estimates of trend, seasonal, cyclical, and irregular factors can be made. These estimates are then used to forecast future values of the time series. You saw that both point forecasts and intuitive confidence interval forecasts can be found using these estimates. In Chapter 7 you studied a technique that is known as Winters' Method. This method is an exponential smoothing procedure that is best used to estimate the trend and seasonal factors of a time series with a linear trend and multiplicative seasonal variation. Again, you learned how to use these estimates to calculate both point forecasts and intuitive confidence interval forecasts.

In Chapter 8 you studied two techniques that are often used to analyze and forecast time series possessing additive seasonal variation. The first technique was a regression technique that employs variables called "dummy variables." These dummy variables insure that the appropriate seasonal parameter is included in the regression model in each time period. Since this technique is a special case of the general time series regression model, it can be used to obtain both point forecasts and "statistically correct" confidence interval forecasts. The second technique was called Additive Winters' Method. This method is an exponential smoothing procedure that is best used to forecast a time series with a linear trend and additive seasonal variation. Again, you learned how to use the estimates obtained using Additive Winters' Method to calculate point forecasts and intuitive confidence interval forecasts.

In Part III we concentrated on forecasting two time series, the Rola-Cola sales time series and the seminar demand time series. We forecasted each of these time series using several of the methods discussed in Chapters 6, 7, and 8. In Chapter 9 we summarized our modeling and forecasting of these time series. You saw that the Rola-Cola sales time series, which approximately possesses multiplicative seasonal variation, was most accurately forecasted using multiplicative models, while the seminar demand time series, which approximately possesses additive seasonal variation, was most accurately forecasted using additive models. In Chapter 9 you also studied two more regression approaches that can be used to forecast seasonal time series. The first of these approaches involves the use of regression models employing trigonometric terms. Such models can be used to forecast time series possessing either additive or multiplicative seasonal variation. The second approach involves using regression models that employ both causal variables and mathematical functions of time. You learned that, since both these approaches are special cases of the general time series regression model, they can be used to calculate point forecasts and "statistically correct" confidence interval forecasts.

EXERCISES

1. Describe the behavior of a time series having multiplicative seasonal variation and a decreasing trend.

2. Describe the behavior of a time series having additive seasonal variation.

3. Describe the behavior of a time series having multiplicative seasonal variation and an increasing trend.

4. Explain why the procedure described in Section 6–2 is called the multiplicative decomposition method.

5. In the multiplicative decomposition method, moving averages are calculated by adding the observations for a number of periods and dividing the result by the number of periods included in the average. Why is the number of periods included in each average equal to the number of periods in a year?

6. Explain the reason for calculating centered moving averages.

7. Consider analyzing a time series of registration figures for a course that is offered on a trimester basis. In using the multiplicative decomposition method, is the calculation of centered moving averages necessary? Why or why not?

8. Explain why the deseasonalized data is used to estimate the trend component in the multiplicative decomposition method.

9. Give two reasons why the estimates of the cyclical component often are not used in forecasting a time series by the multiplicative decomposition method.

10. Briefly explain the intuitive method for obtaining confidence interval forecasts when using the multiplicative decomposition method.

11. Intuitively explain how the permanent component in Winters' Method is updated.

12. Intuitively explain how the trend component in Winters' Method is updated.

13. Intuitively explain how the seasonal component in Winters' Method is updated.

14. Explain how dummy variables are used to forecast a time series with additive seasonal variation.

15. What is the advantage of using time series models that are special cases of the general time series regression model?

16. Suppose that an additive time series model has been used to forecast a time series with a decreasing trend and multiplicative seasonal variation. Describe the pattern of forecast errors that would most likely be produced in this situation.

17. Suppose that a multiplicative time series model has been used to forecast a time series with an increasing trend and additive seasonal variation. Describe the pattern of forecast errors that would most likely be produced in this situation.

18. Describe a situation you are familiar with in which a regression model using both causal variables and mathematical functions of time might be used as a forecasting tool.

19. Briefly explain the intuitive method for obtaining confidence interval forecasts when using Winters' Method.

20. Consider the regression approach to forecasting a time series with additive seasonal variation. Explain the meaning of β_{Sj}, the parameter corresponding to the dummy variable for season j.

VOCABULARY LIST

Additive seasonal variation

Multiplicative seasonal variation

Multiplicative decomposition method

Moving average

Centered moving average

Deseasonalized data

Winters' Method

Permanent component

Trend component

Dummy variables

Trigonometric models

QUIZ

Answer TRUE if the statement is always true. If the statement is not true, replace the underlined word(s) with a word(s) that makes the statement true.

1. If a time series displays <u>multiplicative</u> seasonal variation, the magnitude of the seasonal swing remains the same as time advances.

2. If a time series displays an increasing trend and additive seasonal variation, the magnitude of the seasonal swing <u>increases</u> as time advances.

3. If a time series displays a decreasing trend and multiplicative seasonal variation, the magnitude of the seasonal swing <u>decreases</u> as time advances.

4. In general, <u>the same time series model should</u> be used to forecast time series with additive or multiplicative seasonal variation.

5. In the multiplicative decomposition method, the number of periods included in each moving average is the number of periods in a <u>year</u>.

6. In the multiplicative decomposition method, the centered moving average represents the combined effects of the <u>seasonal</u> and <u>cyclical</u> factors.

7. In the multiplicative decomposition method, the estimate of the trend factor is obtained by applying regression analysis to the <u>original</u> time series observations.

8. If an observation corresponds to a season with a seasonal factor <u>greater than</u> one, the deseasonalized observation is larger than the actual time series observation.

9. In the multiplicative decomposition method, forecasts are made using the estimates of the <u>trend</u> <u>and</u> <u>seasonal</u> or <u>trend</u>, <u>seasonal</u>, <u>and</u> <u>cyclical</u> <u>factors</u>.

10. In the intuitive method for calculating confidence interval forecasts when using the multiplicative decomposition method, the error in a $(100 - \alpha)$ percent confidence interval for a future value of a time series is the error in a $(100 - \alpha)$ percent confidence interval for the corresponding <u>future</u> <u>deseasonalized</u> <u>observation</u>.

11. Winters' Method employs <u>a</u> <u>single</u> <u>smoothing</u> <u>constant</u> to obtain updated estimates of the permanent, trend, and seasonal components.

12. Winters' Method updates the trend component by smoothing the previous estimate of the trend with the difference between the estimate of the <u>seasonal</u> component made in the current period and the estimate of the <u>seasonal</u> component made in the last period.

13. When using Winters' Method, the <u>multiplicative</u> <u>decomposition</u> <u>method</u> is one of the techniques that can be used to determine the initial estimates needed to forecast a time series with multiplicative seasonal variation.

14. A dummy variable for the fourth season in a year is defined to be equal to <u>4</u> if the time period being considered is in the fourth season and is defined to be equal to <u>0</u> otherwise.

15. When the regression approach is used to forecast a time series with additive seasonal variation, confidence interval forecasts obtained are <u>statistically</u> <u>correct</u> rather than <u>intuitive</u>.

16. When Winters' Method is used to forecast a time series with additive or multiplicative seasonal variation, confidence interval forecasts obtained are <u>statistically</u> <u>correct</u> rather than <u>intuitive</u>.

17. The Winters' Method updating and forecasting equations that are appropriate when forecasting a time series with additive seasonal variation can be obtained from the Winters' Method updating and forecasting equations that are appropriate when forecasting a time series with multiplicative seasonal variation by replacing <u>division</u> operations with <u>subtraction</u> operations and <u>addition</u> operations with <u>multiplication</u> operations.

18. When using Winters' Method, <u>the</u> <u>simulation</u> <u>approach</u> is one of the techniques that can be used to determine the initial estimates needed to forecast a time series with additive seasonal variation.

19. Two advantages of the use of regression models having trigonometric terms

are that they can be used to forecast time series having <u>additive</u> or <u>multiplicative</u> seasonal variation and that they yield <u>intuitive</u> confidence interval forecasts.

20. It is <u>impossible</u> to use a forecasting model which is a combination of a causal model and a time series model.

REFERENCES

1. Box, G. E. P., and G. M. Jenkins, *Time Series Analysis, Forecasting and Control,* Holden-Day, Inc., San Francisco, 1970.

2. Brown, R. G., *Statistical Forecasting for Inventory Control,* McGraw-Hill Book Company, New York, 1959.

3. Brown, R. G., *Smoothing, Forecasting and Prediction of Discrete Time Series,* Prentice-Hall, Inc., Englewood Cliffs, New Jersey, 1962.

4. Brown, R. G., *Decision Rules for Inventory Management,* Holt, Rinehart and Winston, Inc., New York, 1967.

5. Chow, W. M., "Adaptive Control of the Exponential Smoothing Constant," *Journal of Industrial Engineering,* vol. 16, no. 5, pp. 314–317, 1965.

6. Draper, N. R., and H. Smith, *Applied Regression Analysis,* John Wiley & Sons, Inc., New York, 1968.

7. Fuller, Wayne A., *Introduction to Statistical Time Series,* John Wiley & Sons, Inc., New York, 1976.

8. Johnson, L. A., and D. C. Montgomery, *Forecasting and Time Series Analysis,* McGraw-Hill Book Company, New York, 1976.

9. Nelson, C. R., *Applied Time Series Analysis for Managerial Forecasting,* Holden-Day, Inc., San Francisco, 1973.

10. Neter, John, and William Wasserman, *Applied Linear Statistical Models,* R. D. Irwin, Homewood, Illinois, 1974.

11. Wheelright, S. C., and S. Makridakis, *Forecasting Methods for Management,* John Wiley & Sons, Inc., New York, 1973.

12. Winters, P. R., "Forecasting Sales by Exponentially Weighted Moving Averages," *Management Science,* vol. 6, no. 3, pp. 324–342, 1960.

PART
IV

The Box-Jenkins Methodology

> The Box-Jenkins methodology is very useful in forecasting future values of a time series having values that are statistically dependent upon or related to each other.

> The Box-Jenkins methodology consists of four basic steps. These steps are called identification, estimation, diagnostic checking, and forecasting.

> The Box-Jenkins methodology chooses a particular time series model to be used from a class of stationary time series models. Models in this class are either autoregressive models, moving-average models, or mixed autoregressive–moving-average models. These models are capable of representing both seasonal and nonseasonal time series.

> The Box-Jenkins methodology is capable of accurately forecasting time series with complicated error structures. We demonstrate this capability by forecasting room occupancy for the Nite's Rest Inc. hotel chain.

CONTENTS

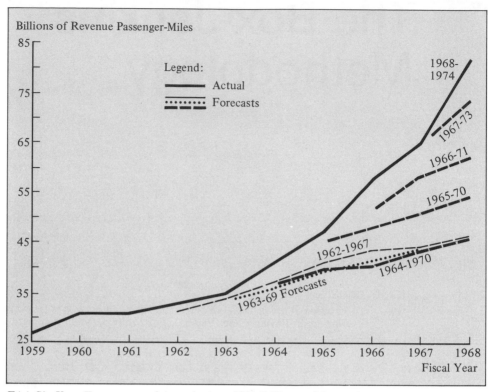

FAA Six-Year Forecasts and Actual, since 1961 (billions of domestic revenue passenger-miles)

Source: Reprinted from "Forecasting: Issues and Challenges for Marketing Management" by Spyros Makridakis and Steven C. Wheelwright, *Journal of Marketing*, vol. 41, no. 4, October 1977, published by the American Marketing Association.

OBJECTIVES FOR PART IV

The regression and exponential smoothing approaches to forecasting that we have previously studied in this book assume that the values of the time series being forecasted are statistically independent of or not related to each other. If the values of the time series being forecasted are statistically dependent upon or related to each other, then the Box-Jenkins methodology uses the dependency to produce forecasts that are likely to be more accurate than the forecasts produced by the regression and exponential smoothing approaches. The objective of Part IV is to discuss the Box-Jenkins methodology. As examples of using the techniques to be discussed, we will forecast sales of Soak-All Paper Towels, which are made by the Big Tree Paper Company, Inc., and the viscosity of Chemical Product XB-77-5, which is produced by Chemo, Inc. In a case study we will also forecast room occupancy for a small hotel chain, Nite's Rest Inc.

INTRODUCTION TO PART IV

All of the regression and exponential smoothing approaches to forecasting that we study in this book assume that the random error components in the time series model

$$y_t = f(\beta_0, \beta_1, \ldots, \beta_p; t) + \epsilon_t$$

are statistically independent of each other. In such a model, if the successive error terms are statistically independent, then so are the successive observations of the time series. However, if the successive error terms are statistically dependent upon each other, then so are the successive observations, and we then call the observations *autocorrelated*. It then follows that accurate forecasting of the time series can probably best be accomplished by employing a model that expresses y_t as a function of present and prior random error components. Thus we consider the model

$$\boxed{y_t = \mu + \psi_0\epsilon_t + \psi_1\epsilon_{t-1} + \psi_2\epsilon_{t-2} + \cdots}$$

where $\epsilon_t, \epsilon_{t-1}, \ldots$ are the present and prior random error components and where $\mu, \psi_0, \psi_1, \psi_2, \ldots$ are the parameters of the model. In this model it is assumed that the error terms $\epsilon_t, \epsilon_{t-1}, \epsilon_{t-2}, \ldots$ are independent and normally distributed with mean 0 and equal variances. However, it is clear that successive values of the time series in this model are dependent, because they are functions of some common error terms. Box and Jenkins [1] have developed a systematic procedure for analyzing and forecasting time series using this model. This procedure is often called the *Box-Jenkins methodology*.

It can be shown that there exist many useful time series models that are special cases of the model

$$y_t = \mu + \psi_0\epsilon_t + \psi_1\epsilon_{t-1} + \psi_2\epsilon_{t-2} + \cdots$$

The Box-Jenkins methodology first develops an appropriate time series model for use in forecasting. This development consists of a three-step iterative procedure. The first step is called *identification*. In this step a tentative model is identified by analysis of the historical data. The second step is called *estimation*. In this step the unknown parameters of the tentative model are estimated. The third step is called *diagnostic checking*. In this step diagnostic checks are performed to test the adequacy of the model, and, if need be, to suggest potential improvements. Once a time series model has been developed, a fourth step, called *forecasting*, generates predictions of future values of the time series. These four steps, identification, estimation, diagnostic checking, and forecasting, are discussed in Chapter 10.

Throughout Chapter 10 one particular time series model will be used to illustrate these concepts. In Chapter 11 some of the other more common models used by the Box-Jenkins methodology will be discussed. We will present models that can be used to analyze and forecast both time series not possessing seasonal variation and time series possessing seasonal variation. In Chapter 12 we present a case study that further illustrates use of the Box-Jenkins methodology.

For other discussions of the Box-Jenkins methodology including analyzing and forecasting seasonal time series, the interested reader is referred to Box and Jenkins [1], which is fairly mathematical and gives an excellent theoretical development with very good examples; to Nelson [12], which is less mathematical and contains excellent computer examples; or to Johnson and Montgomery [8], which gives a brief, usable account of the Box-Jenkins methodology. As will be noticed, the calculations involved in effectively using the Box-Jenkins methodology are very cumbersome. For this reason they are done using the computer. There are many computer packages that do these calculations. For descriptions of two such computer packages the reader is referred to Meeker [10] and Nelson [12]. The following discussion is based upon a critique of the Box-Jenkins methodology given in Johnson and Montgomery [8, pages 235–236].

In practice, regression and exponential smoothing models are frequently applied with good results to forecasting time series with dependent or autocorrelated observations. However, since the Box-Jenkins methodology uses the dependency in the observations more effectively than do regression and exponential smoothing, the Box-Jenkins methodology is likely to produce more accurate forecasts. Moreover, the Box-Jenkins methodology offers a more systematic approach to building, analyzing, and forecasting with time series models. However, the Box-Jenkins methodology has several drawbacks. First, at least 50 and preferably 100 observations are needed to build a good Box-Jenkins model. Hence, Box-Jenkins models are best used to analyze time series in which the sampling interval is small, so that a fairly long history of data can be accumulated. The Box-Jenkins methodology has thus frequently been applied to time series that generate hourly, daily, or weekly observations. For example, readings of the yield of an industrial process might be taken hourly. When monthly or yearly data is collected, the sampling interval may be fairly long and there may not be enough historical data (50 to 100 observations) available to develop a good Box-Jenkins model; or, if the history of data is long enough to provide sufficient data for the model, the underlying process generating the time series may be changing. Thus the first part of the process history may be useless in determining the current nature of the underlying process. For time series exhibiting seasonal variation, the problem is more serious, because each complete year of history is essentially a single observation. That is, in three years of monthly data, we have only three observations for each month. Of course, when an adequate set of historical data is not available one can use the data that does exist to find a preliminary model and then update the preliminary model as new data are observed. Another disadvantage of the Box-Jenkins methodology is that there are no available automatic procedures to either update the estimates of the model parameters as new data are observed or to determine and adapt to changes in the underlying process generating the time series. Such automatic procedures exist for exponential smoothing, but in using the Box-Jenkins methodology one must occasionally repeat the model-building process. A final disadvantage of the Box Jenkins methodology is that, since the model-building process can be rather complex and since a large amount (50 to 100 observations) of reliable data must be obtained, it can be somewhat time consuming and costly to use. We are not implying that a forecasting methodology should not be used because it is somewhat expensive and time consuming. However, it should be remembered from Chapter 1 that *the forecasting method that should be used is one that meets the needs of the forecasting situation at the least cost and inconvenience.* Very frequently the Box-Jenkins methodology is used when only a small number of time series need to be forecasted. Many situations, however, require the forecasting of many different time series. Consider, for example, a distributor who stocks 120,000 different items and who wishes to forecast demand for each of these items. In such a situation it would be particularly important to determine whether the increased accuracy to be obtained by using the Box-Jenkins methodology would justify the time and resources needed for its effective use.

Before proceeding further, it should be noted that the approach we are taking in our discussion of the Box-Jenkins methodology differs from the traditional approach taken in most textbooks, for example by Box and Jenkins [1] and by Nelson [12]. The traditional approach begins with a lengthy, detailed discussion of the theoretical properties of the many models used by the Box-Jenkins methodology and then progresses to a discussion of how the models are used. We have found that many students get bogged down in the discussion of the theoretical properties of the models, because they do not understand why the properties are being discussed. Hence, in order to motivate the reader, we begin by combining discussion of the properties of one model with discussion of how the properties of this model relate to the steps—identification, estimation, diagnostic checking, and forecasting—that are taken in building and then forecasting with an appropriate Box-Jenkins model.

CHAPTER 10

The Basic Steps of the Box-Jenkins Methodology

10–1 INTRODUCTION

The Box-Jenkins methodology consists of several steps. The first step involves using historical observations of a time series to identify a tentative model to be used in forecasting future values of the time series. This identification process is discussed in Section 10–2. We will find that in order to carry out this identification, the concepts of autocorrelation, partial autocorrelation, stationary time series, and nonstationary time series must be understood. The second step of the Box-Jenkins methodology involves estimating the unknown parameters of the tentatively identified model. Estimation of these parameters is discussed in Section 10–3. The third step involves testing the adequacy of the tentatively identified model, and, if need be, suggesting ways to improve the model. This step, called diagnostic checking, is discussed in Section 10–4. When a final time series model has been developed, it is used to forecast future values of the time series. We will show how this forecasting is done in Section 10–5.

10–2 IDENTIFICATION

Identification of a tentative model to be used in forecasting a time series is done through the analysis of historical data. At least 50 and preferably 100 observations are required to identify a good model. In Section 10–2.1 we will discuss the concepts of stationary and nonstationary time series. In Section 10–2.2 we will discuss the concepts of autocorrelation and partial autocorrelation. In Section 10–2.3 we will discuss how the sample autocorrelation function and the sample partial autocorrelation function can be used to identify the particular stationary time series model that adequately describes an observed time series.

10–2.1   STATIONARY TIME SERIES AND NONSTATIONARY TIME SERIES

Consider a time series that can be assumed to be described by the model

$$y_t = \mu + \psi_0 \epsilon_t + \psi_1 \epsilon_{t-1} + \psi_2 \epsilon_{t-2} + \cdots$$

It can be shown that this model can effectively represent both *stationary* and *nonstationary* time series. If a time series is *stationary,* its values fluctuate around a constant mean μ. If a time series is *nonstationary,* it has no constant mean. We will soon see how to determine whether a time series is stationary or nonstationary. This identification is accomplished through the analysis of n observed values y_1, y_2, . . . , y_n of the time series. If the identification indicates that the time series is nonstationary, it is important to transform the nonstationary time series values y_1, y_2, . . . , y_n into values that can be assumed to be described by a stationary time series. If the time series involved does not possess seasonal variation, stationary time series values can frequently be produced by a transformation that involves taking the *first differences* of the original time series values $y_1, y_2, . . . , y_n$. That is, we let

$$z_t = \nabla y_t = y_t - y_{t-1} \qquad \text{for } t = 2, . . . , n$$

Below we list the original values and the first-differenced transformed values of the time series.

Original Values	*First Differences*
y_1	
y_2	$z_2 = \nabla y_2 = y_2 - y_1$
y_3	$z_3 = \nabla y_3 = y_3 - y_2$
.	.
.	.
.	.
y_{n-1}	.
y_n	$z_n = \nabla y_n = y_n - y_{n-1}$

When dealing with business and economic data not possessing seasonal variation, if the original time series values $y_1, y_2, . . . , y_n$ are nonstationary, the transformed values $\nabla y_2, \nabla y_3, . . . , \nabla y_n$ are usually stationary. However, if the transformed values $\nabla y_2, \nabla y_3, . . . , \nabla y_n$ are still nonstationary, taking the *second differences* of the original time series values $y_1, y_2, . . . , y_n$ will usually produce stationary time series values. That is, we let

$$z_t = \nabla^2 y_t = y_t - 2y_{t-1} + y_{t-2} \qquad \text{for } t = 3, . . . , n$$

Below we list the original values and the second-differenced transformed values of the time series.

Original Values	*Second Differences*
y_1	
y_2	
y_3	$z_3 = \nabla^2 y_3 = y_3 - 2y_2 + y_1$
y_4	$z_4 = \nabla^2 y_4 = y_4 - 2y_3 + y_2$
.	.
.	.
.	.
y_{n-2}	.
y_{n-1}	
y_n	$z_n = \nabla^2 y_n = y_n - 2y_{n-1} + y_{n-2}$

On occasion, in order to transform the original time series values into stationary time series values, it is necessary to take either the natural logarithms, or the first or second differences of the natural logarithms, of the original time series values. This is, however, an infrequent occurrence.

10–2.2   AUTOCORRELATION AND PARTIAL AUTOCORRELATION

Assume that we have values $z_a, z_{a+1}, \ldots, z_n$, which can be assumed to have been generated by a stationary time series. Note that we do not write the first value as z_1, because the values $z_a, z_{a+1}, \ldots, z_n$ might be first or second differences of a nonstationary time series $y_1, y_2, \ldots, y_n$. If the values $z_a, z_{a+1}, \ldots, z_n$ are first differences, then

$$z_a = z_2 = \nabla y_2 = y_2 - y_1$$

in which case $a = 2$; whereas if the values $z_a, z_{a+1}, \ldots, z_n$ are second differences, then

$$z_a = z_3 = \nabla^2 y_3 = y_3 - 2y_2 + y_1$$

in which case $a = 3$. Of course, if the original time series values $y_1, y_2, \ldots, y_n$ are stationary, then $z_a = y_1$, in which case $a = 1$.

An important implication of the assumption that the time series values z_a, $z_{a+1}, \ldots, z_n$ are stationary is that the statistical properties of the time series are unaffected by a shift of the time origin. This means, for example, that the statistical relationships between n observations at origin t, say $z_t, z_{t+1}, \ldots, z_{t+n-1}$, are the same as the statistical relationships between n observations at origin $t + j$, say z_{t+j}, $z_{t+j+1}, \ldots, z_{t+j+n-1}$. One of these important relationships is measured by ρ_k, which is the *autocorrelation* between any two time series observations separated by a ''lag'' of k time units. We will not give the precise, mathematical definition of ρ_k; it suffices to say that ρ_k measures the relationship between any two time series observations separated by a lag of k time units. It can be shown that ρ_k is dimensionless, that $-1 \leq \rho_k \leq 1$, and that $\rho_k = \rho_{-k}$, which implies it is necessary to consider only positive lags. It can also be shown that a value of ρ_k close to 1 indicates that observations separated by a lag of k time units have a strong tendency to move together in a linear fashion with a positive slope, whereas a value of ρ_k close to -1 indicates that observations separated by a lag of k time units have a strong tendency to move together in a linear fashion with a negative slope. Although ρ_k is a parameter that cannot be known with certainty, ρ_k can be estimated using the sample observations z_a, z_{a+1}, . . . , z_n. The estimate of ρ_k is called the "sample autocorrelation at lag k," is denoted by the symbol r_k, and is given by the formula

$$ r_k = \frac{\sum_{t=a}^{n-k} (z_t - \bar{z})(z_{t+k} - \bar{z})}{\sum_{t=a}^{n} (z_t - \bar{z})^2} $$

where $\bar{z}$ is the average of the observations z_a, z_{a+1}, . . . , z_n and is given by the formula

$$ \bar{z} = \frac{\sum_{t=a}^{n} z_t}{n - a + 1} $$

The theoretical autocorrelation function is defined to be a listing, or graph, of ρ_k for lags of $k = 1, 2, \ldots$. The sample autocorrelation function is a listing, or graph, of r_k for lags of $k = 1, 2, \ldots$. The theoretical autocorrelation function of a *stationary* time series tends either to *die down* with increasing lag k or to *cut off* after a particular lag $k = q$. When we say the theoretical autocorrelation function tends to "cut off" after a particular lag $k = q$ we mean that

$$ \rho_k = 0 \qquad \text{for } k > q $$

We can determine when the theoretical autocorrelation function cuts off by using the sample autocorrelation function. However, even though the theoretical autocorrelation function may be such that

$$ \rho_k = 0 \qquad \text{for } k > q $$

because of sampling variation, the sample autocorrelation r_k will probably be small, but *not* equal to 0, for $k > q$. The question that then arises is: How small does r_k

have to be to conclude that $\rho_k = 0$? A rough rule of thumb that is frequently used is to conclude that

$$\rho_k = 0 \qquad \text{for } k > q$$

if

$$|r_k| \le 2 \frac{1}{(n - a + 1)^{1/2}} \left(1 + 2 \sum_{j=1}^{q} r_j^2 \right)^{1/2} \qquad \text{for } k > q$$

The reader is referred to Box and Jenkins [1] for a discussion of the statistical theory behind this rule of thumb.

Another method that is often used in deciding whether or not it can be concluded that $\rho_k = 0$ is to calculate the "*t*-like" statistic

$$t_{r_k} = \frac{r_k}{s_{r_k}}$$

where

$$s_{r_k} = \frac{1}{(n - a + 1)^{1/2}} \left(1 + 2 \sum_{j=1}^{k-1} r_j^2 \right)^{1/2}$$

Although this statistic does not have a *t*-distribution, it is asymptotically normal and is often used as a *t*-statistic is used in regression analysis. As a rough rule of thumb, it can be concluded that

$$\rho_k = 0 \qquad \text{if } |t_{r_k}| = \left| \frac{r_k}{s_{r_k}} \right| \le 2$$

Conversely, as a rule of thumb, if $|t_{r_k}| > 2$ it is reasonable to conclude that $\rho_k \ne 0$. Thus, for example, if $|t_{r_k}| > 2$ for lags $1, 2, \ldots, q$ and if $|t_{r_k}| \le 2$ for lags $q + 1$, $q + 2, \ldots$, it is reasonable to conclude that the theoretical autocorrelation function cuts off after lag q.

It can be shown that if a time series is nonstationary, then the sample autocorrelation function will neither cut off nor die down quickly, but rather will *die down extremely slowly*. Hence, if we are given a set of n observed time series values $y_1, y_2, \ldots, y_n$, we first compute the sample autocorrelation function by setting

$$z_a = z_1 = y_1$$

$$z_{a+1} = z_2 = y_2$$

$$\cdot$$
$$\cdot$$
$$\cdot$$

$$z_n = y_n$$

If the sample autocorrelation function of the original observations either dies down or cuts off fairly quickly, we can assume the original time series is stationary. However, if it dies down extremely slowly, then we assume the original time series is nonstationary. The precise meanings of the terms "fairly quickly" and "extremely slowly" are somewhat arbitrary and can best be determined through experience.

   If the original time series is nonstationary and does not possess seasonal variation, we can frequently produce stationary time series values by taking the first differences of the original time series values. That is, we let

$$z_a = z_2 = \nabla y_2 = y_2 - y_1$$

$$z_{a+1} = z_3 = \nabla y_3 = y_3 - y_2$$

$$\cdot \qquad \cdot$$
$$\cdot \qquad \cdot$$
$$\cdot \qquad \cdot$$
$$\cdot \qquad \cdot$$
$$\cdot \qquad \cdot$$

$$z_n = \nabla y_n = y_n - y_{n-1}$$

   If the sample autocorrelation function of these first-differenced transformed values either dies down or cuts off fairly quickly, we can assume the first-differenced values represent a stationary time series. However, if it dies down extremely slowly, then we assume the first differences are nonstationary. In this case we can probably produce stationary time series values by taking the second differences of the original time series values. That is, we let

$$z_a = z_3 = \nabla^2 y_3 = y_3 - 2y_2 + y_1$$

$$z_{a+1} = z_4 = \nabla^2 y_4 = y_4 - 2y_3 + y_2$$

$$\cdot$$
$$\cdot$$
$$\cdot$$

$$z_n = \nabla^2 y_n = y_n - 2y_{n-1} + y_{n-2}$$

If the sample autocorrelation function of these second-differenced transformed values either dies down or cuts off fairly quickly, we can assume the second-differenced values represent a stationary time series. However, if it dies down extremely slowly, then we assume the second differences are nonstationary. If the time series does not possess seasonal variation, it is rarely necessary to do more than take second differences of the original time series in order to achieve stationarity.

Once the original time series values $y_1, y_2, \ldots, y_n$ have been found to be or have been transformed into stationary time series values $z_a, z_{a+1}, \ldots, z_n$, it is important to identify the particular stationary time series model that can be assumed to have generated the values $z_a, z_{a+1}, \ldots, z_n$. In Section 10–2.3 we will see that this can partially be accomplished by determining whether the theoretical autocorrelation function dies down or cuts off, and, if it cuts off, when it cuts off, as determined by the integer q such that

$$\rho_k = 0 \qquad \text{for } k > q$$

Also useful in determining the particular stationary time series model that can be assumed to have generated the observations $z_a, z_{a+1}, \ldots, z_n$ is ρ_{kk}, which is the *partial autocorrelation* between any two time series observations separated by a lag of k time units. We will not give the precise, mathematical definition of ρ_{kk}; it suffices to say that the partial autocorrelation ρ_{kk} may be thought of as the autocorrelation of any two observations, z_t and z_{t+k}, separated by a lag of k time units, with the effects of the intervening observations $z_{t+1}, z_{t+2}, \ldots, z_{t+k-1}$ eliminated. It may be shown that $\rho_{11} = \rho_1$. Although ρ_{kk} is a parameter that cannot be known with certainty, it can be estimated using the sample observations $z_a, z_{a+1}, \ldots, z_n$. The estimate of ρ_{kk} is called the "sample partial autocorrelation at lag k," is denoted by the symbol r_{kk}, and is given by the formula

$$r_{kk} = \begin{cases} r_1 & \text{if } k = 1 \\[2em] \dfrac{r_k - \sum\limits_{j=1}^{k-1} r_{k-1,j}\, r_{k-j}}{1 - \sum\limits_{j=1}^{k-1} r_{k-1,j}\, r_j} & \text{if } k = 2, 3, \ldots \end{cases}$$

where

$$r_{kj} = r_{k-1,j} - r_{kk} r_{k-1,k-j}, \qquad \text{for } j = 1, 2, \ldots, k-1$$

Note that in the above formulas r_k is the sample autocorrelation at lag k. Calculation of r_{kk} using the above formulas will be demonstrated in Example 10.1.

The theoretical partial autocorrelation function is defined to be a listing, or graph, of ρ_{kk} for lags $k = 1, 2, \ldots$. The sample partial autocorrelation function is a listing, or graph, of r_{kk} for lags $k = 1, 2, \ldots$. The theoretical partial autocorrelation function of a stationary time series tends to either die down with increasing lag k or to cut off after a particular lag $k = q$. In Section 10–2.3 we will see that determination of whether the theoretical partial autocorrelation function dies down or cuts off, and, if it cuts off, when it cuts off, as determined by the integer q such that

$$\rho_{kk} = 0 \qquad \text{for } k > q$$

is important in identifying the particular stationary time series model that can be assumed to have generated the values $z_a, z_{a+1}, \ldots, z_n$. We can determine when the theoretical partial autocorrelation function cuts off by using the sample partial autocorrelation function. However, even though the theoretical partial autocorrelation function may be such that

$$\rho_{kk} = 0 \qquad \text{for } k > q$$

because of sampling variation the sample partial autocorrelation r_{kk} will probably be small, but *not* equal to 0, for $k > q$. The question that then arises is: How small does r_{kk} have to be to conclude that $\rho_{kk} = 0$? A rough rule of thumb that is frequently used is to conclude that

$$\rho_{kk} = 0 \qquad \text{for } k > q$$

if

$$|r_{kk}| \leqq 2\, \frac{1}{(n - a + 1)^{1/2}} \qquad \text{for } k > q$$

The reader is referred to Box and Jenkins [1] for a discussion of the statistical theory behind this rule of thumb.

Another method often used in deciding whether or not it can be concluded that $\rho_{kk} = 0$ is to calculate the "t-like" statistic

$$t_{r_{kk}} = \frac{r_{kk}}{\dfrac{1}{(n - a + 1)^{1/2}}}$$

Although this statistic does not have a t-distribution, it is asymptotically normal and is often used as a *t*-statistic is used in regression analysis. As a rough rule of thumb, it can be concluded that

$$
\rho_{kk} = 0 \qquad \text{if} \qquad |t_{r_{kk}}| = \left| \frac{r_{kk}}{\dfrac{1}{(n-a+1)^{1/2}}} \right| \leq 2
$$

Conversely, as a rule of thumb, if $|t_{r_{kk}}| > 2$ it is reasonable to conclude that $\rho_{kk} \neq 0$. Thus, for example, if $|t_{r_{kk}}| > 2$ for lags 1, 2, . . . , q and if $|t_{r_{kk}}| \leq 2$ for lags $q + 1$, $q + 2$, . . . , it is reasonable to conclude that the theoretical partial autocorrelation function cuts off after lag q.

Before continuing to Examples 10.1 and 10.2, it should be noticed that if a particular stationary time series does not possess seasonal variation, then, if either the autocorrelation function or the partial autocorrelation function cuts off, it generally will do so after a lag q that is less than or equal to 2. That is, if

$$
\rho_k = 0 \qquad \text{for } k > q
$$

then, generally speaking, q will be less than or equal to 2. Moreover, if

$$
\rho_{kk} = 0 \qquad \text{for } k > q
$$

then, generally speaking, q will also be less than or equal to 2. If either of these functions does not cut off after a lag q that is less than or equal to 2, it is usually reasonable to assume that the function dies down.

Example 10.1   The Big Tree Paper Company, Inc., makes Soak-All Paper Towels. The company would like to develop a prediction model that can be used to give point forecasts and confidence interval forecasts of weekly sales over 100,000 rolls, in units of 10,000 rolls, of Soak-All Paper Towels. With a reliable model, The Big Tree Paper Company, Inc., can more effectively plan its production schedule, plan its budget, and estimate requirements for producing and storing this product. For the past 120 weeks the company has recorded weekly sales over 100,000 rolls, in units of 10,000 rolls, of Soak-All Paper Towels. The 120 sales figures, $y_1, y_2, \ldots, y_{120}$, are given in Table 10.1 and are plotted in Figure 10.1a. It should be noticed from Figure 10.1a that the original values of the time series do not seem to fluctuate around a constant mean, and hence it would seem that these values are nonstationary. To more precisely determine whether the original time series values are stationary or nonstationary, we will compute the sample autocorrelation function of

TABLE 10.1 *Weekly Sales over 100,000 Rolls of Soak-All Paper Towels (in units of 10,000 rolls)*

t	y_t	t	y_t	t	y_t	t	y_t
1	15.0000	31	10.7752	61	-1.3173	91	10.5502
2	14.4064	32	10.1129	62	-0.6021	92	11.4741
3	14.9383	33	9.9330	63	0.1400	93	11.5568
4	16.0374	34	11.7435	64	1.4030	94	11.7986
5	15.6320	35	12.2590	65	1.9280	95	11.8867
6	14.3975	36	12.5009	66	3.5626	96	11.2951
7	13.8959	37	11.5378	67	1.9615	97	12.7847
8	14.0765	38	9.6649	68	4.8463	98	13.9435
9	16.3750	39	10.1043	69	6.5454	99	13.6859
10	16.5342	40	10.3452	70	8.0141	100	14.1136
11	16.3839	41	9.2835	71	7.9746	101	13.8949
12	17.1006	42	7.7219	72	8.4959	102	14.2853
13	17.7876	43	6.8300	73	8.4539	103	16.3867
14	17.7354	44	8.2046	74	8.7114	104	17.0884
15	17.0010	45	8.5289	75	7.3780	105	15.8861
16	17.7485	46	8.8733	76	8.1905	106	14.8227
17	18.1888	47	8.7948	77	9.9720	107	15.9479
18	18.5997	48	8.1577	78	9.6930	108	15.0982
19	17.5859	49	7.9128	79	9.4506	109	13.8770
20	15.7389	50	8.7978	80	11.2088	110	14.2746
21	13.6971	51	9.0775	81	11.4986	111	15.1682
22	15.0059	52	9.3234	82	13.2778	112	15.3818
23	16.2574	53	10.4739	83	13.5910	113	14.1863
24	14.3506	54	10.6943	84	13.4297	114	13.9996
25	11.9515	55	9.8367	85	13.3125	115	15.2463
26	12.0328	56	8.1803	86	12.7445	116	17.0179
27	11.2142	57	7.2509	87	11.7979	117	17.2929
28	11.7023	58	5.0814	88	11.7319	118	16.6366
29	12.5905	59	1.8313	89	11.6523	119	15.3410
30	12.1991	60	-0.9127	90	11.3718	120	15.6453

$$\bar{y} = 11.58$$

these original values. To do this, we first calculate the mean of the 120 original time series values, which is

$$\bar{y} = \frac{\sum_{t=1}^{120} y_t}{120} = \frac{15.0000 + 14.4064 + \cdots + 15.6453}{120} = 11.58$$

Table 10.2 gives the sample autocorrelation function for lags $k = 1, 2, 3, \ldots, 12$. This function is plotted in Figure 10.2. As an example of the calculations involved, r_3 would be calculated as shown below, where $z_1 = y_1$, $z_2 = y_2$, $\ldots$, $z_{120} = y_{120}$, and $\bar{z} = \bar{y}$.

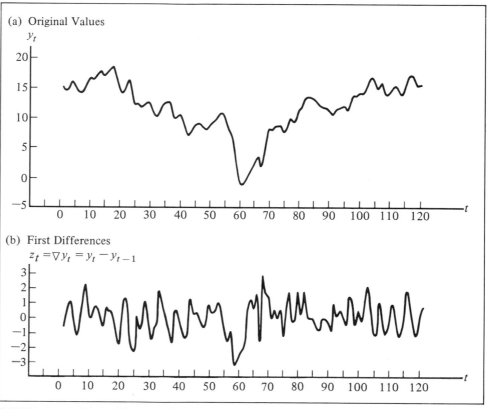

FIGURE 10.1   *Original Values of and First Differences of Weekly Sales over 100,000 Rolls, in Units of 10,000 Rolls, of Soak-All Paper Towels*

$$r_3 = \frac{\displaystyle\sum_{t=1}^{117} (z_t - \bar{z})(z_{t+3} - \bar{z})}{\displaystyle\sum_{t=1}^{120} (z_t - \bar{z})^2}$$

$$= \frac{(z_1 - \bar{z})(z_4 - \bar{z}) + (z_2 - \bar{z})(z_5 - \bar{z}) + \cdots + (z_{117} - \bar{z})(z_{120} - \bar{z})}{(z_1 - \bar{z})^2 + (z_2 - \bar{z})^2 + \cdots + (z_{120} - \bar{z})^2}$$

$$= [(15.0000 - 11.58)(16.0374 - 11.58)$$

$$+ (14.4064 - 11.58)(15.6320 - 11.58) + \cdots$$

$$+ (17.2929 - 11.58)(15.6453 - 11.58)]$$

$$\div [(15.0000 - 11.58)^2 + (14.4064 - 11.58)^2 + \cdots$$

$$+ (15.6453 - 11.58)^2]$$

$$= .85$$

TABLE 10.2   *Sample Autocorrelation Function for the Observations in Table 10.1*

lag k	1	2	3	4	5	6	7	8	9	10	11	12
r_k	.96	.91	.85	.80	.74	.68	.63	.58	.53	.50	.47	.44

Since it is obvious from Table 10.2 and Figure 10.2 that the sample autocorrelation function dies down extremely slowly, we assume that the original time series is nonstationary. Hence, we will attempt to produce stationary time series values by taking the first differences of the original time series values. That is, we let

$$z_a = z_2 = \nabla y_2 = y_2 - y_1 = 14.4064 - 15.0000 = -.5936$$

$$z_{a+1} = z_3 = \nabla y_3 = y_3 - y_2 = 14.9383 - 14.4064 = .5319$$

$$\cdot$$
$$\cdot$$
$$\cdot$$

$$z_{120} = \nabla y_{120} = y_{120} - y_{119} = 15.6453 - 15.3410 = .3043$$

The first differences, $z_2 = \nabla y_2$, $z_3 = \nabla y_3$, . . . , $z_{120} = \nabla y_{120}$, are given in Table 10.3 and plotted in Figure 10.1b. Notice from Figure 10.1 that, whereas the original values of the time series do not seem to fluctuate around a constant mean, the first differences of these original values do seem to fluctuate around a constant mean that is roughly equal to 0. Hence, it would seem that the first differences are stationary. To more precisely determine whether the first differences of the original time series values are stationary or nonstationary, we will compute the sample autocorrelation function of these first differences. To do this, we first calculate the mean of the 119 first differences, which is

$$\bar{z} = \frac{\sum\limits_{t=2}^{120} z_t}{119} = \frac{-.5936 + .5319 + \cdots + .3043}{119} = .005423$$

Table 10.4 gives the sample autocorrelation function of the first differences for lags $k = 1, 2, . . . , 12$. This function is plotted in Figure 10.3. As an example of the calculations involved, r_3 would be calculated as shown below, where $z_2 = \nabla y_2$, $z_3 = \nabla y_3$, . . . , $z_{120} = \nabla y_{120}$.

$$r_3 = \frac{\sum\limits_{t=2}^{117} (z_t - \bar{z})(z_{t+3} - \bar{z})}{\sum\limits_{t=2}^{120} (z_t - \bar{z})^2}$$

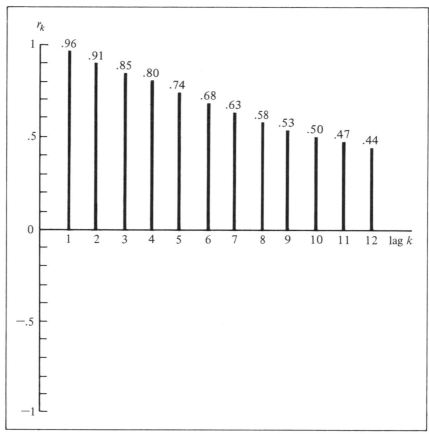

FIGURE 10.2   *Sample Autocorrelation Function for the Observations in Table 10.1*

$$= \frac{(z_2 - \bar{z})(z_5 - \bar{z}) + \cdots + (z_{117} - \bar{z})(z_{120} - \bar{z})}{(z_2 - \bar{z})^2 + \cdots + (z_{120} - \bar{z})^2}$$

$$= [(-.5936 - .005423)(-.4054 - .005423) + \cdots$$

$$+ (.2750 - .005423)(.3043 - .005423)]$$

$$\div [(-.5936 - .005423)^2 + \cdots + (.3043 - .005423)^2]$$

$$= -.07$$

Examination of Table 10.4 and Figure 10.3 indicates that r_k is small for $k > 1$. This would lead us to hypothesize that the theoretical autocorrelation function cuts off after lag 1. That is, we hypothesize that

$$\rho_k = 0 \qquad \text{for } k > 1$$

TABLE 10.3   *First Differences of the Observations in Table 10.1*

t	$z_t = \nabla y_t$ $= y_t - y_{t-1}$	t	$z_t = \nabla y_t$ $= y_t - y_{t-1}$	t	$z_t = \nabla y_t$ $= y_t - y_{t-1}$	t	$z_t = \nabla y_t$ $= y_t - y_{t-1}$
2	−.5936	32	−.6623	62	.7152	92	.9238
3	.5319	33	−.1798	63	.7421	93	.08268
4	1.099	34	1.810	64	1.263	94	.2418
5	−.4054	35	.5154	65	.5249	95	.08809
6	−1.235	36	.2419	66	1.635	96	−.5916
7	−.5015	37	−.9631	67	−1.601	97	1.490
8	.1805	38	−1.873	68	2.885	98	1.159
9	2.298	39	.4395	69	1.699	99	−.2576
10	.1593	40	.2409	70	1.469	100	.4277
11	−.1503	41	−1.062	71	−.03953	101	−.2186
12	.7167	42	−1.562	72	.5213	102	.3903
13	.6871	43	−.8918	73	−.04202	103	2.101
14	−.05226	44	1.375	74	.2575	104	.7016
15	−.7344	45	.3243	75	−1.333	105	−1.202
16	.7475	46	.3444	76	.8124	106	−1.063
17	.4403	47	−.07841	77	1.782	107	1.125
18	.4109	48	−.6371	78	−.2790	108	−.8497
19	−1.014	49	−.2449	79	−.2424	109	−1.221
20	−1.847	50	.8850	80	1.758	110	.3976
21	−2.042	51	.2797	81	.2898	111	.8936
22	1.309	52	.2459	82	1.779	112	.2136
23	1.251	53	1.150	83	.3132	113	−1.195
24	−1.907	54	.2204	84	−.1613	114	−.1867
25	−2.399	55	−.8575	85	−.1173	115	1.247
26	.08132	56	−1.656	86	−.5680	116	1.772
27	−.8186	57	−.9294	87	−.9465	117	.2750
28	.4881	58	−2.170	88	−.06604	118	−.6564
29	.8882	59	−3.250	89	−.07964	119	−1.296
30	−.3914	60	−2.744	90	−.2804	120	.3043
31	−1.424	61	−.4046	91	−.8216		

$$\bar{z} = .005423$$

TABLE 10.4   *Sample Autocorrelation Function for the Observations in Table 10.3*

lag k	1	2	3	4	5	6	7	8	9	10	11	12
r_k	.31	−.06	−.07	.10	.08	.02	−.13	−.12	−.17	−.12	−.05	.02

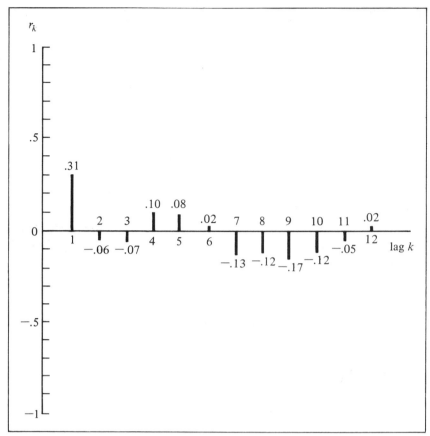

FIGURE 10.3   *Sample Autocorrelation Function for the Observations in Table 10.3*

Letting $q = 1$ and computing

$$2 \frac{1}{(n - a + 1)^{1/2}} \left(1 + 2 \sum_{j=1}^{q} r_j^2 \right)^{1/2} = 2 \frac{1}{(120 - 2 + 1)^{1/2}} (1 + 2r_1^2)^{1/2}$$

$$= 2 \frac{1}{(119)^{1/2}} [1 + 2(.31)^2]^{1/2}$$

$$\approx .20$$

we see that

$$|r_k| \leq .20 \qquad \text{for } k > 1$$

Thus

$$|r_k| \leq 2 \frac{1}{(n - a \cdot + 1)^{1/2}} \left(1 + 2 \sum_{j=1}^{q} r_j^2 \right)^{1/2} \qquad \text{for } k > 1$$

and using our rule of thumb, we conclude that

$$\rho_k = 0 \qquad \text{for } k > 1$$

or that the theoretical autocorrelation function cuts off after lag 1. Alternatively, we can calculate t_{r_k} for lag 1 as follows.

$$s_{r_k} = \frac{1}{(n - a + 1)^{1/2}} \left(1 + 2 \sum_{j=1}^{k-1} r_j^2 \right)^{1/2}$$

$$s_{r_1} = \frac{1}{(120 - 2 + 1)^{1/2}} [1 + 2\,(0)]^{1/2}$$

$$= \frac{1}{(119)^{1/2}}$$

$$= .092$$

and

$$t_{r_1} = \frac{r_1}{s_{r_1}} = \frac{.31}{.092}$$

$$= 3.37$$

Since $|t_{r_1}| > 2$, we conclude that $\rho_1 \neq 0$. We can also calculate t_{r_k} for lag 2 as follows.

$$s_{r_k} = \frac{1}{(n - a + 1)^{1/2}} \left(1 + 2 \sum_{j=1}^{k-1} r_j^2 \right)^{1/2}$$

$$s_{r_2} = \frac{1}{(120 - 2 + 1)^{1/2}} \left(1 + 2 r_1^2 \right)^{1/2}$$

$$= \frac{1}{(119)^{1/2}} [1 + 2\,(.31)^2]^{1/2}$$

$$= .10$$

and

$$t_{r_2} = \frac{r_2}{s_{r_2}} = \frac{-.06}{.10}$$

$$= -.60$$

Since $|t_{r_2}| < 2$, we conclude that $\rho_2 = 0$. Similar calculations reveal that $|t_{r_k}| < 2$ for lags $k = 3, 4, \ldots, 12$. Thus we conclude that $\rho_k = 0$ for $k = 3, 4, \ldots, 12$. So, it is reasonable to conclude that the theoretical autocorrelation function cuts off after lag 1.

This conclusion, which indicates that the first differences are stationary, will be important in Section 10–2.3, in which we identify the particular stationary time series model that can be assumed to have generated these first differences. Also important in this identification are conclusions concerning the theoretical partial autocorrelation function. The sample partial autocorrelation function is given in Table 10.5 and is plotted in Figure 10.4 on page 357. Let us illustrate the calculation of r_{11}, r_{22}, r_{33}, and r_{44} using the formula

$$r_{kk} = \begin{cases} r_1 & \text{if } k = 1 \\ \\ \dfrac{r_k - \displaystyle\sum_{j=1}^{k-1} r_{k-1,j} r_{k-j}}{1 - \displaystyle\sum_{j=1}^{k-1} r_{k-1,j} r_j} & \text{if } k = 2, 3, \ldots \end{cases}$$

where

$$r_{kj} = r_{k-1,j} - r_{kk} r_{k-1,k-j} \qquad \text{for } j = 1, 2, \ldots, k-1$$

$$r_{11} = r_1 = .31$$

$$r_{22} = \frac{r_2 - \displaystyle\sum_{j=1}^{2-1} r_{2-1,j} r_{2-j}}{1 - \displaystyle\sum_{j=1}^{2-1} r_{2-1,j} r_j}$$

$$= \frac{r_2 - r_{11} r_1}{1 - r_{11} r_1} = \frac{-.06 - (.31)(.31)}{1 - (.31)(.31)} = -.18$$

$$r_{21} = r_{11} - r_{22} r_{11} = .31 - (-.18)(.31) = .37$$

$$r_{33} = \frac{r_3 - \displaystyle\sum_{j=1}^{3-1} r_{3-1,j} r_{3-j}}{1 - \displaystyle\sum_{j=1}^{3-1} r_{3-1,j} r_j} = \frac{r_3 - (r_{21} r_2 + r_{22} r_1)}{1 - (r_{21} r_1 + r_{22} r_2)}$$

$$= \frac{-.07 - [(.37)(-.06) + (-.18)(.31)]}{1 - [(.37)(.31) + (-.18)(-.06)]} = .01$$

$$r_{31} = r_{21} - r_{33}r_{22} = .37 - (.01)(-.18) = .37$$

$$r_{32} = r_{22} - r_{33}r_{21} = -.18 - (.01)(.37) = -.18$$

$$r_{44} = \frac{r_4 - \sum_{j=1}^{4-1} r_{4-1,j}r_{4-j}}{1 - \sum_{j=1}^{4-1} r_{4-1,j}r_j} = \frac{r_4 - (r_{31}r_3 + r_{32}r_2 + r_{33}r_1)}{1 - (r_{31}r_1 + r_{32}r_2 + r_{33}r_3)}$$

$$= \frac{.10 - [(.37)(-.07) + (-.18)(-.06) + (.01)(.31)]}{1 - [(.37)(.31) + (-.18)(-.06) + (.01)(-.07)]} = .13$$

We now attempt to describe the behavior of the partial autocorrelation function. Calculating

$$2\frac{1}{(n - a + 1)^{1/2}} = 2\frac{1}{(120 - 2 + 1)^{1/2}} = .18$$

we see that

$$|r_{kk}| \le .18 \qquad \text{for } k > 1$$

Hence, using our rule of thumb we might be tempted to conclude that

$$\rho_{kk} = 0 \qquad \text{for } k > 1$$

However, we see from Table 10.5 and Figure 10.4 that at lags $k = 2, 4, 7,$ and 9 r_{kk} is fairly large, and hence we also might be tempted to conclude that the theoretical partial autocorrelation function dies down.

Alternatively, we can compute $t_{r_{kk}}$ for lags 1 and 2 as follows. For lag 1 we have

$$t_{r_{11}} = \frac{r_{11}}{\dfrac{1}{(n - a + 1)^{1/2}}} = \frac{r_{11}}{\dfrac{1}{(120 - 2 + 1)^{1/2}}} = \frac{r_{11}}{.092} = \frac{.31}{.092} = 3.37$$

For lag 2 we have

$$t_{r_{22}} = \frac{r_{22}}{.092} = \frac{-.18}{.092} = -1.96$$

TABLE 10.5   *Sample Partial Autocorrelation Function for the Observations in Table 10.3*

lag k	1	2	3	4	5	6	7	8	9	10	11	12
r_{kk}	.31	−.18	.01	.13	−.01	.02	−.14	−.04	−.18	−.05	−.01	.03

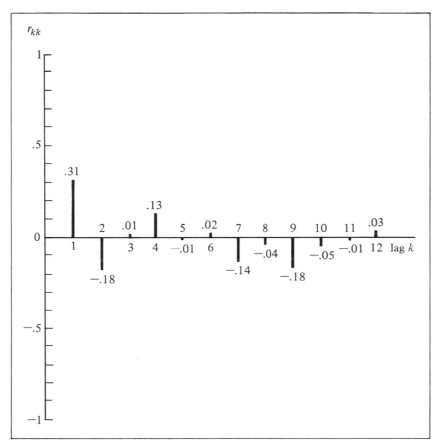

FIGURE 10.4   *Sample Partial Autocorrelation Function for the Observations in Table 10.3*

Since for lag 1 we find that $|t_{r_{11}}| > 2$, we conclude that $\rho_{11} \neq 0$. For lag 2 we find that $|t_{r_{22}}| < 2$, and we might be tempted to conclude that $\rho_{22} = 0$. Furthermore, for lags $k = 3, 4, \ldots, 12$ we note that $|r_{kk}| \leq .18$ and thus for lags $k = 3, 4, \ldots, 12$ we see that $|t_{r_{kk}}|$ will be less than or equal to 1.96 ($|t_{r_{kk}}|$ for lag 2). So we might be tempted to conclude that $\rho_{kk} = 0$ for lags $k = 3, 4, \ldots, 12$, and, hence, conclude that the theoretical partial autocorrelation function cuts off after lag 1. However, we see that for lags $k = 2, 4, 7,$ and $9, |t_{r_{kk}}|$ is reasonably close to 2 (for example, 1.96 for lags $k = 2$ and 9). Thus we might also be tempted to conclude that the theoretical partial autocorrelation function dies down.

To summarize our findings, then, we conclude that, for the first differences, the theoretical autocorrelation function cuts off after lag 1 and the theoretical partial autocorrelation function either cuts off after lag 1 or dies down. This information will be crucial in Section 10–2.3, in which we will identify the stationary time series model that can be assumed to have generated these first differences.

TABLE 10.6   *Daily Readings of the Viscosity of Chemical Product XB-77-5*

t	y_t	t	y_t	t	y_t
1	25.0000	33	34.4337	65	32.2754
2	27.0000	34	35.4844	66	33.2214
3	33.5142	35	33.2381	67	34.5786
4	35.4962	36	36.1684	68	32.3448
5	36.9029	37	34.4116	69	31.5316
6	37.8359	38	33.7668	70	37.8044
7	34.2654	39	33.4246	71	36.0536
8	31.8978	40	33.5719	72	35.7297
9	33.7567	41	35.9222	73	36.7991
10	36.6298	42	33.2125	74	34.9502
11	36.3518	43	37.1668	75	33.5246
12	40.0762	44	35.8138	76	35.1012
13	38.0928	45	33.6847	77	35.9774
14	34.5412	46	33.2761	78	38.0977
15	34.8567	47	38.8163	79	33.4598
16	34.5316	48	42.0838	80	32.9278
17	32.3851	49	40.0069	81	36.5121
18	32.6058	50	33.4514	82	37.4243
19	34.8913	51	30.8413	83	35.1550
20	38.2418	52	30.0655	84	34.4797
21	36.8926	53	37.0544	85	33.2898
22	33.8942	54	39.0982	86	33.9252
23	34.1710	55	37.9075	87	36.1036
24	35.4268	56	36.2393	88	36.7351
25	38.5831	57	34.9535	89	35.4576
26	34.6184	58	33.2061	90	37.5924
27	33.9741	59	34.4261	91	34.4895
28	30.2072	60	37.4511	92	39.1692
29	30.5429	61	37.3335	93	35.8242
30	34.8686	62	38.4679	94	32.3875
31	35.8892	63	33.0976	95	31.2846
32	35.2035	64	32.9285		

$$\bar{y} = 34.93$$

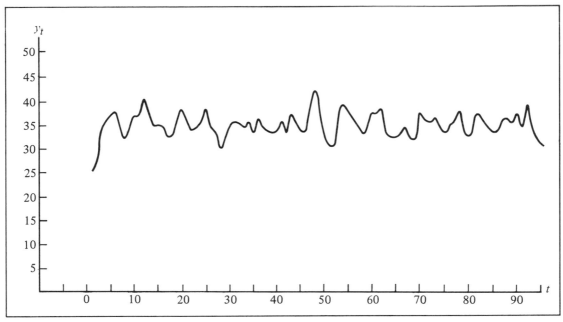

FIGURE 10.5   *Daily Readings of the Viscosity of Chemical Product XB-77-5*

Example
10.2   Chemo, Inc., produces Chemical Product XB-77-5, a product that must have a
rather precisely controlled viscosity. In order to develop a control scheme for their
production process, Chemo, Inc., needs to develop a forecasting model that will
give point forecasts and confidence interval forecasts of the daily viscosity readings
of Chemical Product XB-77-5. For the past 95 days, Chemo, Inc., has recorded the
daily readings of the viscosity of Chemical Product XB-77-5. The 95 daily readings,
$y_1, y_2, \ldots, y_{95}$, are given in Table 10.6 and are plotted in Figure 10.5. It should be
noticed from Figure 10.5 that the original values of the time series seem to fluctuate
around a constant mean, and therefore seem to be stationary. To more precisely
determine whether the original time series values are stationary or nonstationary,
we will compute the sample autocorrelation function of these original values. It can
be shown that the mean of the 95 original time series values is $\bar{y} = 34.93$. The
sample autocorrelation function is calculated as in Example 10.1. Table 10.7 gives
the sample autocorrelation function for lags $k = 1, 2, \ldots, 12$. This function is
plotted in Figure 10.6. Letting $q = 2$ and computing

$$2\,\frac{1}{(n - a + 1)^{1/2}}\left(1 + 2\sum_{j=1}^{q} r_j^2\right)^{1/2} = 2\,\frac{1}{(95 - 1 + 1)^{1/2}}\,[1 + 2\,(r_1^2 + r_2^2)]^{1/2}$$

$$= 2\,\frac{1}{(95)^{1/2}}\,(1 + 2\,[(.44)^2 + (-.11)^2])^{1/2}$$

$$= .24$$

TABLE 10.7   *Sample Autocorrelation Function for the Observations in Table 10.6*

lag k	1	2	3	4	5	6	7	8	9	10	11	12
r_k	.44	−.11	−.34	−.24	.00	.19	.11	−.05	−.14	−.11	−.03	.02

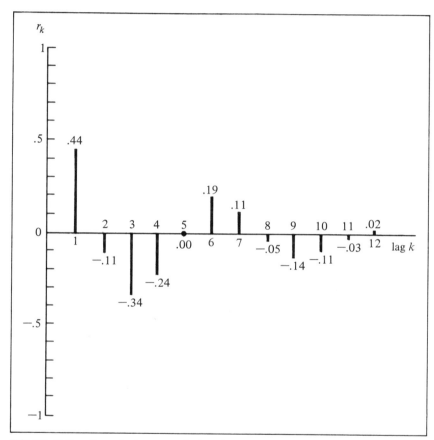

FIGURE 10.6   *Sample Autocorrelation Function for the Observations in Table 10.6*

we see that

$$|r_3| = .34 > .24$$

Since one cannot conclude that the theoretical autocorrelation function cuts off after a lag that is less than or equal to 2, it seems reasonable to conclude that this function dies down.

Alternatively, we can calculate t_{r_k} for lag 3 as follows.

$$s_{r_k} = \frac{1}{(n - a + 1)^{1/2}} \left(1 + 2 \sum_{j=1}^{k-1} r_j^2\right)^{1/2}$$

$$s_{r_3} = \frac{1}{(95 - 1 + 1)^{1/2}} (1 + 2 [(.44)^2 + (-.11)^2])^{1/2}$$

$$= .122$$

and

$$t_{r_3} = \frac{r_3}{s_{r_3}} = \frac{-.34}{.122}$$

Since for lag 3 we have $|t_{r_3}| > 2$, we conclude that $\rho_3 \neq 0$. Thus, we cannot conclude that the theoretical autocorrelation function cuts off after a lag that is less than or equal to 2. So it is reasonable to conclude that this function dies down.

Since, from Table 10.7 and Figure 10.6, it is clear that the function dies down fairly quickly, it is reasonable to assume that the original time series values y_1, y_2, . . . , y_{95} are stationary. For future reference, notice that the sample autocorrelation function dies down in a "damped sine-wave fashion."

The sample partial autocorrelation function of the original time series values can also be calculated as in Example 10.1. Table 10.8 gives this function for lags $k = 1, 2, \ldots, 12$. This function is plotted in Figure 10.7. Examination of Table 10.8 and Figure 10.7 indicates that r_{kk} is small for $k > 2$. Computing

$$2 \frac{1}{(n - a + 1)^{1/2}} = 2 \frac{1}{(95 - 1 + 1)^{1/2}} = .20$$

we see that

$$|r_{kk}| \leq .20 \qquad \text{for } k > 2$$

Hence, using our rule of thumb we conclude that

$$\rho_{kk} = 0 \qquad \text{for } k > 2$$

TABLE 10.8   *Sample Partial Autocorrelation Function for the Observations in Table 10.6*

lag k	1	2	3	4	5	6	7	8	9	10	11	12
r_{kk}	.44	−.38	−.16	−.03	.04	.08	−.10	−.03	−.03	−.03	−.04	−.05

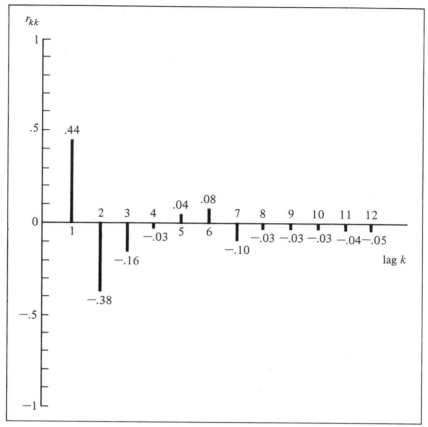

FIGURE 10.7   *Sample Partial Autocorrelation Function for the Observations in*
Table 10.6

Alternatively, we can compute $t_{r_{kk}}$ for lags 1, 2, and 3 as follows. For lag 1 we have

$$t_{r_{11}} = \frac{r_{11}}{\dfrac{1}{(n-a+1)^{1/2}}} = \frac{r_{11}}{\dfrac{1}{(95-1+1)^{1/2}}} = \frac{r_{11}}{.103} = \frac{.44}{.103} = 4.27$$

For lag 2 we have

$$t_{r_{22}} = \frac{r_{22}}{.103} = \frac{-.38}{.103} = -3.69$$

For lag 3 we have

$$t_{r_{33}} = \frac{r_{33}}{.103} = \frac{-.16}{.103} = -1.55$$

TABLE 10.9   *A Summary of Conclusions: Soak-All Paper Towel Sales and Viscosity of Chemical Product XB-77-5*

Conclusions Concerning . . .	Time Series	
	First Differences of Soak-All Paper Towels Sales	Original Values of Viscosity of Chemical Product XB-77-5
Theoretical Autocorrelation Function	Cuts off after lag 1	Dies down
Theoretical Partial Autocorrelation Function	Either cuts off after lag 1 or dies down	Cuts off after lag 2

Since for lags 1 and 2 we have $|t_{r_{kk}}| > 2$, we conclude that $\rho_{11} \neq 0$ and $\rho_{22} \neq 0$. For lag 3 we have $|t_{r_{33}}| < 2$, so we conclude $\rho_{33} = 0$. Furthermore, since for lags $k = 4, 5, \ldots, 12$ we have $|r_{kk}| < .16$, $|t_{r_{kk}}|$ for these lags will be less than 1.55 ($|t_{r_{33}}|$). So we conclude that $\rho_{kk} = 0$ for $k = 4, 5, \ldots, 12$. Thus we conclude that the theoretical partial autocorrelation function cuts off after lag 2.

To summarize our findings, then, we conclude that, for the original time series, the theoretical autocorrelation function dies down in a damped sine-wave fashion, and the theoretical partial autocorrelation function cuts off after lag 2. This information will be crucial in Chapter 11, in which we will identify the stationary time series model that can be assumed to have generated the original time series. The conclusions we have made concerning the theoretical autocorrelation and theoretical partial autocorrelation functions in Examples 10.1 and 10.2 are summarized in Table 10.9.

10–2.3   IDENTIFYING A PARTICULAR STATIONARY TIME SERIES MODEL

Once conclusions have been reached concerning the theoretical autocorrelation function and the theoretical partial autocorrelation function of a stationary time series, these conclusions may be used to identify the particular stationary time series model that can be assumed to have generated the stationary time series under consideration. The Box-Jenkins methodology chooses a particular stationary time series model from a class of stationary time series models. This particular model is then used to forecast future values of the original time series. Models in the aforementioned class of stationary time series models are either autoregressive models, moving-average models, or mixed autoregressive–moving-average models. The precise meaning of these terms will be explained in Chapter 11. The main point, however, is that *each particular model in the class of models is characterized by the behavior of its theoretical autocorrelation function and its theoretical partial autocorrelation function.* If we conclude that the behavior of the theoretical auto-correlation and theoretical partial autocorrelation functions of a given observed stationary time series is identical to the behavior of the theoretical autocorrelation and theoretical partial autocorrelation functions of a given particular stationary time series model, then it is reasonable to tentatively assume that that particular station-ary time series model has generated the given observed stationary time series. It is also reasonable to tentatively assume that that particular model can be used to forecast future values of the given observed time series.

We have previously discussed the property of *stationarity*. There is another property called *invertibility*, that we have not yet discussed. The Box-Jenkins methodology requires that a model to be used in describing and forecasting a stationary time series be both stationary and invertible. We will not formally define the meaning of invertibility. But it can be shown that the conditions for both stationarity and invertibility of a given time series model can be expressed in terms of the parameter(s) of the model.

To illustrate the ideas involved, consider the time series model

$$z_t = \mu + \epsilon_t - \theta_1 \epsilon_{t-1}$$

which is called a *first-order moving-average model*. It can be shown that, for any value of the parameter θ_1, this model describes a stationary time series. Hence, we say there are no conditions that must be imposed on the parameter θ_1 in order to make the above model stationary. However, it can be shown that there is a condition that must be imposed on θ_1 in order to make this model invertible. This condition is that

$$|\theta_1| < 1$$

It will be seen in Section 10–3 that knowledge of the stationarity and invertibility condition(s) on the parameter(s) of a particular time series model is of great importance in estimating the parameter(s) of the model.

It can be shown that the mean of the time series model

$$z_t = \mu + \epsilon_t - \theta_1 \epsilon_{t-1}$$

is μ, that the theoretical partial autocorrelation function of this model dies down in a fashion dominated by "damped exponential decay," and that the theoretical auto-correlation function cuts off after lag 1. In particular, it can be shown that

$$\rho_k = \begin{cases} \dfrac{-\theta_1}{1 + \theta_1^2} & \text{for } k = 1 \\ \\ 0 & \text{for } k > 1 \end{cases}$$

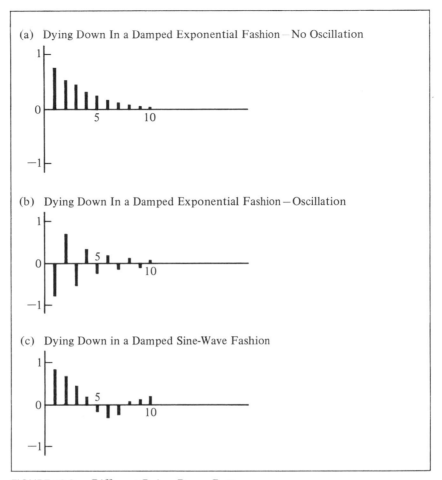

FIGURE 10.8   *Different Dying-Down Patterns*

It will be seen in Section 10–3 that knowledge of the theoretical relationship between ρ_k and the parameter(s) of a given time series model can be of importance in estimating the parameter(s) of the model. It should also be noticed that, in describing the behavior of the partial autocorrelation function of the first-order moving-average process, we use the words "dies down in a fashion dominated by damped exponential decay." In portions of Part IV we will describe the dying down of autocorrelation functions and partial autocorrelation functions with words similar to "in a damped exponential fashion" or "in a damped sine-wave fashion." The intuitive meanings of these terms are illustrated in Figure 10.8. Note that dying down in a damped exponential fashion may involve no oscillation or oscillation of the exponential decay. Until Section 11–5 the reader need only note whether autocorrelation functions and partial autocorrelation functions die down or cut off. Then, the significance of the fashions in which such functions die down will be more fully discussed.

Example
10.3

   Consider the viscosity time series of Table 10.6. In Example 10.2 we concluded that, for this time series, the theoretical autocorrelation function dies down and the theoretical partial autocorrelation function cuts off after lag 2. This behavior is not similar to the behavior of the theoretical autocorrelation and partial autocorrelation functions of the first-order moving-average process model

$$z_t = \mu + \epsilon_t - \theta_1 \epsilon_{t-1}$$

So it is not reasonable to assume that the first-order moving-average process model has generated the time series of Table 10.6.

   Next, consider the sales time series of Table 10.1. In Example 10.1 we concluded that, for the first differences of this time series, the theoretical autocorrelation function cuts off after lag 1 and the theoretical partial autocorrelation function either cuts off after lag 1 or dies down. If we assume that the theoretical partial autocorrelation function dies down, then the behavior of the theoretical autocorrelation and partial autocorrelation functions of the first differences of the time series of Table 10.1 is identical to the behavior of these functions for the first-order moving-average model

$$z_t = \mu + \epsilon_t - \theta_1 \epsilon_{t-1}$$

Hence, it is reasonable to conclude that this model has generated the first differences of the time series of Table 10.1. That is, letting

$$z_t = \nabla y_t = y_t - y_{t-1}$$

it is reasonable to conclude that the model

$$z_t = \mu + \epsilon_t - \theta_1 \epsilon_{t-1}$$

or

$$\nabla y_t = \mu + \epsilon_t - \theta_1 \epsilon_{t-1}$$

or

$$y_t - y_{t-1} = \mu + \epsilon_t - \theta_1 \epsilon_{t-1}$$

or

$$y_t = y_{t-1} + \mu + \epsilon_t - \theta_1 \epsilon_{t-1}$$

has generated the time series of Table 10.1. In Section 10–3 we will discuss estimating the parameter of this model, θ_1, and in Section 10–5 we will discuss forecasting future values of the time series using this model.

10–3 ESTIMATION OF MODEL PARAMETERS

After an appropriate time series model has been identified, estimates of the model parameters are obtained. We will not discuss the theory behind obtaining these estimates. It suffices to say that these estimates are obtained by using a least squares criterion similar to the criterion discussed in Chapter 2. We, therefore, refer to these estimates as least squares estimates. Standard computer programs can be used to find the least squares estimates. Most of the computer programs used apply an iterative search procedure to a sum of squares function and require that the user specify *preliminary estimates* of the unknown parameters as inputs to the computer programs. These preliminary estimates can be obtained through the relationships between the theoretical autocorrelation function and the model parameters. These relationships are used to solve for the model parameters in terms of the theoretical autocorrelations, which are estimated by the corresponding sample autocorrelations, thus producing a set of preliminary estimates of the model parameters. Sometimes this solution process yields more than one set of preliminary estimates. In such a case *exactly one* set of preliminary estimates will satisfy the stationarity and invertibility conditions on the model parameters. The one set of preliminary estimates that does satisfy the stationarity and invertibility conditions on the model parameters is the set that should be chosen.

For example, we have seen in Section 10–2.3 that the relationship between ρ_k and θ_1 in the first-order moving-average model

$$z_t = \mu + \epsilon_t - \theta_1 \epsilon_{t-1}$$

is

$$\rho_k = \begin{cases} \dfrac{-\theta_1}{1 + \theta_1^2} & \text{for } k = 1 \\[2ex] 0 & \text{for } k > 1 \end{cases}$$

That is, we have

$$\rho_1 = \frac{-\theta_1}{1 + \theta_1^2}$$

Solving for θ_1 in terms of ρ_1 we obtain

$$\theta_1 = -\frac{1}{2\rho_1} \pm \left[\frac{1}{(2\rho_1)^2} - 1 \right]^{1/2}$$

Replacing the theoretical autocorrelation ρ_1 with the sample autocorrelation r_1, we obtain a preliminary estimate of θ_1, which is

$$\hat{\theta}_1 = -\frac{1}{2r_1} \pm \left[\frac{1}{(2r_1)^2} - 1 \right]^{1/2}$$

Notice that there are two values for the preliminary estimate $\hat{\theta}_1$. These values are

$$-\frac{1}{2r_1} + \left[\frac{1}{(2r_1)^2} - 1 \right]^{1/2} \quad \text{and} \quad -\frac{1}{2r_1} - \left[\frac{1}{(2r_1)^2} - 1 \right]^{1/2}$$

In order to determine which of these preliminary estimates should be used, recall that the invertibility condition on the first-order moving-average model is

$$|\theta_1| < 1$$

Thus it makes sense to choose the preliminary estimate $\hat{\theta}_1$ that satisfies the condition

$$|\hat{\theta}_1| < 1$$

Since it has been previously stated that μ is the mean of the process, it makes intuitive sense to use the mean of the observed time series values as the preliminary estimate of μ. This mean is

$$\bar{z} = \frac{\sum\limits_{t=a}^{n} z_t}{n - a + 1}$$

However, if $\bar{z}$ is small relative to the time series values $z_a, z_{a+1}, \ldots, z_n$, it is common procedure to assume that μ is 0. Hence, in this situation we use the first-order moving-average model

$$z_t = \epsilon_t - \theta_1 \epsilon_{t-1}$$

If $z_a, z_{a+1}, \ldots, z_n$ in the model

$$z_t = \mu + \epsilon_t - \theta_1 \epsilon_{t-1}$$

are the original time series values $y_1, y_2, \ldots, y_n$, then the assumption that $\mu = 0$ implies that these original time series values are fluctuating around a mean equal to 0. However, if $z_a, z_{a+1}, \ldots, z_n$ are the *first differences* $\nabla y_2, \nabla y_3, \ldots, \nabla y_n$ of the original time series values, then it can be shown that assuming that $\mu = 0$ is equivalent to assuming that there is no deterministic drift (or trend) in the *original* time series values, whereas assuming that $\mu \neq 0$ is equivalent to assuming that there is a deterministic drift (or trend) in the original time series values. Similar comments apply to dealing with the *constant terms* in the other time series models we will study in Chapter 11.

Once preliminary estimates of the model parameters are obtained, these estimates are used as inputs to standard computer programs which, as we have said, find *final estimates* of the model parameters through an iterative procedure. These final estimates are frequently substantially different from the preliminary estimates; and it is, therefore, very dangerous to simply use the preliminary estimates as final estimates of the model parameters.

Notice that, for the first-order moving-average model, it is fairly easy to solve

$$\rho_1 = \frac{-\theta_1}{1 + \theta_1^2}$$

for θ_1 in terms of ρ_1 to give

$$\theta_1 = -\frac{1}{2\rho_1} \pm \left(\frac{1}{(2\rho_1)^2} - 1\right)^{1/2}$$

which is used to obtain the preliminary estimate

$$\hat{\theta}_1 = -\frac{1}{2r_1} \pm \left(\frac{1}{(2r_1)^2} - 1\right)^{1/2}$$

of θ_1. In Chapter 11 we will study some common models, other than the first-order moving-average model, used by the Box-Jenkins methodology to model time series. For each of these models there is a theoretical relationship between the theoretical autocorrelation function and the parameter(s) of the model. However, sometimes it is quite complicated or time consuming to solve such a relationship for the model parameters in terms of the theoretical autocorrelations, and it is important for the reader to know that there exist easy-to-use tables which, for many of the models used by the Box-Jenkins methodology, express preliminary estimates of model parameters in terms of the sample autocorrelations. For example, Box and Jenkins [1] gives such tables.

If, as is sometimes the case with various seasonal models employed by the Box-Jenkins methodology, it is very difficult or time consuming to solve for the model parameters in terms of the autocorrelations and if no tables as described above exist, then one very simple procedure is to use the number .1 as the preliminary estimate of any and every unknown model parameter except μ. We exclude μ because, as explained previously, the preliminary estimate of μ is simply

$$\bar{z} = \frac{\sum\limits_{t=a}^{n} z_t}{n - a + 1}$$

or 0 if $\bar{z}$ is small relative to the time series values $z_a, z_{a+1}, \ldots, z_n$. Using .1 as the preliminary estimate of parameters other than μ is reasonable for two reasons. First, using .1 will generally mean that the preliminary estimate(s) will satisfy the stationarity and invertibility conditions for the model under consideration. For example, we have seen that the invertibility condition for the first-order moving-average model is

$$|\theta_1| < 1$$

Obviously, the preliminary estimate $\hat{\theta}_1 = .1$ satisfies the condition

$$|\hat{\theta}_1| < 1$$

For more complicated models the stationarity and invertibility conditions are more complicated. For example, the second-order moving-average model, to be discussed in Chapter 11, employs parameters denoted by θ_1 and θ_2, and the invertibility conditions for this model are

$$\theta_1 + \theta_2 < 1 \qquad \theta_2 - \theta_1 < 1 \qquad |\theta_2| < 1$$

Again, the preliminary estimates $\hat{\theta}_1 = .1$ and $\hat{\theta}_2 = .1$ satisfy the conditions

$$\hat{\theta}_1 + \hat{\theta}_2 < 1 \qquad \hat{\theta}_2 - \hat{\theta}_1 < 1 \qquad |\hat{\theta}_2| < 1$$

A second reason for using .1 is that the final estimates obtained by computer programs are not very sensitive to the preliminary estimates used as inputs.

In Chapter 11 we will study autoregressive models and mixed autoregressive–moving-average models. These models employ a parameter denoted by δ, which is a function of the mean μ of the process and other parameters in the model. For example, the "first-order mixed autoregressive–moving-average model" employs parameters denoted by δ, θ_1, and ϕ_1, and it is true that

$$\delta = \mu(1 - \phi_1)$$

Simple preliminary estimates of δ, θ_1, and ϕ_1 would then be

$$\hat{\theta}_1 = .1 \qquad \hat{\phi}_1 = .1 \qquad \hat{\delta} = \bar{z}(1 - \hat{\phi}_1) = \bar{z}(1 - .1)$$

where

$$\bar{z} = \frac{\sum\limits_{t=a}^{n} z_t}{n - a + 1}$$

It should be noticed that $\hat{\theta}_1 = .1$ and $\hat{\phi}_1 = .1$ satisfy the stationarity and invertibility conditions for the model, which can be shown to be

$$|\hat{\theta}_1| < 1 \qquad \text{and} \qquad |\hat{\phi}_1| < 1$$

We will not again mention the use of .1 as a preliminary estimate until Section 11–6, which discusses seasonal models. In Sections 11–2, 11–3, and 11–4, which discuss nonseasonal models, we will emphasize using the relationship between the theoretical autocorrelation function and the parameter(s) of a model to obtain preliminary estimates. However, the reader should remember that using .1 as a preliminary estimate will generally be satisfactory.

Example
10.4
We concluded in Example 10.3 that the first differences, given in Table 10.3, of the sales time series of Table 10.1, are adequately described using the first-order moving-average model

$$z_t - \mu + \epsilon_t - \theta_1 \epsilon_{t-1}$$

Since the mean of the first differences is $\bar{z} = .00543$, and since this mean is small compared to the first differences of Table 10.3, we will conclude that $\mu = 0$, and hence that the first-order moving-average model

$$z_t = \epsilon_t - \theta_1 \epsilon_{t-1}$$

describes the first differences.

We now wish to estimate the parameter θ_1. We require a preliminary estimate of θ_1 as input to a computer program that will find the least squares estimate of θ_1. Remembering from Table 10.4 that $r_1 = .31$ and computing

$$-\frac{1}{2r_1} + \left[\frac{1}{(2r_1)^2} - 1 \right]^{1/2} = -1.6129 + 1.2655 = -.3474$$

and

$$-\frac{1}{2r_1} - \left[\frac{1}{(2r_1)^2} - 1 \right]^{1/2} = -1.6129 - 1.2655 = -2.8784$$

we see that the preliminary estimate $\hat{\theta}_1 = -.3474$ satisfies the condition that

$$|\hat{\theta}_1| < 1$$

whereas the preliminary estimate $\hat{\theta}_1 = -2.8784$ does not satisfy this condition. Hence, we use the preliminary estimate

$$\hat{\theta}_1 = -.3474$$

as input to the computer program. After several iterations, this program calculates the final estimate

$$\hat{\theta}_1 = -.3542$$

Hence, we estimate that the model

$$z_t = \epsilon_t - (-.3542)\epsilon_{t-1}$$
$$= \epsilon_t + .3542\epsilon_{t-1}$$

describes the first differences of the time series of Table 10.1. Letting

$$z_t = \nabla y_t = y_t - y_{t-1}$$

we thus estimate that the model

$$z_t = \epsilon_t + .3542\epsilon_{t-1}$$
$$\nabla y_t = \epsilon_t + .3542\epsilon_{t-1}$$
$$y_t - y_{t-1} = \epsilon_t + .3542\epsilon_{t-1}$$
$$y_t = y_{t-1} + \epsilon_t + .3542\epsilon_{t-1}$$

describes the time series of Table 10.1.

10–4 DIAGNOSTIC CHECKING

After a tentative model has been fit to the data, it is important to perform diagnostic checks to test the adequacy of the model and, if need be, to suggest potential improvements. One way this can be accomplished is through the analysis of residuals. That is, we examine the differences between the observed data and the predictions given by the tentatively fit model. Specifically, it has been found that an effective way to measure the overall adequacy of the tentative model is to examine a quantity that determines whether the first K *autocorrelations of the residuals,* considered together, indicate adequacy of the model. This quantity is called the Box-Pierce Chi-Square Statistic, is denoted by the symbol Q, and is computed using the formula

$$Q = (n - d) \sum_{l=1}^{K} r_l^2(\epsilon)$$

where   1. n is the number of observations in the original time series
        2. d is the degree of differencing that was used to transform the original time series into a stationary time series
        3. $r_l^2(\epsilon)$ is the square of $r_l(\epsilon)$, the sample autocorrelation of the residuals at lag l—that is, the sample autocorrelation of residuals separated by a lag of l time units

The modeling process is supposed to account for the relationships between the observations. If it does account for these relationships, the residuals should be unrelated, and hence the autocorrelations of the residuals should be small. Hence, Q should be small. The larger Q is, the larger are the autocorrelations between the residuals, and the more related are the residuals. Thus a large value of Q indicates that the model is inadequate. It is common practice to accept the adequacy of the model if the calculated value of Q is less than $\chi_5^2(K - n_p)$, which is defined to be the point on the scale of the chi-square distribution having $K - n_p$ degrees of freedom such that there is an area of .05 under the curve of this distribution above this point. Here, n_p is the number of parameters that must be estimated in the model under consideration. The choice of K, the number of residual autocorrelations used in the calculation of Q, is arbitrary, but common practice is to compute Q for $K = 12$ (and possibly 24 and 36). A table of chi-square areas is given in Appendix B.

If the model is deemed inadequate, improvements in the model must be made. A good discussion of making such improvements can be found in Nelson [12] and Box and Jenkins [1]. We will give an example of making such improvements in Section 11–6, which discusses seasonal models.

It should also be noted that the adequacy of a Box-Jenkins model can be judged by considering the quantity

$$s = \sqrt{\frac{\text{SSE}}{n - n_p}} = \sqrt{\frac{\sum_{t=1}^{n}(y_t - \hat{y}_t)^2}{n - n_p}}$$

where n = the number of observations in the original time series, and n_p = the number of parameters that must be estimated in the model. As in regression analysis, s measures the overall fit of the model. The smaller s is, the better the overall fit is considered to be. In particular, it should be noted that a smaller value of s usually yields shorter confidence interval forecasts when a Box-Jenkins model is being used.

Example
10.5   In Example 10.4 we tentatively concluded that the model

$$y_t = y_{t-1} + \epsilon_t + .3542\epsilon_{t-1}$$

adequately describes the sales time series of Table 10.1. To examine the adequacy of this model, we first compute the residuals. No residual can be computed for $t = 1$. However, for $t = 2$ we compute the estimate

$$\hat{y}_2 = y_1 + \epsilon_2 + .3542\epsilon_1$$

Since the error terms ϵ_1 and ϵ_2 have zero means, and since we have no other information concerning their values, the best estimate of both ϵ_1 and ϵ_2 is 0. From Table 10.1, $y_1 = 15$. Thus we have

$$\hat{y}_2 = y_1 = 15$$

Since, from Table 10.1 the observed time series value $y_2 = 14.4064$, we have a residual of

$$y_2 - \hat{y}_2 = 14.4064 - 15 = -.5936$$

We can now use this value as an estimate of ϵ_2. Similarly, for $t = 3$ we compute the estimate

$$\hat{y}_3 = y_2 + \epsilon_3 + .3542\epsilon_2$$

Since the best estimate of ϵ_3 is 0, we have

$$\hat{y}_3 = 14.4064 + .3542(-.5936) = 14.1962$$

Since the observed time series value $y_3 = 14.9383$, we have a residual of

$$y_3 - \hat{y}_3 = 14.9383 - 14.1962 = .7421$$

We can use this value as an estimate of ϵ_3. The remaining residuals for periods $t = 4, 5, \ldots, 120$ may be calculated in a similar manner. For future reference it should be noticed that $\hat{y}_{120}$ can be calculated to be

$$\hat{y}_{120} = 14.9553$$

Since $y_{120} = 15.6453$, this gives a residual of

$$y_{120} - \hat{y}_{120} = 15.6453 - 14.9553 = .6900$$

We can use this value as an estimate of ϵ_{120}. Next, the sample autocorrelations of the residuals are calculated. We choose $K = 12$ and list the first 12 residual auto-correlations below.

l	1	2	3	4	5	6	7	8	9	10	11	12
$r_l(\epsilon)$	.01	−.04	−.10	.13	.03	.06	−.14	−.03	−.15	−.06	−.04	.03

We next compute

$$Q = (n - d) \sum_{l=1}^{K} r_l^2(\epsilon)$$

$$= (120 - 1) \sum_{l=1}^{12} r_l^2(\epsilon)$$

$$= (119)[(.01)^2 + (-.04)^2 + \cdots + (.03)^2]$$

$$= 9.6$$

where we have used $d = 1$ since first differencing was required to transform the original time series values into stationary values. Since θ_1 is the one parameter estimated in the first-order moving-average model, $n_p = 1$. Hence

$$\chi_5^2(K - n_p = 12 - 1 = 11) = 19.675$$

Since

$$Q = 9.6 < 19.675 = \chi_5^2(11)$$

we conclude that the fitted model adequately describes the time series of Table 10.1.

10-5 FORECASTING

Once an appropriate model has been adopted, it is used to forecast future values of the time series. Assume that we are at time origin T and wish to forecast $y_{T+\tau}$, which is the value of the time series to be observed at time $T + \tau$. Generally, as will be demonstrated in the next example, the forecast of $y_{T+\tau}$ must be built up recursively from the forecasts of y_{T+1}, y_{T+2}, ..., $y_{T+\tau-1}$. In this procedure, let $\hat{y}_{T+t}(T)$ denote the forecast for y_{T+t} that is made at time origin T. It will be found that $\hat{y}_{T+t}(T)$ will be a function of y_{T+j}, which is a value of the time series, and error terms ϵ_{T+j}. First, consider y_{T+j}. If $j \leqslant 0$, then y_{T+j}, the actual value of the time series, has been observed and we use this observed value. If $j > 0$, then y_{T+j}, the actual value of the time series, has not been observed, and we use $\hat{y}_{T+j}(T)$, the estimate of the observed value. Next, consider the error term ϵ_{T+j}. If $j \leq 0$, we estimate ϵ_{T+j} by the one-period-ahead forecast error

$$y_{T+j} - \hat{y}_{T+j}(T + j - 1)$$

If $j > 0$, we set ϵ_{T+j} equal to 0.

Since there is a theoretical statistical basis behind the Box-Jenkins methodology, statistically correct confidence intervals for future values of the time series may be computed. We will not discuss the technical details involved in computing these confidence intervals. It suffices to say that there are standard computer programs that yield both point forecasts and confidence intervals for future values of the time series.

Example
10.6

We have concluded that the model

$$y_t = y_{t-1} + \epsilon_t + .3542\epsilon_{t-1}$$

adequately describes the sales time series of Table 10.1. Moreover, in Example 10.5 we saw that $\hat{y}_{120} = 14.9553$, and since $y_{120} = 15.6453$, we have a residual for period 120 of

$$y_{120} - \hat{y}_{120} = 15.6453 - 14.9553 = .6900$$

At time origin 120 we may find forecasts for future values of the time series. For example, a forecast of y_{121} is

$$\hat{y}_{121}(120) = y_{120} + \epsilon_{121} + .3542\epsilon_{120}$$

$$= 15.6453 + 0 + .3542(.69)$$

$$= 15.8898$$

Note that in making this forecast we set ϵ_{121} equal to 0, and since the one-period-ahead forecast error is not available, we have estimated ϵ_{120} by .69, which is the residual for period 120. Furthermore, it can be shown that a 95 percent confidence interval for y_{121} is

$$[15.8898 - 2.0362, 15.8898 + 2.0362] \quad \text{or} \quad [13.8536, 17.9260]$$

A forecast of y_{122} made at time origin 120 is

$$\hat{y}_{122}(120) = y_{121} + \epsilon_{122} + .3542c_{121}$$

$$= \hat{y}_{121}(120) + 0 + 0$$

$$= 15.8898$$

Note that we have used the previously generated forecast, $\hat{y}_{121}(120) = 15.8898$, as our estimate of y_{121}. Furthermore, it can be shown that a 95 percent confidence interval for y_{122} is

$$[15.8898 - 3.4278, 15.8898 + 3.4278] \quad \text{or} \quad [12.4620, 19.3176]$$

In general, a point forecast of $y_{120+\tau}$, for $\tau \geq 2$, is

$$\hat{y}_{120+\tau}(120) = y_{120+\tau-1} + \epsilon_{120+\tau} + .3542\epsilon_{120+\tau-1}$$

$$= \hat{y}_{120+\tau-1}(120)$$

$$= 15.8898$$

and the larger τ is, the wider is the 95 percent confidence interval for $y_{120+\tau}$.

Assume now that the actual observation at time 121 is $y_{121} = 16.1099$. Then, although the only way to update the estimate of θ_1 is to refit the entire model, we can obtain a new forecast of y_{122} by using the current estimate of θ_1. Since the one-period-ahead forecast error is

$$y_{121} - \hat{y}_{121}(120) = 16.1099 - 15.8898 = .2201$$

a forecast of y_{122} made at time 121 is

$$\hat{y}_{122}(121) = y_{121} + \epsilon_{122} + .3542\epsilon_{121}$$

$$= 16.1099 + .3542(.2201)$$

$$= 16.1879$$

Moreover, a forecast of y_{123} made at time 121 is

$$\hat{y}_{123}(121) = y_{122} + \epsilon_{123} + .3542\epsilon_{122}$$

$$= \hat{y}_{122}(121)$$

$$= 16.1879$$

PROBLEMS

1. Consider the following tables. Table 1 gives 90 values $y_1, y_2, \ldots, y_{90}$ of weekly sales, in units of 1000 tubes, of Decay-Away Toothpaste. The mean of these 90 values is $\bar{y} = 674.3$. Table 2 gives the sample autocorrelation function for the values $y_1, y_2, \ldots, y_{90}$ for lags $k = 1, 2, \ldots, 12$. Table 3 gives the first differences $z_2, \ldots, z_{90}$ of the original time series values. The mean of these values is $\bar{z} = 8.927$. Finally, Tables 4 and 5 give, respectively, the sample

TABLE 1   *Weekly Sales of Decay-Away Toothpaste (in units of 1000 tubes)*

t	y_t	t	y_t	t	y_t
1	235.000	31	551.925	61	846.962
2	239.000	32	557.929	62	853.830
3	244.090	33	564.285	63	860.840
4	252.731	34	572.164	64	871.075
5	264.377	35	582.926	65	877.792
6	277.934	36	595.295	66	881.143
7	286.687	37	607.028	67	884.226
8	295.629	38	617.541	68	890.208
9	310.444	39	622.941	69	894.966
10	325.112	40	633.436	70	901.288
11	336.291	41	647.371	71	913.138
12	344.459	42	658.230	72	922.511
13	355.399	43	670.777	73	930.786
14	367.691	44	685.457	74	941.306
15	384.003	45	690.992	75	950.305
16	398.042	46	693.557	76	952.373
17	412.969	47	700.675	77	960.042
18	422.901	48	712.710	78	968.100
19	434.960	49	726.513	79	972.477
20	445.853	50	736.429	80	977.408
21	455.929	51	743.203	81	977.602
22	465.584	52	751.227	82	979.505
23	477.894	53	764.265	83	982.934
24	491.408	54	777.852	84	985.833
25	507.712	55	791.070	85	991.350
26	517.237	56	805.844	86	996.291
27	524.349	57	815.122	87	1003.100
28	532.104	58	822.905	88	1010.320
29	538.097	59	830.663	89	1018.420
30	544.948	60	839.600	90	1029.480

$$\bar{y} = 674.3$$

TABLE 2   *Sample Autocorrelation Function for the Observations in Table 1*

lag k	1	2	3	4	5	6	7	8	9	10	11	12
r_k	.97	.94	.90	.87	.84	.81	.77	.74	.71	.68	.64	.61

TABLE 3 *First Differences of the Observations in Table 1*

t	$z_t = \nabla y_t$ $= y_t - y_{t-1}$	t	$z_t = \nabla y_t$ $= y_t - y_{t-1}$	t	$z_t = \nabla y_t$ $= y_t - y_{t-1}$
2	4.000	31	6.977	61	7.362
3	5.090	32	6.004	62	6.868
4	8.641	33	6.356	63	7.010
5	11.65	34	7.879	64	10.235
6	13.56	35	10.76	65	6.717
7	8.753	36	12.37	66	3.351
8	8.942	37	11.73	67	3.083
9	.14.81	38	10.51	68	5.982
10	14.67	39	5.400	69	4.758
11	11.18	40	10.49	70	6.322
12	8.168	41	13.94	71	11.85
13	10.94	42	10.86	72	9.373
14	12.29	43	12.55	73	8.275
15	16.31	44	14.68	74	10.52
16	14.04	45	5.535	75	8.999
17	14.93	46	2.565	76	2.068
18	9.932	47	7.118	77	7.669
19	12.06	48	12.04	78	8.058
20	10.89	49	13.80	79	4.347
21	10.08	50	9.916	80	4.961
22	9.655	51	6.774	81	.1938
23	12.31	52	8.024	82	1.903
24	13.51	53	13.04	83	3.429
25	16.30	54	13.59	84	2.899
26	9.525	55	13.22	85	5.517
27	7.112	56	14.77	86	4.941
28	7.755	57	9.278	87	6.809
29	5.993	58	7.783	88	7.220
30	6.851	59	7.758	89	8.100
		60	8.937	90	11.06

$$\bar{z} = 8.927$$

TABLE 4 *Sample Autocorrelation Function for the Observations in Table 3*

lag k	1	2	3	4	5	6	7	8	9	10	11	12
r_k	.64	.32	.25	.24	.26	.26	.17	.09	.04	.04	.04	.07

TABLE 5   *Sample Partial Autocorrelation Function for the
Observations in Table 3*

lag k	1	2	3	4	5	6	7	8	9	10	11	12
r_{kk}	.64	−.16	.19	.04	.12	.06	−.09	.01	−.06	.03	−.03	.07

autocorrelation function and the sample partial autocorrelation function for the first differences $z_2, \ldots, z_{90}$ for lags $k = 1, 2, \ldots, 12$.

a. Demonstrate how r_2 of Table 2 is calculated.
b. Demonstrate how z_2, z_3, and z_{90} of Table 3 are calculated.
c. Demonstrate how r_4 of Table 4 is calculated.
d. Demonstrate how r_{11}, r_{22}, r_{33}, and r_{44} of Table 5 are calculated.
e. Using the rules of thumb and the sample autocorrelation and sample partial autocorrelation functions, draw appropriate conclusions about stationarity and about the behavior of the theoretical autocorrelation and theoretical partial autocorrelation functions.

2.   In the context of Example 10.6, assume that the actual observation at time 122 is $y_{122} = 16.1526$.

a. Find a forecast that is made at time 122 for y_{123}.
b. Find a forecast that is made at time 122 for y_{124}.

CHAPTER 11

Common Models of the Box-Jenkins Methodology

11–1 INTRODUCTION

The Box-Jenkins methodology chooses a particular time series model from a class of stationary time series models. This particular model is then used to forecast future values of a time series. In this chapter we discuss several of the common models used by the Box-Jenkins methodology. We begin by presenting models that are useful in forecasting nonseasonal time series. In Section 11–2 we discuss a class of models called *moving-average models,* and in Section 11–3 we discuss a class of models called *autoregressive models.* Then, in Section 11–4 we present models that contain both moving-average terms and autoregressive terms. These models are called *mixed autoregressive–moving-average models.* Section 11–5 summarizes our discussion of nonseasonal Box-Jenkins models. Finally, in Section 11–6 we will discuss several Box-Jenkins models that are useful in forecasting time series that possess seasonal variation.

11–2 MOVING-AVERAGE MODELS

The model

$$z_t = \mu + \epsilon_t - \theta_1 \epsilon_{t-1} - \theta_2 \epsilon_{t-2} - \cdots - \theta_q \epsilon_{t-q}$$

is called a *moving-average model of order q.* The minus signs are introduced by convention. It can be shown that there are no conditions that must be imposed on the parameters θ_1, θ_2, . . . , θ_q to make this model stationary, but there are conditions that must be imposed on these parameters to make the model invertible. It can be shown that the theoretical partial autocorrelation function of this model dies down, and the theoretical autocorrelation function cuts off after lag q.

One useful moving-average process model is the *first-order moving-average model*

$$z_t = \mu + \epsilon_t - \theta_1 \epsilon_{t-1}$$

which has been discussed in Sections 10–2.3 through 10–5.

Another useful moving-average process model is the *second-order moving-average model*

$$z_t = \mu + \epsilon_t - \theta_1 \epsilon_{t-1} - \theta_2 \epsilon_{t-2}$$

It can be shown that the conditions that must be imposed on the parameters θ_1 and θ_2 to make this model invertible are

$$\theta_1 + \theta_2 < 1 \qquad \theta_2 - \theta_1 < 1 \qquad |\theta_2| < 1$$

It can be shown that the mean of the time series model

$$z_t = \mu + \epsilon_t - \theta_1 \epsilon_{t-1} - \theta_2 \epsilon_{t-2}$$

is μ, that the theoretical partial autocorrelation function of this model dies down according to a mixture of damped exponentials and/or damped sine waves, and that the theoretical autocorrelation function cuts off after lag 2. In particular, it can be shown that

$$\rho_1 = \frac{-\theta_1(1 - \theta_1)}{1 + \theta_1^2 + \theta_2^2}$$

$$\rho_2 = \frac{-\theta_2}{1 + \theta_1^2 + \theta_2^2}$$

$$\rho_k = 0 \qquad \text{for } k > 2$$

These relationships can be used to solve for the parameters θ_1 and θ_2 in terms of ρ_1 and ρ_2, which are estimated by r_1 and r_2, thus producing a set of preliminary estimates $\hat{\theta}_1$ and $\hat{\theta}_2$ that satisfies the invertibility conditions

$$\hat{\theta}_1 + \hat{\theta}_2 < 1 \qquad \hat{\theta}_2 - \hat{\theta}_1 < 1 \qquad |\hat{\theta}_2| < 1$$

Also, since the mean of the second-order moving-average model is μ, a reasonable preliminary estimate of μ is

$$\bar{z} = \frac{\displaystyle\sum_{t=a}^{n} z_t}{n - a + 1}$$

Example 11.1   Assume that, for a particular observed time series $y_1, y_2, \ldots, y_n$, both the sample autocorrelation function of the original time series values and the sample autocorrelation function of the first differences of these original values die down extremely slowly. However, assume that the sample autocorrelation function of the second differences of the original time series values cuts off after lag 2, and that the sample partial autocorrelation function of these second differences dies down in a damped sine-wave fashion. It is then reasonable to conclude that the second differences are stationary, and that the second-order moving-average model

$$z_t = \mu + \epsilon_t - \theta_1 \epsilon_{t-1} - \theta_2 \epsilon_{t-2}$$

has generated the second differences of the original time series. That is, letting

$$z_t = \nabla^2 y_t = y_t - 2y_{t-1} + y_{t-2}$$

it is reasonable to conclude that the model

$$z_t = \mu + \epsilon_t - \theta_1 \epsilon_{t-1} - \theta_2 \epsilon_{t-2}$$

$$\nabla^2 y_t = \mu + \epsilon_t - \theta_1 \epsilon_{t-1} - \theta_2 \epsilon_{t-2}$$

$$y_t - 2y_{t-1} + y_{t-2} = \mu + \epsilon_t - \theta_1 \epsilon_{t-1} - \theta_2 \epsilon_{t-2}$$

$$y_t = \mu + \epsilon_t - \theta_1 \epsilon_{t-1} - \theta_2 \epsilon_{t-2} + 2y_{t-1} - y_{t-2}$$

can be used to forecast future values of the original time series.

11–3 AUTOREGRESSIVE MODELS

The model

$$z_t = \delta + \phi_1 z_{t-1} + \phi_2 z_{t-2} + \cdots + \phi_p z_{t-p} + \epsilon_t$$

is called an *autoregressive process of order p*. The term "autoregressive" is used because z_t, the current value of the time series, is "regressed," or expressed as a function of, $z_{t-1}, z_{t-2}, \ldots, z_{t-p}$, which are the previous values of the same time series. It can be shown that there are no conditions that must be imposed on the parameters $\phi_1, \phi_2, \ldots, \phi_p$ to make this model invertible, but there are conditions that must be imposed on these parameters to make this model stationary. It can be shown that the theoretical partial autocorrelation function of this model cuts off after lag p, and that the theoretical autocorrelation function dies down. In passing, it should be noticed that just the opposite is true for the moving-average process of order q, since the theoretical autocorrelation function of the moving-average process cuts off after lag q, and the theoretical partial autocorrelation function of the moving-average process dies down.

One useful autoregressive process model is the *first-order autoregressive model*

$$z_t = \delta + \phi_1 z_{t-1} + \epsilon_t$$

It can be shown that the condition that must be imposed on the parameter ϕ_1 to make this model stationary is

$$|\phi_1| < 1$$

It can be shown that the mean of the model

$$z_t = \delta + \phi_1 z_{t-1} + \epsilon_t$$

is

$$\mu = \frac{\delta}{1 - \phi_1}$$

that the theoretical partial autocorrelation function of this model cuts off after lag 1, and that the theoretical autocorrelation function dies down in a damped exponential fashion. In particular, it can be shown that

$$\rho_k = \phi_1^k \qquad \text{for } k \geqslant 1$$

Since this equation implies that

$$\rho_1 = \phi_1$$

we solve for ϕ_1 in terms of ρ_1, which is estimated by r_1, thus producing the preliminary estimate of ϕ_1, which is

$$\hat{\phi}_1 = r_1$$

This preliminary estimate must satisfy the stationarity condition

$$\boxed{|\hat{\phi}_1| < 1}$$

Also, the mean of the first-order autoregressive model is

$$\mu = \frac{\delta}{1 - \phi_1}$$

This implies that

$$\delta = \mu(1 - \phi_1)$$

and thus a reasonable preliminary estimate of δ is

$$\boxed{\hat{\delta} = \bar{z}(1 - \hat{\phi}_1)}$$

where

$$\bar{z} = \frac{\displaystyle\sum_{t=a}^{n} z_t}{n - a + 1}$$

Another useful autoregressive process model is the *second-order autoregressive model*

$$\boxed{z_t = \delta + \phi_1 z_{t-1} + \phi_2 z_{t-2} + \epsilon_t}$$

It can be shown that the conditions that must be imposed on the parameters ϕ_1 and ϕ_2 to make this model stationary are

$$\phi_1 + \phi_2 < 1 \qquad \phi_2 - \phi_1 < 1 \qquad |\phi_2| < 1$$

It can also be shown that the mean of this model is

$$\mu = \frac{\delta}{1 - \phi_1 - \phi_2}$$

that the theoretical partial autocorrelation function cuts off after lag 2, and that the theoretical autocorrelation function dies down according to a mixture of damped exponentials and/or damped sine waves. In particular, it can be shown that the following recursive relationship exists for the theoretical autocorrelation function. We first solve the equations

$$\rho_1 = \phi_1 + \phi_2\rho_1 \qquad \text{and} \qquad \rho_2 = \phi_1\rho_1 + \phi_2$$

for ρ_1 and ρ_2 in terms of ϕ_1 and ϕ_2. This yields

$$\rho_1 = \frac{\phi_1}{1 - \phi_2} \qquad \text{and} \qquad \rho_2 = \frac{\phi_1^2}{1 - \phi_2} + \phi_2$$

Then, we can obtain ρ_k for $k \geqslant 3$ using the equation

$$\rho_k = \phi_1\rho_{k-1} + \phi_2\rho_{k-2}$$

The equations

$$\rho_1 = \phi_1 + \phi_2\rho_1 \qquad \text{and} \qquad \rho_2 = \phi_1\rho_1 + \phi_2$$

are usually called the Yule-Walker equations. They can be solved for ϕ_1 and ϕ_2 in terms of ρ_1 and ρ_2, which are estimated by r_1 and r_2, thus producing the preliminary estimates of ϕ_1 and ϕ_2, which are

$$\hat{\phi}_1 = r_1 \left(\frac{1 - r_2}{1 - r_1^2} \right) \qquad \text{and} \qquad \hat{\phi}_2 = \frac{r_2 - r_1^2}{1 - r_1^2}$$

These preliminary estimates must satisfy the stationarity conditions

$$\hat{\phi}_1 + \hat{\phi}_2 < 1 \qquad \hat{\phi}_2 - \hat{\phi}_1 < 1 \qquad |\hat{\phi}_2| < 1$$

Also, the mean of the second-order autoregressive model is

$$\mu = \frac{\delta}{1 - \phi_1 - \phi_2}$$

This implies that

$$\delta = \mu (1 - \phi_1 - \phi_2)$$

and thus a reasonable preliminary estimate of δ is

$$\hat{\delta} = \bar{z}(1 \quad \hat{\phi}_1 \quad \hat{\phi}_2)$$

where

$$\bar{z} = \frac{\displaystyle\sum_{t=a}^{n} z_t}{n - a + 1}$$

Example 11.2   Consider the viscosity time series of Table 10.6. In Example 10.2 we concluded that, for the original values of this time series, the theoretical autocorrelation function dies down in a damped sine-wave fashion, and the theoretical partial autocorrelation function cuts off after lag 2. Thus, the behavior of the theoretical autocorrelation and theoretical partial autocorrelation functions of this time series is identical to the behavior of these functions for the second-order autoregressive model

$$z_t = \delta + \phi_1 z_{t-1} + \phi_2 z_{t-2} + \epsilon_t$$

Hence, letting $z_t = y_t$, it is reasonable to conclude that the model

$$y_t = \delta + \phi_1 y_{t-1} + \phi_2 y_{t-2} + \epsilon_t$$

has generated the time series of Table 10.6.

Having tentatively identified a particular stationary time series model, we next wish to estimate the parameters δ, ϕ_1, and ϕ_2. As input to a computer program that will find the least squares estimates of δ, ϕ_1, and ϕ_2, we require preliminary estimates of δ, ϕ_1, and ϕ_2. From Tables 10.6 and 10.7 we have $\bar{y} = 34.93$, $r_1 = .44$, and $r_2 = -.11$. Thus we compute the preliminary estimates

$$\hat{\phi}_1 = r_1 \left(\frac{1 - r_2}{1 - r_1^2} \right) = .6057$$

$$\hat{\phi}_2 = \frac{r_2 - r_1^2}{1 - r_1^2} = -.3765$$

$$\hat{\delta} = \bar{y}(1 - \hat{\phi}_1 - \hat{\phi}_2) = 26.93$$

of, respectively, ϕ_1, ϕ_2, and δ. Note that we do not assume that the mean of the time series is 0 because $\bar{y} = 34.93$ is not small compared to the observations of the time series in Table 10.6. Notice also that $\hat{\phi}_1$ and $\hat{\phi}_2$ satisfy the stationarity conditions

$$\hat{\phi}_1 + \hat{\phi}_2 < 1 \qquad \hat{\phi}_2 - \hat{\phi}_1 < 1 \qquad |\hat{\phi}_2| < 1$$

The preliminary estimates are used as inputs to a computer program. After several iterations, the computer program calculates the final estimates of, respectively, δ, ϕ_1, and ϕ_2:

$$\hat{\delta} = 26.34 \qquad \hat{\phi}_1 = .6857 \qquad \hat{\phi}_2 = -.4395$$

Hence, we tentatively conclude that the model

$$y_t = 26.34 + .6857y_{t-1} - .4395y_{t-2} + \epsilon_t$$

adequately describes the time series of Table 10.6.

To examine the adequacy of this model, we first compute the residuals. No residuals can be computed for $t = 1$ or $t = 2$. However, for $t = 3$ we compute the estimate of y_3, which is

$$\begin{aligned}
\hat{y}_3 &= 26.34 + .6857y_2 - .4395y_1 + \epsilon_3 \\
&= 26.34 + .6857(27) - .4395(25) \\
&= 33.8696
\end{aligned}$$

Note that we have estimated ϵ_3 as zero and have used the observed values $y_2 = 27$ and $y_1 = 25$ from Table 10.6. Since from Table 10.6 the observed time series value $y_3 = 33.5142$, we have a residual of

$$y_3 - \hat{y}_3 = 33.5142 - 33.8696 = -.3554$$

The remaining residuals for periods $t = 4, 5, \ldots, 95$ may be calculated in a similar manner. Next, the sample autocorrelation function of the residuals may be calculated. We choose $K = 12$ and list the first 12 residual autocorrelations.

l	1	2	3	4	5	6	7	8	9	10	11	12
$r_l(\epsilon)$	$-.14$	.09	$-.06$	$-.07$	$-.03$	.10	$-.02$	$-.04$	$-.05$	$-.06$	.03	$-.02$

We next compute

$$\begin{aligned}
Q &= (n - d) \sum_{l=1}^{K} r_l^2(\epsilon) \\
&= (95 - 0) \sum_{l=1}^{12} r_l^2(\epsilon) \\
&= (95)[(-.14)^2 + (.09)^2 + \cdots + (-.02)^2] \\
&= 5.5
\end{aligned}$$

where we have used $d = 0$, since the original time series values are stationary. Since δ, ϕ_1, and ϕ_2 are the parameters that must be estimated in the second-order autoregressive model, $n_p = 3$, and hence

$$\chi^2_5(K - n_p = 12 - 3 = 9) = 16.92$$

Since

$$Q = 5.5 < 16.92 = \chi^2_5(9)$$

we conclude that the fitted model adequately describes the viscosity time series of Table 10.6.

We now consider using the model

$$y_t = 26.34 + .6857y_{t-1} - .4395y_{t-2} + \epsilon_t$$

to forecast future values of the time series. For example, forecasts of y_{96}, y_{97}, and y_{98} at time origin 95 are given below.

$$\hat{y}_{96}(95) = 26.34 + .6857y_{95} - .4395y_{94} + \epsilon_{96}$$
$$= 26.34 + .6857(31.2846) - .4395(32.3875)$$
$$= 33.5576$$

$$\hat{y}_{97}(95) = 26.34 + .6857y_{96} - .4395y_{95} + \epsilon_{97}$$
$$= 26.34 + .6857\hat{y}_{96}(95) - .4395y_{95}$$
$$= 26.34 + .6857(33.5576) - .4395(31.2846)$$
$$= 35.6008$$

$$\hat{y}_{98}(95) = 26.34 + .6857y_{97} - .4395y_{96} + \epsilon_{98}$$
$$= 26.34 + .6857\hat{y}_{97}(95) - .4395\hat{y}_{96}(95)$$
$$= 26.34 + .6857(35.6008) - .4395(33.5576)$$
$$= 36.0029$$

Moreover, it can be shown that 95 percent confidence intervals for y_{96}, y_{97}, and y_{98} are as follows.

For y_{96}: $[33.5576 - 4.1568, 33.5576 + 4.1568]$ or $[29.4008, 37.7144]$

For y_{97}: $[35.6008 - 5.0401, 35.6008 + 5.0401]$ or $[30.5607, 40.6409]$

For y_{98}: $[36.0029 - 5.0417, 36.0029 + 5.0417]$ or $[30.9612, 41.0446]$

*11–3 USING REGRESSION TO ANALYZE AUTOREGRESSIVE MODELS

The autoregressive model of order p,

$$z_t = \delta + \phi_1 z_{t-1} + \phi_2 z_{t-2} + \ldots + \phi_p z_{t-p} + \epsilon_t$$

may be considered as a regression model in which z_t is the dependent variable and $z_{t-1}, z_{t-2}, \ldots, z_{t-p}$ are the independent variables. Hence, regression analysis may be used to find the least squares estimates of $\delta, \phi_1, \phi_2, \ldots, \phi_p$ and to find confidence intervals for future values of the time series. For example, the least squares estimates of δ, ϕ_1, and ϕ_2 in the second-order autoregressive model

$$z_t = \delta + \phi_1 z_{t-1} + \phi_2 z_{t-2} + \epsilon_t$$

are given by the matrix algebra equation

$$\begin{bmatrix} \hat{\delta} \\ \hat{\phi}_1 \\ \hat{\phi}_2 \end{bmatrix} = (\mathbf{Z}'\mathbf{Z})^{-1}\mathbf{Z}'\mathbf{z}$$

where, if $z_a, z_{a+1}, \ldots, z_n$ are the observed values of the time series, then

$$\mathbf{z} = \begin{bmatrix} z_{a+2} \\ z_{a+3} \\ \cdot \\ \cdot \\ \cdot \\ \cdot \\ z_n \end{bmatrix} \quad \text{and} \quad \mathbf{Z} = \begin{bmatrix} 1 & z_{a+1} & z_a \\ 1 & z_{a+2} & z_{a+1} \\ \cdot & \cdot & \cdot \\ \cdot & \cdot & \cdot \\ \cdot & \cdot & \cdot \\ 1 & z_{n-1} & z_{n-2} \end{bmatrix}$$

This regression approach will seldom yield results identical to those obtained by using standard computer programs that perform the calculations of the Box-Jenkins methodology. This is because these standard computer programs use an algorithm that differs from the calculations used in the regression approach. The regression approach cannot be used to find the least squares estimates of the parameters in moving-average models or in mixed autoregressive–moving-average models, because these models are not linear functions of the unknown parameters and hence are not "regression models." However, the algorithm used by standard computer programs used for the Box-Jenkins methodology is more general and will calculate the least squares estimates of the unknown parameters in moving-average models or in mixed autoregressive–moving-average models.

Example
11.3   Given the viscosity data of Table 10.6, the least squares estimates of the parameters δ, ϕ_1, and ϕ_2 in the second-order autoregressive model can be found using the regression approach. These estimates are

$$\begin{bmatrix} \hat{\delta} \\ \hat{\phi}_1 \\ \hat{\phi}_2 \end{bmatrix} = (\mathbf{Z}'\mathbf{Z})^{-1}\mathbf{Z}'\mathbf{z} = \begin{bmatrix} 25.82 \\ .6891 \\ -.4283 \end{bmatrix}$$

where we have used the matrices

$$\mathbf{z} = \begin{bmatrix} z_{a+2} \\ z_{a+3} \\ \cdot \\ \cdot \\ \cdot \\ z_n \end{bmatrix} = \begin{bmatrix} y_3 \\ y_4 \\ \cdot \\ \cdot \\ \cdot \\ y_{95} \end{bmatrix} = \begin{bmatrix} 33.5142 \\ 35.4962 \\ \cdot \\ \cdot \\ \cdot \\ 31.2846 \end{bmatrix}$$

$$\mathbf{Z} = \begin{bmatrix} 1 & z_{a+1} & z_a \\ 1 & z_{a+2} & z_{a+1} \\ \cdot & \cdot & \cdot \\ \cdot & \cdot & \cdot \\ \cdot & \cdot & \cdot \\ 1 & z_{n-1} & z_{n-2} \end{bmatrix} = \begin{bmatrix} 1 & y_2 & y_1 \\ 1 & y_3 & y_2 \\ \cdot & \cdot & \cdot \\ \cdot & \cdot & \cdot \\ \cdot & \cdot & \cdot \\ 1 & y_{94} & y_{93} \end{bmatrix}$$

$$= \begin{bmatrix} 1 & 27 & 25 \\ 1 & 33.5142 & 27 \\ \cdot & \cdot & \cdot \\ \cdot & \cdot & \cdot \\ \cdot & \cdot & \cdot \\ 1 & 32.3875 & 35.8242 \end{bmatrix}$$

Notice that these regression estimates do indeed differ somewhat from the estimates

$$\hat{\delta} = 26.34 \qquad \hat{\phi}_1 = .6857 \qquad \hat{\phi}_2 = -.4395$$

which are obtained using standard computer programs of the Box-Jenkins methodology.

11–4 MIXED AUTOREGRESSIVE–MOVING-AVERAGE MODELS

The model

$$z_t = \delta + \phi_1 z_{t-1} + \phi_2 z_{t-2} + \cdots + \phi_p z_{t-p}$$
$$- \theta_1 \epsilon_{t-1} - \theta_2 \epsilon_{t-2} - \cdots - \theta_q \epsilon_{t-q} + \epsilon_t$$

is called a mixed *autoregressive–moving-average model of order (p,q)*. It can be shown that the stationarity condition for the autoregressive process of order p and the invertibility condition for the moving-average process of order q are the respective stationarity and invertibility conditions for the mixed model of order (p,q). It can also be shown that the theoretical partial autocorrelation function of this mixed model dies down, and that the theoretical autocorrelation function also dies down.

It has been found that one useful mixed model is the *mixed autoregressive–moving-average model of order (1,1)* which is

$$z_t = \delta + \phi_1 z_{t-1} + \epsilon_t - \theta_1 \epsilon_{t-1}$$

This model is stationary if

$$|\phi_1| < 1$$

and is invertible if

$$|\theta_1| < 1$$

It can be shown that the mean of this model is

$$\mu = \frac{\delta}{1 - \phi_1}$$

that its theoretical partial autocorrelation function dies down in a fashion dominated by damped exponential decay, and that its theoretical autocorrelation function also dies down in a damped exponential fashion. In particular, it can be shown that

$$\rho_1 = \frac{(1 - \phi_1 \theta_1)(\phi_1 - \theta_1)}{1 + \theta_1^2 - 2\theta_1 \phi_1}$$

$$\rho_2 = \phi_1 \rho_1$$

$$\rho_k = \phi_1 \rho_{k-1} \qquad \text{for } k \geqslant 3$$

These relationships can be used to solve for the parameters ϕ_1 and θ_1 in terms of ρ_1 and ρ_2, which are estimated by r_1 and r_2, thus producing a set of preliminary estimates $\hat{\phi}_1$ and $\hat{\theta}_1$ that satisfy the stationarity and invertibility conditions

$$|\hat{\phi}_1| < 1 \qquad \text{and} \qquad |\hat{\theta}_1| < 1$$

Also, since the mean of the mixed model,

$$\mu = \frac{\delta}{1 - \phi_1}$$

implies that

$$\delta = \mu(1 - \phi_1)$$

a reasonable preliminary estimate of δ is

$$\hat{\delta} = \bar{z}(1 - \hat{\phi}_1)$$

where

$$\bar{z} = \frac{\displaystyle\sum_{t=a}^{n} z_t}{n - a + 1}$$

Example 11.4   Assume that for a particular observed time series $y_1, y_2, \ldots, y_n$, the sample autocorrelation function of the original time series values dies down extremely slowly. However, assume that the sample autocorrelation function of the first differences of these original values dies down fairly quickly in a damped exponential fashion, and that the sample partial autocorrelation function of these first differences also dies down fairly quickly in a fashion dominated by damped exponential decay. It is then reasonable to conclude that the first differences are stationary, and that the mixed model

$$z_t = \delta + \phi_1 z_{t-1} + \epsilon_t - \theta_1 \epsilon_{t-1}$$

has generated the first differences of the original time series. That is, letting

$$z_t = \nabla y_t = y_t - y_{t-1}$$

it is reasonable to conclude that the model

$$z_t = \delta + \phi_1 z_{t-1} + \epsilon_t - \theta_1 \epsilon_{t-1}$$

$$\nabla y_t = \delta + \phi_1 \nabla y_{t-1} + \epsilon_t - \theta_1 \epsilon_{t-1}$$

$$y_t - y_{t-1} = \delta + \phi_1(y_{t-1} - y_{t-2}) + \epsilon_t - \theta_1 \epsilon_{t-1}$$

$$y_t = \delta + (\phi_1 + 1)y_{t-1} - \phi_1 y_{t-2} + \epsilon_t - \theta_1 \epsilon_{t-1}$$

can be used to forecast future values of the original time series.

11–5 A SUMMARY OF NONSEASONAL BOX-JENKINS MODELS

Thus far in Chapter 11, we have discussed some of the common models used by the Box-Jenkins methodology to forecast time series that do not possess seasonal variation. Before proceeding to a discussion of models that can be used to forecast time series possessing seasonal variation, we will summarize the nonseasonal models and their properties. Table 11.1 summarizes the discussion of the general moving-average, autoregressive, and mixed autoregressive–moving-average models we have presented. The table indicates whether or not each model requires conditions for stationarity and invertibility, and includes a description of the behavior of the theoretical autocorrelation function and theoretical partial autocorrelation function for each model. Table 11.2 summarizes the discussion of the specific models presented to this point. The table gives the stationarity and invertibility conditions required by each model, and includes a description of the behavior of the theoretical autocorrelation and partial autocorrelation functions for each model. Note that use of .1 as the preliminary estimate of model parameters satisfies the stationarity and invertibility conditions for each of the models.

Examination of Tables 11.1 and 11.2 indicates that none of these basic models have *both* a theoretical autocorrelation function that cuts off *and* a theoretical partial autocorrelation function that cuts off. This fact is often of help in the identification of an appropriate Box-Jenkins model. For example, consider the identification of a Box-Jenkins model to be used in forecasting the Soak-All Paper Towels sales time series presented in Table 10.1 of Example 10.1. In Example 10.3 we concluded that the first-order moving-average model has generated the first differences of this time series. We reached this conclusion because, for the first differences of this time series, the theoretical autocorrelation function cuts off after

TABLE 11.1 *General Models*

Model	Theoretical Partial Autocorrelation Function	Theoretical Autocorrelation Function	Stationarity Conditions?	Invertibility Conditions?
Moving-average of order q $$z_t = \mu + \epsilon_t - \theta_1 \epsilon_{t-1} - \theta_2 \epsilon_{t-2} - \cdots - \theta_q \epsilon_{t-q}$$	Dies down	Cuts off after lag q	No	Yes
Autoregressive of order p $$z_t = \delta + \phi_1 z_{t-1} + \phi_2 z_{t-2} + \cdots + \phi_p z_{t-p} + \epsilon_t$$	Cuts off after lag p	Dies down	Yes	No
Mixed autoregressive–moving-average of order (p,q) $$z_t = \delta + \phi_1 z_{t-1} + \phi_2 z_{t-2} + \cdots + \phi_p z_{t-p} - \theta_1 \epsilon_{t-1} - \theta_2 \epsilon_{t-2} - \cdots - \theta_q \epsilon_{t-q} + \epsilon_t$$	Dies down	Dies down	Yes	Yes

TABLE 11.2   *Specific Models*

Model	Theoretical Partial Autocorrelation Function	Theoretical Autocorrelation Function	Stationarity Conditions	Invertibility Conditions
First-order moving-average $z_t = \mu + \epsilon_t - \theta_1 \epsilon_{t-1}$	Dies down in a fashion dominated by damped exponential decay	Cuts off after lag 1	None	$|\theta_1| < 1$
Second-order moving-average $z_t = \mu + \epsilon_t - \theta_1 \epsilon_{t-1} - \theta_2 \epsilon_{t-2}$	Dies down according to a mixture of damped exponentials and/or damped sine waves	Cuts off after lag 2	None	$\theta_1 + \theta_2 < 1$ $\theta_2 - \theta_1 < 1$ $|\theta_2| < 1$
First-order autoregressive $z_t = \delta + \phi_1 z_{t-1} + \epsilon_t$	Cuts off after lag 1	Dies down in a damped exponential fashion	$|\phi_1| < 1$	None
Second-order autoregressive $z_t = \delta + \phi_1 z_{t-1} + \phi_2 z_{t-2} + \epsilon_t$	Cuts off after lag 2	Dies down according to a mixture of damped exponentials and/or damped sine waves	$\phi_1 + \phi_2 < 1$ $\phi_2 - \phi_1 < 1$ $|\phi_2| < 1$	None
Mixed autoregressive–moving-average of order $(1,1)$ $z_t - \delta + \phi_1 z_{t-1} + \epsilon_t - \theta_1 \epsilon_{t-1}$	Dies down in a fashion dominated by damped exponential decay	Dies down in a damped exponential fashion	$|\phi_1| < 1$	$|\theta_1| < 1$

lag 1, and the theoretical partial autocorrelation function either cuts off after lag 1 or dies down. Since the theoretical autocorrelation function cuts off after lag 1, and since none of the basic Box-Jenkins models have both a theoretical autocorrelation function that cuts off and a theoretical partial autocorrelation function that cuts off, it is reasonable to assume that the theoretical partial autocorrelation function must die down. We conclude, therefore, that the first-order moving-average model has generated the first differences of this time series.

It should be noticed that in Table 11.1 we have summarized the fashions or patterns in which the autocorrelation functions and partial autocorrelation functions of the various models die down. Recall that the intuitive meanings of the expressions "damped exponential fashion" and "damped sine-wave fashion" are illustrated in Figure 10.8. Let us first consider the autoregressive process. The difference between the fashions in which the theoretical autocorrelation functions of the first-order autoregressive process and the second-order autoregressive process die down can sometimes be used to tell, without examining the theoretical partial autocorrelation function, whether a given observed time series has been generated by a first-order autoregressive process or a second-order autoregressive process. For example, it seems from Table 10.7 and Figure 10.6 that the sample autocorrelation function of the viscosity time series dies down in a damped sine-wave fashion. Since there is no first-order autoregressive process for which the theoretical autocorrelation function dies down in a damped sine-wave fashion, whereas there is such a second-order autoregressive process, it is reasonable to believe that a second-order autoregressive process has generated the viscosity time series.

Notice also that in Table 11.2 it is clear that both the theoretical autocorrelation function of a first-order autoregressive process and the theoretical autocorrelation function of a second-order autoregressive process can die down in a similar damped exponential fashion. For this reason, it is sometimes very difficult or impossible to tell the difference between a first-order autoregressive process and a second-order autoregressive process by looking at only the sample autocorrelation function. Hence, it is important to also look at the sample partial autocorrelation function and determine whether it seems to cut off after lag 1 or lag 2, as we did in Example 10.2 in identifying that a second-order autoregressive process generated the viscosity time series.

Similarly, the difference between the fashions in which the theoretical partial autocorrelation functions of the first-order moving-average process and the second-order moving-average process die down can sometimes be used to tell, without examining the theoretical autocorrelation function, whether a given observed time series has been generated by a first-order moving-average process or a second-order moving-average process. In Table 11.2 we say the theoretical autocorrelation function of the first-order autoregressive process dies down ''in a damped exponential fashion,'' and the theoretical partial autocorrelation function of a first-order moving-average process dies down ''in a fashion dominated by damped exponential decay.'' The word ''dominated'' implies that exponential decay may be difficult to recognize in the sample partial autocorrelation function. For example, consider the sample partial autocorrelation function of the first differences of the Soak-All Paper Towels sales time series, which we have identified as a first-order moving-average process. In Table 10.5 and Figure 10.4, where these differences are shown, exponential decay is somewhat difficult to recognize. Since, as expressed in Table 11.2, both the theoretical partial autocorrelation function of a first-order moving-average process and the theoretical partial autocorrelation function of a second-order moving-average process may die down in a similar damped exponential fashion, it is sometimes very difficult or impossible to tell the difference between a first-order and a second-order moving-average process by looking at only the sample partial autocorrelation function. Hence, it is important to also look at the sample autocorrelation function and determine whether it seems to cut off after lag 1 or lag 2, as we did in Example 10.1 in identifying that a first-order moving-average process generated the first differences of the Soak-All Paper Towels sales time series.

Notice that the models used by the Box-Jenkins methodology contain relatively few parameters. For example, none of the models summarized in Table 11.2

contain more than three parameters. In general, an important objective of the Box-Jenkins methodology is to develop *parsimonious models,* that is, models that adequately describe time series yet employ relatively few parameters. In general, if two different time series models seem to describe a particular observed time series equally well, and if one model contains fewer parameters than the other, then the time series modeler is probably well advised to choose the model with fewer parameters. One reason for choosing such a model is that it contains fewer parameters to estimate. An example of such a choice is given in Section 11–6, which discusses Box-Jenkins seasonal models.

11–6 SEASONAL MODELS

Box and Jenkins have developed models that are useful in describing and forecasting time series having seasonal variation. The general approach, which involves identification, estimation, diagnostic checking, and forecasting, is the same for seasonal models as it is for the nonseasonal models we have previously discussed. In this section we will discuss modeling seasonal time series using the Box-Jenkins methodology. Other discussions and examples of modeling seasonal time series are given in Box and Jenkins [1], Nelson [12], Johnson and Montgomery [8], Mabert [9], and Miller and Wichern [11].

To simplify our discussion of seasonal models, we need to become familiar with the symbol B, which is called the backshift operator and which shifts the subscript of a time series observation or error term backward in time by one period. That is,

$$By_t = y_{t-1}; \qquad \text{for example, } By_{50} = y_{49}$$

$$Bz_t = z_{t-1}; \qquad \text{for example, } Bz_{50} = z_{49}$$

$$B\epsilon_t = \epsilon_{t-1}; \qquad \text{for example, } B\epsilon_{50} = \epsilon_{49}$$

Next, the symbol B^k, which intuitively represents B raised to a power equal to k, shifts the subscript of a time series observation or error term backward in time by k periods. That is,

$$B^k y_t = y_{t-k}; \qquad \text{for example, } B^{12} y_{50} = y_{38}$$

$$B^k z_t = z_{t-k}; \qquad \text{for example, } B^{12} z_{50} = z_{38}$$

$$B^k \epsilon_t = \epsilon_{t-k}; \qquad \text{for example, } B^{12} \epsilon_{50} = \epsilon_{38}$$

In the rest of this section we will perform various algebraic manipulations with the backshift operator B. We will not discuss the theory justifying these manipulations, but the reader should accept the fact that the manipulations are legitimate.

As in the analysis of nonseasonal time series, the first step in analyzing a seasonal time series when using the Box-Jenkins methodology is to find a set of stationary time series values $z_a, z_{a+1}, \ldots, z_n$. Assume we have a set of n observed time series values $y_1, y_2, \ldots, y_n$. We should remember that if these values do not possess seasonal variation and if they are nonstationary, then we can usually transform them into stationary time series values $z_a, z_{a+1}, \ldots, z_n$ using either the transformation

$$z_t = \nabla y_t = y_t - y_{t-1}$$

or the transformation

$$z_t = \nabla^2 y_t = y_t - 2y_{t-1} + y_{t-2}$$

If the time series values $y_1, y_2, \ldots, y_n$ do possess seasonal variation, they may be nonstationary, and it is possible that one of the transformations $z_t = \nabla y_t$ or $z_t = \nabla^2 y_t$ will produce stationary time series values $z_a, z_{a+1}, \ldots, z_n$. However, frequently *seasonal differencing* is required to produce stationary time series values $z_a, z_{a+1}, \ldots, z_n$. To begin discussion of seasonal differencing, we express the *nonseasonal operator* ∇ as

$$\boxed{\nabla = 1 - B}$$

Thus

$$\nabla y_t = (1 - B)y_t = y_t - By_t = y_t - y_{t-1}$$

and

$$\begin{aligned}
\nabla^2 y_t = (1 - B)^2 y_t &= (1 - 2B + B^2)y_t \\
&= y_t - 2By_t + B^2 y_t \\
&= y_t - 2y_{t-1} + y_{t-2}
\end{aligned}$$

In general,

$$\boxed{\nabla^d y_t = (1 - B)^d y_t}$$

where d is any degree of nonseasonal differencing that may be required to produce stationary time series values. Next, let L be the number of seasons in a year. Thus,

$L = 12$ for monthly data and $L = 4$ for quarterly data. We define the *seasonal operator* ∇_L as

$$\nabla_L = (1 - B^L)$$

Thus

$$\nabla_L y_t = (1 - B^L)y_t = y_t - B^L y_t = y_t - y_{t-L}$$

That is,

$$\nabla_L y_t = y_t - y_{t-L}$$

Similarly,

$$\nabla_L^2 y_t = (1 - B^L)^2 y_t = (1 - 2B^L + B^{2L})y_t$$
$$= y_t - 2B^L y_t + B^{2L} y_t$$
$$= y_t - 2y_{t-L} + y_{t-2L}$$

That is,

$$\nabla_L^2 y_t = y_t - 2y_{t-L} + y_{t-2L}$$

In general

$$\nabla_L^D y_t = (1 - B^L)^D y_t$$

where D is any degree of seasonal differencing that may be required to produce stationary time series values.

In general, when using the Box-Jenkins methodology to analyze and forecast a time series that possesses seasonal variation, we should first plot the time series and determine whether its variability increases or decreases as time advances. If the variability of a seasonal time series is constant as time advances, that seasonal time series will appear to approximately possess additive seasonal variation. Hence, constant variability of a seasonal time series may be thought of as being approximately equivalent to additive seasonal variation. If the variability of a seasonal time series is changing (increasing or decreasing) as time advances, that seasonal time series will appear to approximately possess multiplicative seasonal variation. Hence, changing variability of a seasonal time series may be thought of as being approximately equivalent to multiplicative seasonal variation. If the seasonal time series values $y_1, y_2, \ldots, y_n$ seem to exhibit changing variability (multiplicative seasonal variation) as time advances, it is often desirable when using the Box-Jenkins methodology to use the transformation

$$y^*_t = \ln y_t$$

to produce values $y^*_1, y^*_2, \ldots, y^*_n$ which approximately possess constant variability (additive seasonal variation).

We will now discuss ways to transform the seasonal time series values $y^*_1, y^*_2, \ldots, y^*_n$ to stationary time series values $z_a, z_{a+1}, \ldots, z_n$. This discussion will assume that the logarithmic transformation $y^*_t = \ln y_t$ has been used. This is because business and economic time series possessing seasonal variation frequently have variability that changes as time advances (multiplicative seasonal variation), and thus the transformation $y^*_t = \ln y_t$ is frequently used. For time series possessing variability that remains constant as time advances (additive seasonal variation), the following discussion concerning transformations to achieve stationarity applies with y^*_t replaced by y_t. A general transformation that will usually produce stationary time series values is of the form

$$z_t = \nabla_L^D \nabla^d y^*_t = (1 - B^L)^D (1 - B)^d y^*_t$$

Before considering particular forms of this transformation, it should be realized that if the values $z_a, z_{a+1}, \ldots, z_n$ produced by any particular transformations have a sample autocorrelation function that cuts off or dies down fairly quickly, then they may be considered stationary; whereas if they have a sample autocorrelation function that dies down extremely slowly, then they should be considered nonstationary. The difference in meaning between the terms "dies down fairly quickly" and "dies down extremely slowly" is even more arbitrary for seasonal time series than for nonseasonal time series, and must be determined through experience. The sample autocorrelation function and the sample partial autocorrelation function are computed for original values and transformed values $z_a, z_{a+1}, \ldots, z_n$ of seasonal time series exactly as they are computed for nonseasonal time series. These calculations have been explained in Section 10–2.2. Moreover, the rules of thumb, given in Section 10–2.2, for determining when the theoretical autocorrelation function and the theoretical partial autocorrelation function cut off also apply to seasonal time series. In our examples of modeling nonseasonal time series, we computed the sample autocorrelation function and the sample partial autocorrelation function for lags from 1 to 12. One reason for calculating these functions only through a lag of 12 is that if a particular stationary time series does not possess seasonal variation, then, if either the theoretical autocorrelation function or the theoretical partial autocorrelation function cuts off, it will generally do so after a lag q that is less than or equal to 2. If a particular stationary time series does possess seasonal variation, then, if either the theoretical autocorrelation function or the

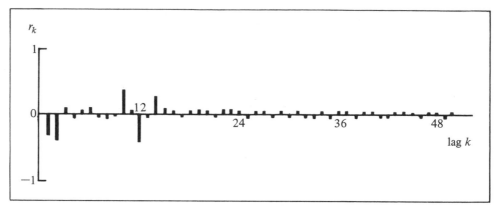

FIGURE 11.1   *A Sample Autocorrelation Function with Spikes at Lags 1, 2, 10, 12, and 14*

theoretical partial autocorrelation function cuts off, it will generally do so after a lag q that is less than or equal to $2L + 2$, and it will frequently do so after a lag q that is less than or equal to $L + 2$. It should be noted that if the theoretical autocorrelation function or theoretical partial autocorrelation function cuts off after lag $L + 2$ (or lag $2L + 2$), some of the theoretical autocorrelations or theoretical partial autocorrelations at lags less than $L + 2$ (or $2L + 2$) will be different from zero while others will equal zero. In this respect, when we say that a sample autocorrelation function (sample partial autocorrelation function) has spikes at a given set of lags, we mean that the sample autocorrelations (sample partial autocorrelations) at this set of lags are larger in absolute value than the sample autocorrelations (sample partial autocorrelations) at other lags. Spikes in the sample autocorrelation function (sample partial autocorrelation function) indicate nonzero autocorrelations (partial autocorrelations) in the theoretical autocorrelation function (theoretical partial autocorrelation function). For example, then, if we are dealing with monthly seasonal data ($L = 12$), and the sample autocorrelation function of a particular set of transformed values $z_a, z_{a+1}, \ldots, z_n$ has spikes at lags 1, 2, 10, 12, and 14, and seems to cut off after lag 14, as illustrated in Figure 11.1, it is reasonable to believe the transformed values $z_a, z_{a+1}, \ldots, z_n$ are stationary.

In order to determine the particular transformation of the form

$$z_t = \nabla_L^D \nabla^d y^*{}_t = (1 - B^L)^D (1 - B)^d y^*{}_t$$

that is required to produce stationary time series values of a seasonal time series, the sample autocorrelation function and the sample partial autocorrelation function are usually computed for all lags from 1 to $4L$. For most business and economic time series possessing seasonal variation, d is either 0 or 1 and D is either 0 or 1. If $d = 0$ and $D = 0$, then

$$z_t = \nabla_L^0 \nabla^0 y^*{}_t = (1 - B^L)^0 (1 - B)^0 y^*{}_t = y^*{}_t$$

That is,

$$z_t = \nabla_L^0 \nabla^0 y*_t = y*_t$$

Hence

$$z_a = z_1 = y*_1$$

$$z_{a+1} = z_2 = y*_2$$

$$\cdot$$
$$\cdot$$
$$\cdot$$

$$z_n = y*_n$$

and the transformed values are the logarithms of the original values. If the sample autocorrelation function of the values $y*_1, y*_2, \ldots, y*_n$ either cuts off or dies down fairly quickly, then the values $y*_1, y*_2, \ldots, y*_n$ should be considered stationary. Sometimes, for seasonal time series the sample autocorrelation function of the values $y*_1, y*_2, \ldots, y*_n$ has spikes at the "seasonal lags" L, $2L$, $3L$, and $4L$. In this case, the time series values should be considered nonstationary, even if, ignoring these spikes, the sample autocorrelation function dies down fairly quickly. Figure 11.2 illustrates a sample autocorrelation function that dies down very slowly and has spikes at lags L, $2L$, $3L$, and $4L$, where $L = 12$. Figure 11.3 illustrates a sample autocorrelation function that dies down fairly quickly and has spikes at lags L, $2L$, $3L$, and $4L$, where $L = 12$.

If the values $y*_1, y*_2, \ldots, y*_n$ are nonstationary, a differencing transformation of the form

$$z_t = \nabla_L^D \nabla^d y*_t = (1 - B^L)^D (1 - B)^d y*_t$$

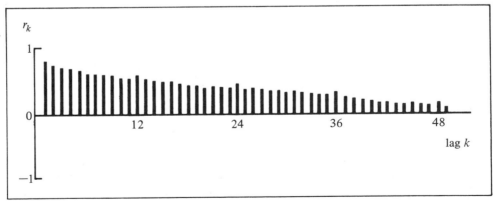

FIGURE 11.2   *A Sample Autocorrelation Function That Dies Down Very Slowly and Has Spikes at Lags 12, 24, 36, and 48*

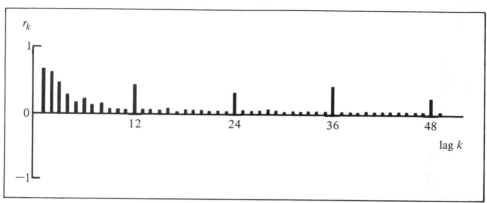

FIGURE 11.3   *A Sample Autocorrelation That Dies Down Fairly Quickly and Has Spikes at Lags 12, 24, 36 and 48*

with $d = 1$ and $D = 0$, or $d = 0$ and $D = 1$, or $d = 1$ and $D = 1$ should be considered. If we set $d = 1$ and $D = 0$, then

$$z_t = \nabla_L^0 \nabla^1 y^*_t = (1 - B^L)^0 (1 - B)^1 y^*_t$$
$$= (1 - B)y^*_t$$
$$= y^*_t - B y^*_t$$
$$= y^*_t - y^*_{t-1}$$

That is,

$$\boxed{z_t = \nabla_L^0 \nabla^1 y^*_t = y^*_t - y^*_{t-1}}$$

Hence

$$z_a = z_2 = y^*_2 - y^*_1$$
$$z_{a+1} = z_3 = y^*_3 - y^*_2$$
$$\cdot$$
$$\cdot$$
$$\cdot$$
$$z_n = y^*_n - y^*_{n-1}$$

If the sample autocorrelation function of these values either dies down or cuts off fairly quickly, then these values should be considered stationary. Sometimes, for seasonal time series the sample autocorrelation function of the above transformed values dies down quickly but has spikes at the seasonal lags L, $2L$, $3L$, and $4L$. Again, see Figure 11.3. This implies that the transformed values should be considered nonstationary. A transformation that will then frequently produce stationary time series values is obtained by setting $d = 1$ and $D = 1$, in which case

$$z_t = \nabla_L^1 \nabla^1 y^*_t = (1 - B^L)^1 (1 - B)^1 y^*_t$$
$$= (1 - B^L)[y^*_t - y^*_{t-1}]$$
$$= [y^*_t - y^*_{t-1}] - B^L(y^*_t - y^*_{t-1})$$
$$= y^*_t - y^*_{t-1} - B^L y^*_t + B^L y^*_{t-1}$$
$$= y^*_t - y^*_{t-1} - y^*_{t-L} + y^*_{t-1-L}$$

That is

$$\boxed{z_t = \nabla_L^1 \nabla^1 y^*_t = y^*_t - y^*_{t-1} - y^*_{t-L} + y^*_{t-1-L}}$$

Hence

$$z_a = z_{L+2} = y^*_{L+2} - y^*_{L+1} - y^*_2 + y^*_1$$
$$z_{a+1} = z_{L+3} = y^*_{L+3} - y^*_{L+2} - y^*_3 + y^*_2$$
$$.$$
$$.$$
$$.$$
$$z_n = y^*_n - y^*_{n-1} - y^*_{n-L} + y^*_{n-1-L}$$

A transformation that sometimes produces stationary time series values is obtained by setting $d = 0$ and $D = 1$, in which case

$$z_t = \nabla_L^1 \nabla^0 y^*_t = (1 - B^L)^1 (1 - B)^0 y^*_t$$
$$= (1 - B^L) y^*_t$$
$$= y^*_t - B^L y^*_t$$
$$= y^*_t - y^*_{t-L}$$

That is,

$$\boxed{z_t = \nabla_L^1 \nabla^0 y^*_t = y^*_t - y^*_{t-L}}$$

Hence

$$z_a = z_{L+1} = y^*_{L+1} - y^*_1$$
$$z_{a+1} = z_{L+2} = y^*_{L+2} - y^*_2$$
$$.$$
$$.$$
$$.$$
$$z_n = y^*_n - y^*_{n-L}$$

If the sample autocorrelation function of these values either dies down or cuts off fairly quickly, then these values should be considered stationary. Sometimes, for seasonal time series the sample autocorrelation function of the above transformed values does not have any spikes at the seasonal lags L, $2L$, $3L$, and $4L$ but dies down very slowly. A transformation that will then frequently produce stationary time series values is obtained by setting $d = 1$ and $D = 1$, in which case

$$z_t = \nabla_L^1 \nabla^1 y^*_t = (1 - B^L)^1 (1 - B)^1 y^*_t$$

This transformation has been previously discussed.

We will now consider two examples of deriving a set of stationary time series values z_a, z_{a+1}, . . . , z_n. The first example was motivated by a similar example discussed by Mabert [9]. The second example was motivated by a similar example discussed by Nelson [12].

Example 11.5   The Tri-State Natural Gas Company wishes to forecast future values of monthly natural gas consumption. The company will use monthly data ($L = 12$) on natural gas consumption in its market for the years 1960 to 1978 to develop a Box-Jenkins forecasting model.

When this historical data, which consists of 228 observations, is plotted, the natural gas consumption displays an upward trend and seasonal variation, with gas consumption being highest in the colder months of the year and lowest in the warmer months. The seasonal time series values y_1, y_2, . . . , y_{228} also seem to exhibit increasing variability (multiplicative seasonal variation) as time advances. Because of this increasing variability, the use of the transformation $y^*_t = \ln y_t$ is appropriate. Thus the values y^*_1, y^*_2, . . . , y^*_{228} are obtained. These values must now be transformed into stationary time series values z_a, z_{a+1}, . . . , z_n.

Several of the differencing transformations that might be considered are listed below.

$z_t = \nabla_{12}^0 \nabla^0 y^*_t$ $= y^*_t$	$z_t = \nabla_{12}^0 \nabla^1 y^*_t$ $= y^*_t - y^*_{t-1}$	$z_t = \nabla_{12}^1 \nabla^0 y^*_t$ $= y^*_t - y^*_{t-12}$	$z_t = \nabla_{12}^1 \nabla^1 y^*_t$ $= y^*_t - y^*_{t-1} - y^*_{t-12} + y^*_{t-1-12}$
$z_1 = y^*_1$ $z_2 = y^*_2$	$z_2 = y^*_2 - y^*_1$ $z_3 = y^*_3 - y^*_2$ . . .		
		$z_{13} = y^*_{13} - y^*_1$ $z_{14} = y^*_{14} - y^*_2$.	$z_{14} = y^*_{14} - y^*_{13} - y^*_2 + y^*_1$ $z_{15} = y^*_{15} - y^*_{14} - y^*_3 + y^*_2$.
$z_{228} = y^*_{228}$	$z_{228} = y^*_{228} - y^*_{227}$	$z_{228} = y^*_{228} - y^*_{216}$	$z_{228} = y^*_{228} - y^*_{227} - y^*_{216} + y^*_{215}$

Suppose that we apply these transformations to the values y^*_1, y^*_2, . . . , y^*_{228} and obtain the following results. The sample autocorrelation function of the values y^*_1, y^*_2, . . . , y^*_{228} dies down extremely slowly and has spikes at the "seasonal lags" 12, 24, 36, and 48. Because of this, it would seem that the values $y^*_1, y^*_2, \ldots, y^*_{228}$ are nonstationary. The sample autocorrelation function of the values $z_2, z_3, \ldots, z_{228}$ produced by the transformation

$$z_t = y^*_t - y^*_{t-1}$$

dies down fairly quickly but has spikes at the seasonal lags 12, 24, 36, and 48. Hence, it would seem that these values are also nonstationary. The sample autocorrelation function of the values $z_{13}, z_{14}, \ldots, z_{228}$ produced by the transformation

$$z_t = y^*_t - y^*_{t-12}$$

dies down fairly quickly in a sine-wave fashion and does not have any spikes. Because of this, it would seem that the values produced by this transformation might be stationary. Last, the sample autocorrelation function of the values $z_{14}, z_{15}, \ldots, z_{228}$ produced by the transformation

$$z_t = y^*_t - y^*_{t-1} - y^*_{t-12} + y^*_{t-1-12}$$

has spikes at lags 1 and 12 and cuts off after lag 12. Thus, it would seem that the values produced by this transformation might also be stationary.

In summary, we conclude that both the transformation

$$z_t = y^*_t - y^*_{t-12}$$

and the transformation

$$z_t = y^*_t - y^*_{t-1} - y^*_{t-12} + y^*_{t-13}$$

can be used to produce a set of stationary time series values $z_a, z_{a+1}, \ldots, z_n$.

Example 11.6

The State License Bureau wishes to forecast monthly registrations of new automobiles in Central County, a large metropolitan area. The Bureau will use monthly data $(L = 12)$ on registrations of new automobiles from 1968 to 1978 to develop a Box-Jenkins forecasting model.

When their historical data of 132 observations is plotted, the registration time series displays an upward trend and seasonal variation, with low numbers of registrations in the late summer during model change over time and high numbers of registrations in the late fall immediately after the introduction of new models. The seasonal time series values $y_1, y_2, \ldots, y_{132}$ also seem to exhibit increasing variability (multiplicative seasonal variation) as time advances. Because of this increasing variability, the use of the transformation $y^*_t = \ln y_t$ is appropriate. Thus the values $y^*_1, y^*_2, \ldots, y^*_{132}$ are obtained. These values must now be transformed into stationary time series values $z_a, z_{a+1}, \ldots, z_n$.

Suppose that we apply several transformations to the values $y^*_1, y^*_2, \ldots, y^*_{132}$ and obtain the following results. The sample autocorrelation function of the values $y^*{}_1, y^*{}_2, \ldots, y^*{}_{132}$ dies down extremely slowly and has spikes at the seasonal lags 12, 24, 36, and 48. Thus it would seem that these values are nonstationary. The sample autocorrelation function of the values produced by the transformation

$$z_t = y^*{}_t - y^*{}_{t-1}$$

dies down fairly quickly but has spikes at the seasonal lags 12, 24, 36, and 48. Hence, it would seem that these transformed values are also nonstationary. The sample autocorrelation function of the values produced by the transformation

$$z_t = y^*{}_t - y^*{}_{t-12}$$

does not have any spikes at lags 12, 24, 36, and 48 but dies down very slowly. Thus it would seem that these values are also nonstationary. Finally, the sample autocorrelation function of the values produced by the transformation

$$z_t = y^*{}_t - y^*{}_{t-1} - y^*{}_{t-12} + y^*{}_{t-1-12}$$

has spikes at lags 1, 2, 10, 12, and 14 and cuts off after lag 14. Because of this, it would seem that these values are stationary. So we conclude that the transformation

$$z_t = y^*{}_t - y^*{}_{t-1} - y^*{}_{t-12} + y^*{}_{t-13}$$

can be used to produce a set of stationary time series values $z_a, z_{a+1}, \ldots, z_n$.

Once the values of a time series possessing seasonal variation have been transformed to stationary time series values $z_a, z_{a+1}, \ldots, z_n$, we must identify the particular stationary time series model that can be assumed to have generated the values $z_a, z_{a+1}, \ldots, z_n$. To begin our discussion of identification, recall that in Section 11–4 we considered the mixed autoregressive–moving-average process of order (p, q), which is

$$z_t = \delta + \phi_1 z_{t-1} + \phi_2 z_{t-2} + \cdots + \phi_p z_{t-p}$$
$$-\theta_1 \epsilon_{t-1} - \theta_2 \epsilon_{t-2} - \cdots - \theta_q \epsilon_{t-q} + \epsilon_t$$

This is equivalent to

$$z_t - \phi_1 z_{t-1} - \phi_2 z_{t-2} - \cdots - \phi_p z_{t-p}$$
$$= \delta + \epsilon_t - \theta_1 \epsilon_{t-1} - \theta_2 \epsilon_{t-2} - \cdots - \theta_q \epsilon_{t-q}$$

or

$$z_t - \phi_1 B z_t - \phi_2 B^2 z_t - \cdots - \phi_p B^p z_t$$
$$= \delta + \epsilon_t - \theta_1 B \epsilon_t - \theta_2 B^2 \epsilon_t - \cdots - \theta_q B^q \epsilon_t$$

or

$$(1 - \phi_1 B - \phi_2 B^2 - \cdots - \phi_p B^p)z_t$$
$$= \delta + (1 - \theta_1 B - \theta_2 B^2 - \cdots - \theta_q B^q)\epsilon_t$$

In this representation we call

$$\phi_p(B) = (1 - \phi_1 B - \phi_2 B^2 - \cdots - \phi_p B^p)$$

the *nonseasonal autoregressive operator of order p,* and we call

$$\theta_q(B) = (1 - \theta_1 B - \theta_2 B^2 - \cdots - \theta_q B^q)$$

the *nonseasonal moving-average operator of order q.*

In order to model stationary time series possessing seasonal variation, it is useful to utilize

$$\phi_P(B^L) = (1 - \phi_{1,L} B^L - \phi_{2,L} B^{2L} - \cdots - \phi_{P,L} B^{PL})$$

which is called the *seasonal autoregressive operator of order P,* and

$$\theta_Q(B^L) = (1 - \theta_{1,L} B^L - \theta_{2,L} B^{2L} - \cdots - \theta_{Q,L} B^{QL})$$

which is called the *seasonal moving-average operator of order Q.*

The Box-Jenkins methodology then models stationary time series values z_a, $z_{a+1}, \ldots, z_n$ possessing seasonal variation by combining the above four operators to obtain the *general multiplicative seasonal model,* which is

$$\phi_p(B)\phi_P(B^L)z_t = \delta + \theta_q(B)\theta_Q(B^L)\epsilon_t$$

or

$$(1 - \phi_1 B - \phi_2 B^2 - \cdots - \phi_p B^p) \times (1 - \phi_{1,L} B^L - \phi_{2,L} B^{2L} - \cdots - \phi_{P,L} B^{PL})z_t$$
$$= \delta + (1 - \theta_1 B - \theta_2 B^2 - \cdots - \theta_q B^q) \times (1 - \theta_{1,L} B^L - \theta_{2,L} B^{2L} - \cdots - \theta_{Q,L} B^{QL})\epsilon_t$$

This model is sometimes called the mixed autoregressive–moving-average model of order (p,P,q,Q). It should be noted that this general model is analogous to the mixed autoregressive–moving-average process of order (p,q), with seasonal operators also included in the model.

Identification of the particular special form of this model that adequately describes a particular stationary time series z_a, z_{a+1}, . . . , z_n possessing seasonal variation involves determining which of the operators $\phi_P(B^L)$, $\phi_p(B)$, $\theta_Q(B^L)$, and $\theta_q(B)$ should be included in the model and the order of each operator that is included. This identification can theoretically be done by comparison of the sample autocorrelation function and the sample partial autocorrelation function of the particular observed stationary time series with the theoretical autocorrelation functions and the theoretical partial autocorrelation functions of various special forms of the general multiplicative seasonal model. Box and Jenkins [1] present a summary of the theoretical autocorrelation functions for several special forms, and this summary is useful in model identification. In practice, however, comparison of the sample and theoretical autocorrelation and partial autocorrelation functions can be difficult. For this reason the approach we will use to develop an appropriate Box-Jenkins seasonal model for a particular seasonal time series is somewhat intuitive and sometimes involves performing several iterations of the modeling process. At each stage we will use the sample autocorrelation and sample partial autocorrelation functions of the observed time series to identify the most obvious components of a suitable model; and then we will use the analysis of the residuals from the fitted model to develop a more appropriate model, if the analysis of these residuals indicates that another model is more appropriate. Frequently, fairly simple special forms of the general multiplicative seasonal model can be used to adequately represent stationary seasonal time series.

We will now attempt to give some general guidelines for identifying which of the components $\phi_P(B^L)$, $\phi_p(B)$, $\theta_Q(B^L)$, and $\theta_q(B)$ should be used in the general multiplicative seasonal model to adequately represent a particular observed stationary time series.

First, if the sample autocorrelation function of the observed stationary time series dies down fairly quickly, and the sample partial autocorrelation function cuts off after a lag that is substantially less than L (perhaps less than or equal to 2), then use of some form of the nonseasonal autoregressive operator

$$\phi_p(B) = (1 - \phi_1 B - \phi_2 B^2 - \cdots - \phi_p B^p)$$

might be appropriate. As will be illustrated in the examples to follow, *the lag at which the sample partial autocorrelation function cuts off suggests the order p of the nonseasonal autoregressive operator that should be used.*

Second, if the sample autocorrelation function of the observed stationary time series dies down fairly quickly, and the sample partial autocorrelation function cuts off after a lag that is nearly equal to L, $L + 1$, $L + 2$, $2L$, $2L + 1$, or $2L + 2$, then use of some form of the seasonal autoregressive operator

$$\phi_P(B^L) = (1 - \phi_{1,L} B^L - \phi_{2,L} B^{2L} - \cdots - \phi_{P,L} B^{PL})$$

might be appropriate if no spikes exist at lags substantially smaller than L (perhaps at lags 1 and/or 2). However, if such spikes do exist, use of some form of the product of the nonseasonal autoregressive operator and the seasonal autoregressive operator

$$\phi_p(B)\phi_P(B^L) = (1 - \phi_1 B - \phi_2 B^2 - \cdots - \phi_p B^p)$$
$$\times (1 - \phi_{1,L}B^L - \phi_{2,L}B^{2L} - \cdots - \phi_{P,L}B^{PL})$$

might be appropriate. As will be illustrated in the examples to follow, *the lags at which spikes occur in the sample partial autocorrelation function help to suggest the order P of the seasonal autoregressive operator or the orders of the operators in the product of operators that should be used.*

Third, if the sample partial autocorrelation function of the observed stationary time series dies down fairly quickly, and the sample autocorrelation function cuts off after a lag that is substantially less than L (perhaps less than or equal to 2), then use of some form of the nonseasonal moving-average operator

$$\theta_q(B) = (1 - \theta_1 B - \theta_2 B^2 - \cdots - \theta_q B^q)$$

might be appropriate. As will be illustrated in the examples to follow, *the lag at which the sample autocorrelation function cuts off suggests the order q of the nonseasonal moving-average operator that should be used.*

Fourth, if the sample partial autocorrelation function of the observed stationary time series dies down fairly quickly, and the sample autocorrelation function cuts off after a lag that is nearly equal to $L, L + 1, L + 2, 2L, 2L + 1,$ or $2L + 2$, then use of some form of the seasonal moving-average operator

$$\theta_Q(B^L) = (1 - \theta_{1,L}B^L - \theta_{2,L}B^{2L} - \cdots - \theta_{Q,L}B^{QL})$$

might be appropriate if no spikes exist at lags substantially smaller than L (perhaps at lags 1 and/or 2). However, if such spikes do exist, use of some form of the product of the nonseasonal moving-average operator and the seasonal moving-average operator

$$\theta_q(B)\theta_Q(B^L) = (1 - \theta_1 B - \theta_2 B^2 - \cdots - \theta_q B^q)$$
$$\times (1 - \theta_{1,L}B^L - \theta_{2,L}B^{2L} - \cdots - \theta_{Q,L}B^{QL})$$

might be appropriate. As will be illustrated in the examples to follow, *the lags at which spikes occur in the sample autocorrelation function help to suggest the order Q of the seasonal moving-average operator or the orders of the operators used in the product of operators that should be used.*

Fifth, if both the sample autocorrelation function and the sample partial autocorrelation function die down fairly quickly, then use of some form of both of the operators $\theta_q(B)\theta_Q(B^L)$ and $\phi_p(B)\phi_P(B^L)$ might be appropriate, although the form probably would be quite simple.

It should be realized that the guidelines just given do not represent a totally systematic method for identifying the obvious components of the general multiplicative seasonal model. Sometimes modified forms of the operators and/or model just discussed might be appropriate in a given situation. For example, consider a situation in which the sample autocorrelation function for a stationary time series can be interpreted as cutting off after a lag nearly equal to L, $L + 1$, $L + 2$, $2L$, $2L + 1$, or $2L + 2$—which suggests that some form of the seasonal moving-average operator should be used—and can also be interpreted as dying down fairly quickly—which suggests that some form of the nonseasonal autoregressive operator might be appropriate. Then, since both patterns seem to be present, use of some form of the general multiplicative model involving both the seasonal moving-average operator and the nonseasonal autoregressive operator might be appropriate. Also, as we have said, it is very possible that the forecaster will have to perform several iterations of the modeling process. The reader will gain insight into the entire modeling process when we present examples of identifying seasonal time series models. Before proceeding to these examples, however, we will briefly discuss estimation, diagnostic checking, and forecasting when using Box-Jenkins seasonal time series models.

Once the original values y_1, y_2, $\ldots$, y_n of a seasonal time series have been transformed to stationary time series values z_a, z_{a+1}, $\ldots$, z_n, and once an appropriate particular form of the general multiplicative seasonal model

$$\phi_p(B)\phi_P(B^L)z_t = \delta + \theta_q(B)\theta_Q(B^L)\epsilon_t$$

has been tentatively identified as the model that has generated z_a, z_{a+1}, $\ldots$, z_n, we need to estimate the parameters in the tentatively identified model. Stationarity and invertibility conditions exist for each special form of the general multiplicative seasonal model. The reader is referred to Box and Jenkins [1] for discussion of ways to determine these conditions. It can be shown that use of .1 as the preliminary estimate of each and every parameter contained in the operators $\phi_P(B^L)$, $\phi_p(B)$, $\theta_Q(B^L)$, and $\theta_q(B)$ will generally satisfy the stationarity and invertibility conditions of the particular model under consideration. It can also be shown that if μ is the mean of the general multiplicative seasonal model, then

$$\delta = (1 - \phi_1 - \phi_2 - \cdots - \phi_p)(1 - \phi_{1,L} - \phi_{2,L} - \cdots - \phi_{P,L})\mu$$

The preliminary estimate of μ is

$$\bar{z} = \frac{\displaystyle\sum_{t=a}^{n} z_t}{n - a + 1}$$

Thus, the preliminary estimate of δ is

$$\hat{\delta} = (1 - \hat{\phi}_1 - \hat{\phi}_2 - \cdots - \hat{\phi}_p)(1 - \hat{\phi}_{1,L} - \hat{\phi}_{2,L} - \cdots - \hat{\phi}_{P,L})\bar{z}$$

where $\hat{\phi}_1$, $\hat{\phi}_2$, . . . , $\hat{\phi}_p$, $\hat{\phi}_{1,L}$, $\hat{\phi}_{2,L}$, . . . , $\hat{\phi}_{P,L}$ are the preliminary estimates of, respectively, $\phi_1, \phi_2, \ldots, \phi_p, \phi_{1,L}, \phi_{2,L}, \ldots, \phi_{P,L}$. Remember, it generally suffices to set each of these preliminary estimates equal to .1. Of course, if $\bar{z}$ is small relative to $z_a, z_{a+1}, \ldots, z_n$, then we assume that μ is 0 and hence that δ is 0. Note that if neither of the autoregressive operators $\phi_P(B^L)$ or $\phi_p(B)$ is involved in the special form of the general multiplicative seasonal model under consideration, then

$$\phi_1 = \phi_2 = \cdots = \phi_p = \phi_{1,L} = \phi_{2,L} = \cdots = \phi_{P,L} = 0$$

and hence

$$\delta = \mu$$

Once preliminary estimates of unknown parameters have been obtained, they are used as inputs to computer programs that produce final least squares estimates of the unknown parameters.

After a tentative model has been fit to the data, it is important to perform diagnostic checks to test the adequacy of the model and, if need be, to suggest potential improvements. As explained in Section 10–4, this can be accomplished through the calculation of residual autocorrelations and the quantity Q. For seasonal time series, it is customary to choose K, the number of residual autocorrelations used in the calculation of Q, to be $3L$ (or 36 if $L = 12$). In the examples to follow we will discuss the use of residual autocorrelations to improve a tentatively identified model.

Once an appropriate model has been adopted, it is used to forecast future values of the time series. Forecasting is accomplished for seasonal time series using the principles discussed in Section 10–5. We will discuss forecasting future values of a seasonal time series in Example 11.7, which follows.

Example
11.7   Consider Example 11.5 in which the Tri-State Natural Gas Company wishes to forecast natural gas consumption. We have observed that the sample autocorrelation function of the values $z_{13}, z_{14}, \ldots, z_{228}$ produced by the transformation

$$z_t = y^*{}_t - y^*{}_{t-12}$$

dies down fairly quickly in a sine-wave fashion and does not have any spikes. We have also observed that the sample autocorrelation function of the values $z_{14}, z_{15},$ . . . , z_{228} produced by the transformation

$$z_t = y^*{}_t - y^*{}_{t-1} - y^*{}_{t-12} + y^*{}_{t-13}$$

has spikes at lags 1 and 12 and cuts off after lag 12. Because of these observations we have concluded that both of these transformations can be used to produce a set of stationary time series values $z_a, z_{a+1}, \ldots, z_n$.

Let us first consider the values $z_{13}, z_{14}, \ldots, z_{228}$ produced by the transformation $z_t = y^*_t - y^*_{t-12}$. Suppose that in addition to the fact that the sample autocorrelation function of these values dies down in a sine-wave fashion, the sample partial autocorrelation function has spikes at lags 1 and 2 and cuts off after lag 2. It would seem from our guidelines that the nonseasonal autoregressive operator $\phi_p(B)$ in the form

$$\phi_2(B) = 1 - \phi_1 B - \phi_2 B^2$$

should be used as a component in the general multiplicative seasonal model. Note that an autoregressive operator of order 2 is used because the sample partial autocorrelation function cuts off after lag 2 (refer to Table 11.2). Suppose that we fit the model

$$(1 - \phi_1 B - \phi_2 B^2)z_t = \epsilon_t$$

to the stationary values $z_{13}, z_{14}, \ldots, z_{228}$ given by the transformation $z_t = y^*_t - y^*_{t-12}$. We have set δ equal to zero in this model because we are assuming that

$$\bar{z} = \frac{\sum\limits_{t=13}^{228} z_t}{216}$$

is small relative to the values $z_{13}, z_{14}, \ldots, z_{228}$ produced by the transformation. Diagnostic checks are performed by calculating the autocorrelations of the residuals obtained from the fitted model and by calculating the quantity Q. Suppose that this procedure indicates that the model is inadequate. We now observe that the autocorrelation function of the residuals has a spike at lag 12 and hence identify a new model to be

$$(1 - \phi_1 B - \phi_2 B^2)z_t = (1 - \theta_{1,12} B^{12})\epsilon_t$$

We identify this model because the spike at lag 12 in the autocorrelation function of the residuals indicates that the residuals, which essentially represent the error terms, are related. In particular, the residuals separated by 12 time periods are related. This suggests that a moving-average operator of order 1 relating error terms separated by a lag of 12 time periods should be included in the model. When this model is fit to the data, assume that diagnostic checks indicate that the model is adequate, and hence it is reasonable to believe that this model has generated the values produced by the transformation

$$z_t = y^*_t - y^*_{t-12}$$

In summary, then, since

$$
\begin{aligned}
(1 - \phi_1 B - \phi_2 B^2)z_t &= (1 - \phi_1 B - \phi_2 B^2)(y^*_t - y^*_{t-12}) \\
&= (1 - \phi_1 B - \phi_2 B^2)y^*_t - (1 - \phi_1 B - \phi_2 B^2)y^*_{t-12} \\
&= y^*_t - \phi_1 B y^*_t - \phi_2 B^2 y^*_t - [y^*_{t-12} - \phi_1 B y^*_{t-12} - \phi_2 B^2 y^*_{t-12}] \\
&= y^*_t - \phi_1 y^*_{t-1} - \phi_2 y^*_{t-2} - y^*_{t-12} + \phi_1 y^*_{t-13} + \phi_2 y^*_{t-14}
\end{aligned}
$$

and

$$
\begin{aligned}
(1 - \theta_{1,12} B^{12})\epsilon_t &= \epsilon_t - \theta_{1,12} B^{12}\epsilon_t \\
&= \epsilon_t - \theta_{1,12}\epsilon_{t-12}
\end{aligned}
$$

it is reasonable to believe that the time series model

$$
y^*_t - \phi_1 y^*_{t-1} - \phi_2 y^*_{t-2} - y^*_{t-12} + \phi_1 y^*_{t-13} + \phi_2 y^*_{t-14} = \epsilon_t - \theta_{1,12}\epsilon_{t-12}
$$

or

$$
y^*_t = \phi_1 y^*_{t-1} + \phi_2 y^*_{t-2} + y^*_{t-12} - \phi_1 y^*_{t-13} - \phi_2 y^*_{t-14} + \epsilon_t - \theta_{1,12}\epsilon_{t-12}
$$

adequately describes the time series values $y^*_1, y^*_2, \ldots, y^*_{228}$.

Let us next consider the values $z_{14}, z_{15}, \ldots, z_{228}$ produced by the transformation $z_t = y^*_t - y^*_{t-1} - y^*_{t-12} + y^*_{t-13}$. Suppose that in addition to the fact that the sample autocorrelation function has spikes at lags 1 and 12 and cuts off after lag 12, the sample partial autocorrelation function dies down quickly. It would seem from our guidelines that the product of the nonseasonal moving-average operator $\theta_q(B)$ and the seasonal moving-average operator $\theta_Q(B^L)$ should be used as a component in the general multiplicative seasonal model. The spike at lag 1 in the sample autocorrelation function suggests that a nonseasonal moving-average operator of order 1 be used, while the spike at lag 12 suggests that a seasonal moving-average operator of order 1 be used. We will, therefore, use the product of these operators, which is

$$
\theta_1(B)\theta_1(B^{12}) = (1 - \theta_1 B)(1 - \theta_{1,12} B^{12})
$$

Assume that the model

$$
z_t = (1 - \theta_1 B)(1 - \theta_{1,12} B^{12})\epsilon_t
$$

is fit to the stationary values $z_{14}, z_{15}, \ldots, z_{228}$ given by the transformation

$$
z_t = y^*_t - y^*_{t-1} - y^*_{t-12} + y^*_{t-13}
$$

and that diagnostic checks indicate that the model is adequate. We have set δ equal to 0 in this model because we are assuming that

$$\bar{z} = \frac{\sum_{t=14}^{228} z_t}{215}$$

is small relative to the values produced by the transformation.

To summarize, then, we find that both the model

$$(1 - \phi_1 B - \phi_2 B^2) z_t = (1 - \theta_{1,12} B^{12}) \epsilon_t$$

where

$$z_t = y^*_t - y^*_{t-1}$$

and the model

$$z_t = (1 - \theta_1 B)(1 - \theta_{1,12} B^{12}) \epsilon_t$$

where

$$z_t - y^*_t - y^*_{t-1} - y^*_{t-12} + y^*_{t-13}$$

adequately represent the natural gas consumption time series.

Let us now consider choosing which of the above two models might be used to forecast future values of natural gas consumption. Suppose that the mean squared error for both of these models is nearly the same. Then the Tri-State Natural Gas Company should probably choose the most parsimonious model—the model with the fewest parameters. That is, since the first model contains three parameters, ϕ_1, ϕ_2, and $\theta_{1,12}$, and the second model contains two parameters, θ_1 and $\theta_{1,12}$, the company should probably choose the second model to forecast future values of natural gas consumption.

We will now demonstrate how the second model

$$z_t = (1 - \theta_1 B)(1 - \theta_{1,12} B^{12}) \epsilon_t$$

where

$$z_t = y^*_t - y^*_{t-1} - y^*_{t-12} + y^*_{t-13}$$

can be used to forecast future values of natural gas consumption. Since

$$(1 - \theta_1 B)(1 - \theta_{1,12} B^{12}) \epsilon_t = (1 - \theta_1 B)[\epsilon_t - \theta_{1,12} B^{12} \epsilon_t]$$

$$= (1 - \theta_1 B)[\epsilon_t - \theta_{1,12} \epsilon_{t-12}]$$

$$= \epsilon_t - \theta_{1,12}\epsilon_{t-12} - \theta_1 B\epsilon_t$$
$$+ \theta_1\theta_{1,12}B\epsilon_{t-12}$$
$$= \epsilon_t - \theta_{1,12}\epsilon_{t-12} - \theta_1\epsilon_{t-1}$$
$$+ \theta_1\theta_{1,12}\epsilon_{t-13}$$

it follows that the second model can be written in the form

$$y^*_t - y^*_{t-1} - y^*_{t-12} + y^*_{t-13} = \epsilon_t - \theta_{1,12}\epsilon_{t-12} - \theta_1\epsilon_{t-1} + \theta_1\theta_{1,12}\epsilon_{t-13}$$

or

$$y^*_t = y^*_{t-1} + y^*_{t-12} - y^*_{t-13} + \epsilon_t - \theta_1\epsilon_{t-1} - \theta_{1,12}\epsilon_{t-12} + \theta_1\theta_{1,12}\epsilon_{t-13}$$

The Tri-State Natural Gas Company finds that the least squares estimates of θ_1 and $\theta_{1,12}$ are, respectively,

$$\hat{\theta}_1 = .258 \quad \text{and} \quad \hat{\theta}_{1,12} = .721$$

Hence, the equation that can be used to forecast future values of natural gas consumption is

$$y^*_t = y^*_{t-1} + y^*_{t-12} - y^*_{t-13} + \epsilon_t - \hat{\theta}_1\epsilon_{t-1} - \hat{\theta}_{1,12}\epsilon_{t-12} + \hat{\theta}_1\hat{\theta}_{1,12}\epsilon_{t-13}$$
$$= y^*_{t-1} + y^*_{t-12} - y^*_{t-13} + \epsilon_t - .258\epsilon_{t-1} - .721\epsilon_{t-12} + .186\epsilon_{t-13}$$

Forecasting of future values of y^*_t is done using the principles discussed in Section 10–5. However, remembering that

$$y^*_t = \ln y_t \quad \text{and hence that} \quad y_t = \exp(y^*_t)$$

where y_t represents the actual natural gas consumption in period t, it follows that if $\hat{y}^*_{T+t}(T)$ is the point forecast of y^*_{T+t} made at time origin T, then the point forecast of y_{T+t} made at time origin T is

$$\hat{y}_{T+t}(T) = \exp[\hat{y}^*_{T+t}(T)]$$

Confidence interval forecasts can also be computed for y^*_t. The calculations are done by computer programs of the Box-Jenkins methodology. If $[a,b]$ is a confidence interval for y^*_t, then

$$[\exp^a, \exp^b]$$

is a confidence interval for y_t.

Since the historical data consists of 228 observations, we will first consider making forecasts for future values of $y*_t$ at time origin 228, which is December of 1978. Using the equation

$$y*_t = y*_{t-1} + y*_{t-12} - y*_{t-13} + \epsilon_t - .258\epsilon_{t-1} - .721\epsilon_{t-12} + .186\epsilon_{t-13}$$

a forecast of $y*_{229}$ is

$$\hat{y}*_{229}(228) = y*_{228} + y*_{217} - y*_{216} + \epsilon_{229} - .258\epsilon_{228} - .721\epsilon_{217} + .186\epsilon_{216}$$

In making this forecast, the actual observed historical values $y*_{228}$, $y*_{217}$, and $y*_{216}$ are used. We set ϵ_{229} equal to 0 since $y*_{229}$ has not been observed. Since $y*_{228}$, $y*_{217}$, and $y*_{216}$ have been observed and since the one-period-ahead forecast errors are not available, we estimate ϵ_{228} by the residual $(y*_{228} - \hat{y}*_{228})$, ϵ_{217} by the residual $(y*_{217} - \hat{y}*_{217})$ and ϵ_{216} by the residual $(y*_{216} - \hat{y}*_{216})$, where $\hat{y}*_{228}$, $\hat{y}*_{217}$, and $\hat{y}*_{216}$ are the predicted values, obtained from fitting the model to the relevant observed historical data. A forecast of $y*_{230}$ is

$$y*_{230}(228) = y*_{229} + y*_{218} - y*_{217} + \epsilon_{230} - .258\epsilon_{229} - .721\epsilon_{218} + .186\epsilon_{217}$$

In making this forecast, the actual observed historical values $y*_{218}$ and $y*_{217}$ are used, but since $y*_{229}$ has not been observed, we estimate $y*_{229}$ by $\hat{y}*_{229}(228)$, which is the previously made forecast of $y*_{229}$. We set both ϵ_{230} and ϵ_{229} equal to 0, since $y*_{230}$ and $y*_{229}$ have not been observed. Since $y*_{218}$ and $y*_{217}$ have been observed and since the one-period-ahead forecast errors are not available, we estimate ϵ_{218} by the residual $(y*_{218} - \hat{y}*_{218})$ and ϵ_{217} by the residual $(y*_{217} - \hat{y}*_{217})$, where $\hat{y}*_{218}$ and $\hat{y}*_{217}$ are the predicted values, obtained from fitting the model to the observed historical data, of the historical values $y*_{218}$ and $y*_{217}$. Forecasts of other future values of $y*_t$ are made in a similar manner.

Assume now that y_{229}, the actual natural gas consumption in period 229, is observed. Then, although the only way to update the estimates of the model parameters θ_1 and $\theta_{1,12}$ is to refit the entire model, we can obtain new forecasts using y_{229} and the current estimates of θ_1 and $\theta_{1,12}$. Since $y*_{229} = \ln y_{229}$, a forecast of $y*_{230}$ made at time origin 229, which is January of 1979, is

$$y*_{230}(229) = y*_{229} + y*_{218} - y*_{217} + \epsilon_{230} - .258\epsilon_{229} - .721\epsilon_{218} + .186\epsilon_{217}$$

In making this forecast, the actual observed historical values $y*_{229}$, $y*_{218}$, and $y*_{217}$ are used. We set ϵ_{230} equal to 0 since $y*_{230}$ has not been observed. Since $y*_{229}$ has been observed and since the one-period-ahead forecast error

$$y*_{229} - \hat{y}*_{229}(228)$$

is available, we estimate ϵ_{229} by this one-period-ahead forecast error. Since $y*_{218}$ and $y*_{217}$ have been observed and since the one-period-ahead forecast errors are not available, we estimate ϵ_{218} by the residual $(y*_{218} - \hat{y}*_{218})$ and ϵ_{217} by the residual $(y*_{217} - \hat{y}*_{217})$, where $\hat{y}*_{218}$ and $\hat{y}*_{217}$ are the predicted values, obtained from the fit made at time origin 228 of the model to the historical data, of, respectively, the historical values y''_{218} and y''_{217}.

Before proceeding to the next example, note that the model

$$z_t = (1 - \theta_1 B)(1 - \theta_{1,12} B^{12})\epsilon_t$$

where

$$z_t = y^*_t - y^*_{t-1} - y^*_{t-12} + y^*_{t-13}$$

used in this example to forecast natural gas consumption has been used frequently in the forecasting literature to forecast future values of seasonal time series. For example, Box and Jenkins [1] use this model to forecast monthly totals of international airline passengers.

As an added note to illustrate the use of the autocorrelation function of the residuals obtained after a tentative Box-Jenkins seasonal model has been fitted to a stationary time series, consider fitting the above model to a different time series. Suppose that the autocorrelation function of the residuals for the tentative model exhibits a spike at lag 6. The spike indicates that the residuals separated by 6 time periods are related. This suggests that a "modified" moving-average operator of the form

$$(1 - \theta_1 B)(1 - \theta_{1,6} B^6 - \theta_{1,12} B^{12})\epsilon_t$$

be used. Thus we would fit the model

$$z_t = (1 - \theta_1 B)(1 - \theta_{1,6} B^6 - \theta_{1,12} B^{12})\epsilon_t$$

where

$$z_t = y^*_t - y^*_{t-1} - y^*_{t-12} + y^*_{t-13}$$

to the stationary time series. The use of such modified operators is common when using seasonal Box-Jenkins models.

Example   Consider Example 11.6 in which the State License Bureau wishes to forecast
11.8   monthly registrations of new automobiles in Central County. We have observed that the sample autocorrelation function of the values $z_{14}, z_{15}, \ldots, z_{132}$ produced by the transformation

$$z_t = y^*_t - y^*_{t-1} - y^*_{t-12} + y^*_{t-13}$$

has spikes at lags 1, 2, 10, 12, and 14 and cuts off after lag 14. Because of this observation we have concluded that this transformation can be used to produce a set of stationary time series values.

Suppose that in addition to the behavior of the sample autocorrelation function, we observe that the sample partial autocorrelation function dies down quickly. It would seem from our guidelines that the product of the nonseasonal moving-average operator $\theta_q(B)$ and the seasonal moving-average operator $\theta_Q(B^L)$ should be used as a component in the general multiplicative seasonal model. The spikes at lags 1 and 2 in the sample autocorrelation function suggest that a nonseasonal moving-average operator of order 2 be used, while the spikes at lags 10, 12, and 14 suggest that a seasonal moving-average operator of order 1 be used. Hence, we will use the product of these operators, which is

$$\theta_2(B)\theta_1(B^{12}) = (1 - \theta_1 B - \theta_2 B^2)(1 - \theta_{1,12}B^{12})$$

Assume that the model

$$z_t = (1 - \theta_1 B - \theta_2 B^2)(1 - \theta_{1,12}B^{12})\epsilon_t$$

is fit to the stationary values $z_{14}, z_{15}, \ldots, z_{132}$ given by the transformation

$$z_t = y^*{}_t - y^*{}_{t-1} - y^*{}_{t-12} + y^*{}_{t-13}$$

and assume that diagnostic checks indicate that the model is adequate. We have set δ equal to 0 in this model because we are assuming that

$$\bar{z} = \frac{\displaystyle\sum_{t=14}^{132} z_t}{119}$$

is small relative to the values $z_{14}, z_{15}, \ldots, z_{132}$ produced by the transformation $z_t = y^*{}_t - y^*{}_{t-1} - y^*{}_{t-12} + y^*{}_{t-13}$.

Since

$$(1 - \theta_1 B - \theta_2 B^2)(1 - \theta_{1,12}B^{12})\epsilon_t = (1 - \theta_1 B - \theta_2 B^2)[\epsilon_t - \theta_{1,12}B^{12}\epsilon_t]$$

$$= (1 - \theta_1 B - \theta_2 B^2)[\epsilon_t - \theta_{1,12}\epsilon_{t-12}]$$

$$= \epsilon_t - \theta_1 B\epsilon_t - \theta_2 B^2\epsilon_t - \theta_{1,12}\epsilon_{t-12}$$

$$+ \theta_1\theta_{1,12}B\epsilon_{t-12} + \theta_2\theta_{1,12}B^2\epsilon_{t-12}$$

$$= \epsilon_t - \theta_1\epsilon_{t-1} - \theta_2\epsilon_{t-2} - \theta_{1,12}\epsilon_{t-12}$$

$$+ \theta_1\theta_{1,12}\epsilon_{t-13} + \theta_2\theta_{1,12}\epsilon_{t-14}$$

it follows that the model can be written in the form

$$y^*{}_t - y^*{}_{t-1} - y^*{}_{t-12} + y^*{}_{t-13} = \epsilon_t - \theta_1\epsilon_{t-1} - \theta_2\epsilon_{t-2} - \theta_{1,12}\epsilon_{t-12}$$

$$+ \theta_1\theta_{1,12}\epsilon_{t-13} + \theta_2\theta_{1,12}\epsilon_{t-14}$$

or

$$y^*_t = y^*_{t-1} + y^*_{t-12} - y^*_{t-13} + \epsilon_t - \theta_1\epsilon_{t-1} - \theta_2\epsilon_{t-2} - \theta_{1,12}\epsilon_{t-12}$$

$$+ \theta_1\theta_{1,12}\epsilon_{t-13} + \theta_2\theta_{1,12}\epsilon_{t-14}$$

After estimating the parameters in the model, the State License Bureau can use this model to forecast future values of monthly registrations of new automobiles.

PROBLEMS

1. Consider the Decay-Away Toothpaste sales time series presented in Table 1 in the problem set for Chapter 10.

   a. Discuss why it is reasonable to conclude that the first-order autoregressive model

   $$z_t = \delta + \phi_1 z_{t-1} + \epsilon_t$$

   can be assumed to have generated the first differences of this time series.

   b. Assume that a standard computer program of the Box-Jenkins methodology yielded least squares estimates

   $$\hat{\delta} = 3.03 \quad \text{and} \quad \hat{\phi}_1 = .6584$$

   Find the preliminary estimates that would be used as inputs to this computer program.

   c. Specify the model that describes the Decay-Away Toothpaste sales time series.

   d. Compute the residual $y_3 - \hat{y}_3$.

   e. Assume that we choose $K = 12$ and compute the first 12 residual autocorrelations to be as follows.

l	1	2	3	4	5	6	7	8	9	10	11	12
$r_l(\epsilon)$	.09	−.21	−.02	.03	.05	.17	.03	−.01	−.06	−.01	−.01	.06

   Assuming that $\chi^2_5(10) = 18.31$, calculate Q and determine whether or not the tentative model specified in Problem 3 is adequate.

   f. Find forecasts of Decay-Away Toothpaste sales for periods 91, 92, and 93. That is, find $\hat{y}_{91}(90)$, $\hat{y}_{92}(90)$, and $\hat{y}_{93}(90)$.

   g. If errors in 95 percent confidence intervals for y_{91}, y_{92}, and y_{93} are, respectively, 5.42, 10.5, and 15.45, find 95 percent confidence intervals for y_{91}, y_{92}, and y_{93}.

---

*h.   Using the regression approach, show how the least squares estimates of δ and ϕ_1 would be computed.

---

2.   In the context of Example 11.7, assume that y^*_{229} has been observed and discuss how a forecast of y^*_{231} would be made at time origin 229.

3. a. Assume that we transform the values of a monthly time series $y_1, y_2, \ldots, y_{120}$ possessing seasonal variation to new values $z_{13}, \ldots, z_{120}$ by the differencing transformation

$$z_t = y_t - y_{t-12}$$

Assume that, for the transformed values $z_{13}, \ldots, z_{120}$, the sample autocorrelation function dies down fairly quickly, and the sample partial autocorrelation function has a spike at lag 1 and cuts off after lag 1. Suggest a tentative model for this time series.

   b. Assume that after fitting the tentative model, the autocorrelation function of the residuals has a spike at lag 12, and the calculation of Q indicates that the model is inadequate. Suggest a new, improved model.

CHAPTER 12

Nite's Rest Inc.—A Box-Jenkins Case Study

We conclude our discussion of the Box-Jenkins methodology with a case study. This case study was developed and written by Dr. William Q. Meeker, Department of Statistics, Iowa State University. The authors are very grateful to Professor Meeker for his contribution to this book.

12–1 INTRODUCTION

Nite's Rest Inc. operates four hotels in Central City. The analysts in the operating division of the corporation were asked to develop a model that could be used to obtain short-term forecasts (up to 1 year) of the number of occupied rooms in the hotels. These forecasts were needed by various personnel to assist in decision making with regard to hiring additional help during the summer months, ordering materials that have long delivery lead times, budgeting of local advertising expenditures, etc.

   The available historical data consisted of the number of occupied rooms during each day for the 15 years from 1963 to 1977. Because it was desired to obtain k-step-ahead monthly forecasts for $k = 1, 2, \ldots, 12$, these data were reduced to monthly averages by dividing each monthly total by the number of days in the month. The monthly room averages for 1963 to 1976 are denoted by $y_1, y_2, \ldots, y_{168}$, are given in Table 12.1, and are plotted in Figure 12.1.

   At the outset, it was decided to perform all analyses with the data from 1963 to 1976 so that forecasts for 1977 could be used as a check on the validity of the model. Further, it was decided to perform an initial analysis using a relatively sim-

TABLE 12.1   *Monthly Hotel Room Averages for 1963–1976*

t	y_t	t	y_t	t	y_t	t	y_t	t	y_t
1	501.	43	785.	85	645.	127	1067.		
2	488.	44	830.	86	602.	128	1038.		
3	504.	45	645.	87	601.	129	812.		
4	578.	46	643.	88	709.	130	790.		
5	545.	47	551.	89	706.	131	692.		
6	632.	48	606.	90	817.	132	782.		
7	728.	49	585.	91	930.	133	758.		
8	725.	50	553.	92	983.	134	709.		
9	585.	51	576.	93	745.	135	715.		
10	542.	52	665.	94	735.	136	788.		
11	480.	53	656.	95	620.	137	794.		
12	530.	54	720.	96	698.	138	893.		
13	518.	55	826.	97	665.	139	1046.		
14	489.	56	838.	98	626.	140	1075.		
15	528.	57	652.	99	649.	141	812.		
16	599.	58	661.	100	740.	142	822.		
17	572.	59	584.	101	729.	143	714.		
18	659.	60	644.	102	824.	144	802.		
19	739.	61	623.	103	937.	145	748.		
20	758.	62	553.	104	994.	146	731.		
21	602.	63	599.	105	781.	147	748.		
22	587.	64	657.	106	759.	148	827.		
23	497.	65	680.	107	643.	149	788.		
24	558.	66	759.	108	728.	150	937.		
25	555.	67	878.	109	691.	151	1076.		
26	523.	68	881.	110	649.	152	1125.		
27	532.	69	705.	111	656.	153	840.		
28	623.	70	684.	112	735.	154	864.		
29	598.	71	577.	113	7?3.	155	717.		
30	683.	72	656.	114	837.	156	813.		
31	774.	73	645.	115	995.	157	811.		
32	780.	74	593.	116	1040.	158	732.		
33	609.	75	617.	117	809.	159	745.		
34	604.	76	686.	118	793.	160	844.		
35	531.	77	679.	119	692.	161	833.		
36	592.	78	773.	120	763.	162	935.		
37	578.	79	906.	121	723.	163	1110.		
38	543.	80	934.	122	655.	164	1124.		
39	565.	81	713.	123	658.	165	868.		
40	648.	82	710.	124	761.	166	860.		
41	615.	83	600.	125	768.	167	762.		
42	697.	84	676.	126	885.	168	877.		

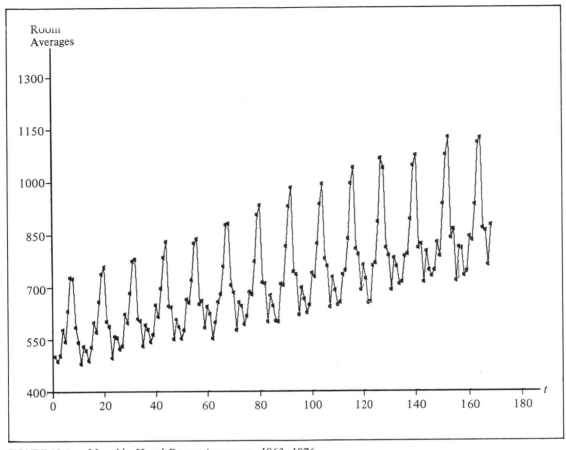

FIGURE 12.1   *Monthly Hotel Room Averages, 1963–1976*

ple multiple regression analysis before proceeding to the more complicated Box-Jenkins models. The Box-Jenkins models were analyzed with the TSERIES computer system (see Meeker [10]).

12–2 INITIAL ANALYSIS

Figure 12.1 shows that the monthly room averages follow a strong trend and that they have a seasonal pattern with one major and several minor peaks during the year. It also appears that the amount of seasonal variation is increasing with the level of the time series, indicating that the use of a transformation (such as the natural logarithms of the observations) might be warranted. The natural logarithms of the room averages are denoted by $y^*_1, y^*_2, \ldots, y^*_{168}$ and are plotted in Figure 12.2. Examination of Figures 12.1 and 12.2 indicates that the log transformation has equalized the amount of seasonal variation over the range of the data.

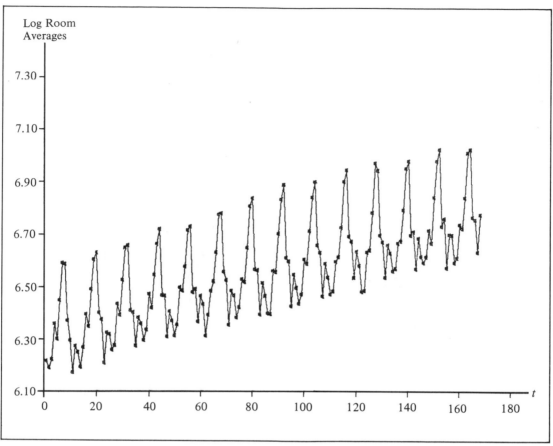

FIGURE 12.2   *Monthly Log Hotel Room Averages, 1963–1976*

In order to obtain more information concerning this time series, it was decided to fit the following dummy-variable multiple regression model to the actual room averages, $y_1, y_2, \ldots, y_{168}$, and to the logs of the room averages, $y^*_1, y^*_2, \ldots, y^*_{168}$.

$$y_t = \beta_0 + \beta_1 t + \beta_{S2} x_{S2,t} + \cdots + \beta_{S12} x_{S12,t} + \epsilon_t$$

where

$$x_{Sj,t} = \begin{cases} 1 & \text{if } t \text{ is in month } j \\ 0 & \text{otherwise} \end{cases} \quad (j = 2, \ldots, 12)$$

Both models fit the data well, confirming our previous observation of strong trend and seasonal components. Closer comparison showed that the fit to the logs of the room averages was superior. For example, a plot of the residuals versus the fitted values for the log model (not shown here) exhibited improved homogeneity of variance and no outliers (that is, no residuals substantially larger in absolute value

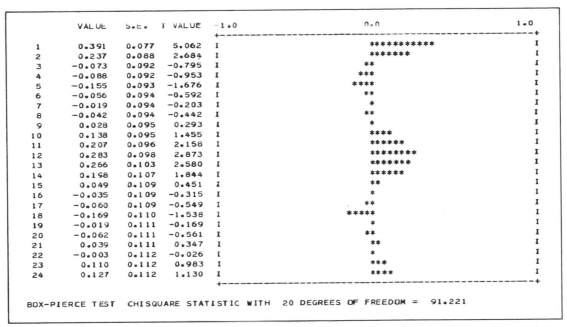

	VALUE	S.E.	T VALUE	-1.0	0.0	1.0
1	0.391	0.077	5.062	I	**********	I
2	0.237	0.088	2.684	I	*******	I
3	-0.073	0.092	-0.795	I	**	I
4	-0.088	0.092	-0.953	I	***	I
5	-0.155	0.093	-1.676	I	****	I
6	-0.056	0.094	-0.592	I	**	I
7	-0.019	0.094	-0.203	I	*	I
8	-0.042	0.094	-0.442	I	**	I
9	0.028	0.095	0.293	I	*	I
10	0.138	0.095	1.455	I	****	I
11	0.207	0.096	2.158	I	******	I
12	0.283	0.098	2.873	I	********	I
13	0.266	0.103	2.580	I	*******	I
14	0.198	0.107	1.844	I	******	I
15	0.049	0.109	0.451	I	**	I
16	-0.035	0.109	-0.315	I	*	I
17	-0.060	0.109	-0.549	I	**	I
18	-0.169	0.110	-1.538	I	*****	I
19	-0.019	0.111	-0.169	I	*	I
20	-0.062	0.111	-0.561	I	**	I
21	0.039	0.111	0.347	I	**	I
22	-0.003	0.112	-0.026	I	*	I
23	0.110	0.112	0.983	I	***	I
24	0.127	0.112	1.130	I	****	I

BOX-PIERCE TEST   CHISQUARE STATISTIC WITH   20 DEGREES OF FREEDOM =   91.221

FIGURE 12.3   *Autocorrelation Function for Residuals of Dummy-Variable Multiple Regression Model Fit to Natural Logarithms of Room Averages*

than the other residuals), correcting problems that existed before the log transformation was used. For these reasons and because the seasonal variation was thought to be more nearly multiplicative, the logs of the room averages were used in all subsequent analyses.

Figure 12.3 shows the autocorrelation function for the residuals of the dummy-variable multiple regression model fit to the log room averages. This figure, which is taken from the output supplied by the TSERIES computer system, gives the autocorrelation for lags $k = 1, 2, \ldots, 24$, the standard errors s_{r_k} (denoted S.E. on the computer output), the "t-like" statistic t_{r_k} (denoted T VALUE on the computer output), and a plot of the autocorrelations for lags $k = 1, 2, \ldots, 24$. (Recall that t_{r_k} and s_{r_k} are discussed in Section 10-2.2.) The autocorrelation function reveals several significant autocorrelations as indicated by T VALUES with absolute values reasonably close to or greater than 2. These significant autocorrelations indicate that observations in the log room average time series may be related. Hence, a Box-Jenkins seasonal model probably will forecast the log room averages more accurately than the dummy-variable multiple regression model we have considered.

This initial analysis has provided us with the following information.

1. A strong trend has persisted over the years. The trend coefficient in the dummy-variable model is highly significant ($t = 80.65$).

2. The room averages follow a strong seasonal pattern.

3. Seasonal variation increases with time. Taking logs of the room averages has equalized this variation.

```
              VALUE   S.E.   T VALUE  -1.0                     0.0                          1.0
                                       +--------------------------------------------------------------------+
      1       0.797   0.077   10.324   I                       ************************I
      2       0.600   0.116    5.162   I                       *****************       I
      3       0.391   0.133    2.929   I                       ***********             I
      4       0.271   0.140    1.938   I                       ********                I
      5       0.137   0.143    0.961   I                       ****                    I
      6       0.131   0.144    0.913   I                       ****                    I
      7       0.134   0.145    0.924   I                       ****                    I
      8       0.253   0.145    1.741   I                       *******                 I
      9       0.356   0.148    2.407   I                       **********              I
     10       0.531   0.153    3.475   I                       ***************         I
     11       0.690   0.164    4.219   I                       *******************     I
     12       0.854   0.180    4.740   I                       ************************* I
     13       0.672   0.203    3.314   I                       *******************     I
     14       0.491   0.216    2.277   I                       **************          I
     15       0.296   0.222    1.334   I                       *********               I
     16       0.187   0.224    0.832   I                       *****                   I
     17       0.058   0.225    0.259   I                       **                      I
     18       0.047   0.225    0.210   I                       **                      I
     19       0.046   0.226    0.206   I                       **                      I
     20       0.156   0.226    0.690   I                       *****                   I
     21       0.251   0.226    1.108   I                       *******                 I
     22       0.413   0.228    1.814   I                       ***********             I
     23       0.561   0.232    2.416   I                       ****************        I
     24       0.714   0.240    2.971   I                       ********************    I
     25       0.550   0.253    2.178   I                       ***************         I
     26       0.383   0.260    1.475   I                       **********              I
     27       0.205   0.263    0.781   I                       ******                  I
     28       0.100   0.264    0.380   I                       ***                     I
     29      -0.020   0.264   -0.075   I                       *                       I
     30      -0.032   0.264   -0.122   I                       *                       I
     31      -0.031   0.264   -0.119   I                       *                       I
     32       0.065   0.264    0.245   I                       **                      I
     33       0.152   0.264    0.575   I                       ****                    I
     34       0.302   0.265    1.141   I                       ********                I
     35       0.443   0.267    1.659   I                       ***********             I
     36       0.584   0.271    2.155   I                       *****************       I
     37       0.436   0.279    1.564   I                       ************            I
     38       0.283   0.283    1.002   I                       ********                I
                                       +--------------------------------------------------------------------+

   BOX-PIERCE TEST   CHISQUARE STATISTIC WITH   20 DEGREES OF FREEDOM = 638.386
```

FIGURE 12.4a   *Sample Autocorrelation Function for Time Series #1*

4. Before taking logs of the room averages, residual variation from the dummy-variable model did not appear to be homogeneous when plotted against the fitted values from the model. Again, analyzing the logs of the room averages seems to have alleviated this problem.

5. Correlations among the residuals from the dummy-variable model tell us that it is likely that better forecasts can be obtained by using a Box-Jenkins seasonal model.

12–3 IDENTIFICATION

This section illustrates how one can obtain information about a time series, leading to identification of a Box-Jenkins model. Several types of regular and seasonal differencing were considered. These are listed as follows.

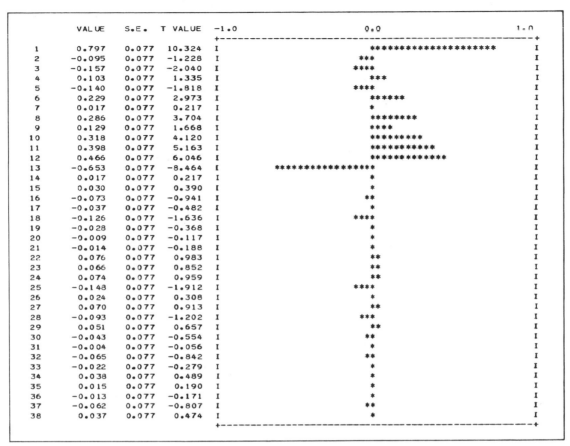

```
        VALUE   S.E.   T VALUE  -1.0                           0.0                              1.0
                                +-----------------------------------------------------------------+
    1    0.797   0.077  10.324  I                              *********************               I
    2   -0.095   0.077  -1.228  I                           ***                                    I
    3   -0.157   0.077  -2.040  I                          ****                                    I
    4    0.103   0.077   1.335  I                              ***                                 I
    5   -0.140   0.077  -1.818  I                          ****                                    I
    6    0.229   0.077   2.973  I                              ******                              I
    7    0.017   0.077   0.217  I                              *                                   I
    8    0.286   0.077   3.704  I                              ********                            I
    9    0.129   0.077   1.668  I                              ****                                I
   10    0.318   0.077   4.120  I                              *********                           I
   11    0.398   0.077   5.163  I                              ***********                         I
   12    0.466   0.077   6.046  I                              *************                       I
   13   -0.653   0.077  -8.464  I          *****************                                       I
   14    0.017   0.077   0.217  I                              *                                   I
   15    0.030   0.077   0.390  I                              *                                   I
   16   -0.073   0.077  -0.941  I                            **                                    I
   17   -0.037   0.077  -0.482  I                             *                                    I
   18   -0.126   0.077  -1.636  I                          ****                                    I
   19   -0.028   0.077  -0.368  I                             *                                    I
   20   -0.009   0.077  -0.117  I                             *                                    I
   21   -0.014   0.077  -0.188  I                             *                                    I
   22    0.076   0.077   0.983  I                             **                                   I
   23    0.066   0.077   0.852  I                             **                                   I
   24    0.074   0.077   0.959  I                             **                                   I
   25   -0.148   0.077  -1.912  I                          ****                                    I
   26    0.024   0.077   0.308  I                             *                                    I
   27    0.070   0.077   0.913  I                             **                                   I
   28   -0.093   0.077  -1.202  I                           ***                                    I
   29    0.051   0.077   0.657  I                             **                                   I
   30   -0.043   0.077  -0.554  I                            **                                    I
   31   -0.004   0.077  -0.056  I                             *                                    I
   32   -0.065   0.077  -0.842  I                            **                                    I
   33   -0.022   0.077  -0.279  I                             *                                    I
   34    0.038   0.077   0.489  I                             *                                    I
   35    0.015   0.077   0.190  I                             *                                    I
   36   -0.013   0.077  -0.171  I                             *                                    I
   37   -0.062   0.077  -0.807  I                            **                                    I
   38    0.037   0.077   0.474  I                             *                                    I
                                +-----------------------------------------------------------------+
```

FIGURE 12.4b   *Sample Partial Autocorrelation Function for Time Series #1*

Time Series #1: $z_t = y^*_t$

Time Series #2: $z_t = y^*_t - y^*_{t-1}$

Time Series #3: $z_t = y^*_t - y^*_{t-12}$

Time Series #4: $z_t = y^*_t - y^*_{t-1} - y^*_{t-12} + y^*_{t-13}$

Figures 12.4, 12.5, 12.6, and 12.7 show the sample autocorrelation and sample partial autocorrelation functions of the log room averages with these different amounts of regular and seasonal differencing. These figures, which are taken from the output supplied by the TSERIES computer system, give the sample autocorrelation or sample partial autocorrelation (denoted VALUE on the computer output) for lags $k = 1, 2, \ldots, 38$, the standard error (denoted S.E. on the computer output), the "t-like" statistic t_{r_k} or $t_{r_{kk}}$ (denoted T VALUE on the computer output), and a plot of the sample autocorrelations or sample partial autocorrelations for lags $k = 1, 2, \ldots, 38$. (Recall that t_{r_k} and $t_{r_{kk}}$ are discussed in Section 10–2.2.)

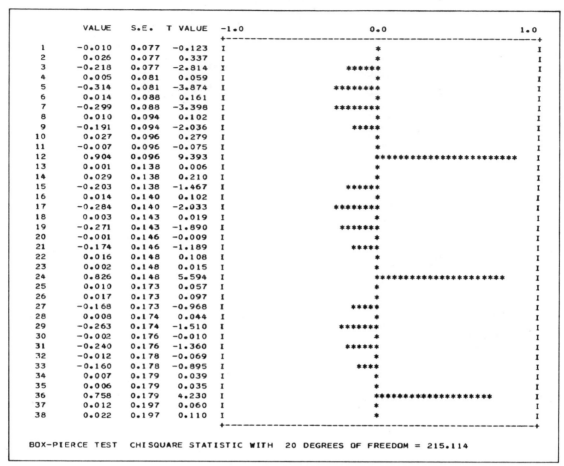

	VALUE	S.E.	T VALUE	-1.0	0.0	1.0
1	-0.010	0.077	-0.123	I	*	I
2	0.026	0.077	0.337	I	*	I
3	-0.218	0.077	-2.814	I	******	I
4	0.005	0.081	0.059	I	*	I
5	-0.314	0.081	-3.874	I	********	I
6	0.014	0.088	0.161	I	*	I
7	-0.299	0.088	-3.398	I	********	I
8	0.010	0.094	0.102	I	*	I
9	-0.191	0.094	-2.036	I	*****	I
10	0.027	0.096	0.279	I	*	I
11	-0.007	0.096	-0.075	I	*	I
12	0.904	0.096	9.393	I	************************	I
13	0.001	0.138	0.006	I	*	I
14	0.029	0.138	0.210	I	*	I
15	-0.203	0.138	-1.467	I	******	I
16	0.014	0.140	0.102	I	*	I
17	-0.284	0.140	-2.033	I	********	I
18	0.003	0.143	0.019	I	*	I
19	-0.271	0.143	-1.890	I	*******	I
20	-0.001	0.146	-0.009	I	*	I
21	-0.174	0.146	-1.189	I	*****	I
22	0.016	0.148	0.108	I	*	I
23	0.002	0.148	0.015	I	*	I
24	0.826	0.148	5.594	I	**********************	I
25	0.010	0.173	0.057	I	*	I
26	0.017	0.173	0.097	I	*	I
27	-0.168	0.173	-0.968	I	*****	I
28	0.008	0.174	0.044	I	*	I
29	-0.263	0.174	-1.510	I	*******	I
30	-0.002	0.176	-0.010	I	*	I
31	-0.240	0.176	-1.360	I	******	I
32	-0.012	0.178	-0.069	I	*	I
33	-0.160	0.178	-0.895	I	****	I
34	0.007	0.179	0.039	I	*	I
35	0.006	0.179	0.035	I	*	I
36	0.758	0.179	4.230	I	********************	I
37	0.012	0.197	0.060	I	*	I
38	0.022	0.197	0.110	I	*	I

BOX–PIERCE TEST CHISQUARE STATISTIC WITH 20 DEGREES OF FREEDOM = 215.114

FIGURE 12.5a   *Sample Autocorrelation Function for Time Series #2*

Examination of the sample autocorrelation functions of time series #1 and #2 indicate nonstationarity since the sample autocorrelation function for time series #1 (Figure 12.4a) dies down extremely slowly and the sample autocorrelation function for time series #2 (Figure 12.5a) has spikes at seasonal lags $k = 12$, 24, and 36. However, inspection of the sample autocorrelation functions for time series #3 and #4 (Figures 12.6a and 12.7a) indicates that these series are stationary. Because time series #3 has a lower order of differencing, it is probably the one to use for further analysis.

Further examination of time series #3 shows that the sample autocorrelation function seems to cut off after lag 12 with a significant autocorrelation ($t_{r_k} = -3.511$) at lag 12. The sample partial autocorrelation function for time series #3 seems to die down at or near seasonal lags with significant partial autocorrelations at lags $k = 12$, 25, and 37, the latter two being considerably smaller than the first. This indicates the need for a seasonal moving-average term of order 1. Another interpretation of the sample autocorrelation function for time series #3 might be

	VALUE	S.E.	T VALUE	−1.0	0.0	1.0
1	−0.010	0.077	−0.123	I	*	I
2	0.026	0.077	0.336	I	*	I
3	−0.218	0.077	−2.812	I	*****	I
4	0.001	0.077	0.018	I	*	I
5	−0.318	0.077	−4.114	I	********	I
6	−0.041	0.077	−0.534	I	**	I
7	−0.357	0.077	−4.608	I	*********	I
8	−0.189	0.077	−2.443	I	*****	I
9	−0.381	0.077	−4.918	I	*********	I
10	−0.507	0.077	−6.555	I	*************	I
11	−0.727	0.077	−9.399	I	*******************	I
12	0.560	0.077	7.239	I	****************	I
13	−0.128	0.077	−1.655	I	****	I
14	−0.181	0.077	−2.339	I	*****	I
15	−0.015	0.077	−0.196	I	*	I
16	−0.068	0.077	−0.878	I	**	I
17	0.114	0.077	1.473	I	***	I
18	−0.072	0.077	−0.929	I	**	I
19	0.071	0.077	0.914	I	**	I
20	−0.008	0.077	−0.099	I	*	I
21	−0.025	0.077	−0.320	I	*	I
22	−0.008	0.077	−0.100	I	*	I
23	0.011	0.077	0.138	I	*	I
24	0.086	0.077	1.117	I	***	I
25	−0.009	0.077	−0.112	I	*	I
26	−0.069	0.077	−0.896	I	**	I
27	0.134	0.077	1.737	I	****	I
28	−0.045	0.077	−0.581	I	**	I
29	−0.035	0.077	−0.453	I	*	I
30	0.051	0.077	0.663	I	**	I
31	0.012	0.077	0.151	I	*	I
32	0.035	0.077	0.448	I	*	I
33	−0.052	0.077	−0.675	I	**	I
34	0.032	0.077	0.415	I	*	I
35	−0.008	0.077	−0.103	I	*	I
36	0.009	0.077	0.111	I	*	I
37	−0.028	0.077	−0.366	I	*	I
38	0.077	0.077	0.990	I	**	I

FIGURE 12.5b *Sample Partial Autocorrelation Function for Time Series #2*

that it dies down fairly quickly with several significant autocorrelations at the lower lags. These significant autocorrelations indicate the need for some low-order autoregressive and/or moving-average terms in the model. Another interpretation of the sample partial autocorrelation function for time series #3 might be that, ignoring lags near the seasonal lags $k = 12$, 24, and 36, it seems to cut off at a low lag with significant partial autocorrelations at the lower lags. While it is difficult to infer much from the pattern of these autocorrelation and partial autocorrelation functions, one interpretation would indicate the need for some autoregressive terms.

For some time series applications, the appropriate Box-Jenkins model can be identified correctly after only a simple examination of the sample autocorrelation and sample partial autocorrelation functions of the time series and/or its differences if the process is nonstationary. However, more often than not, especially with seasonal time series, one will generally have two or more candidate models that require further comparison. This was the case for the hotel room average series. Initial analysis suggested the need for a seasonal moving-average operator and

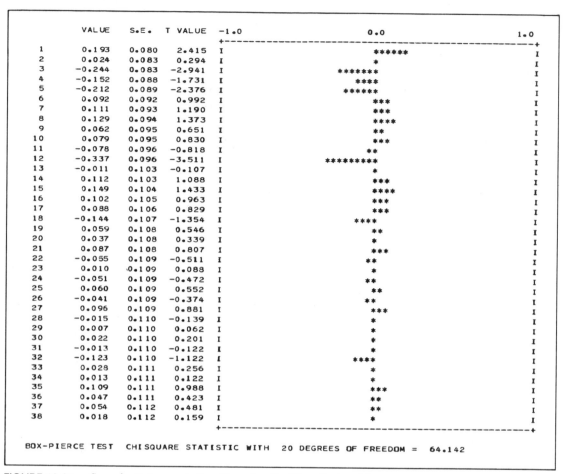

	VALUE	S.E.	T VALUE
1	0.193	0.080	2.415
2	0.024	0.083	0.294
3	−0.244	0.083	−2.941
4	−0.152	0.088	−1.731
5	−0.212	0.089	−2.376
6	0.092	0.092	0.992
7	0.111	0.093	1.190
8	0.129	0.094	1.373
9	0.062	0.095	0.651
10	0.079	0.095	0.830
11	−0.078	0.096	−0.818
12	−0.337	0.096	−3.511
13	−0.011	0.103	−0.107
14	0.112	0.103	1.088
15	0.149	0.104	1.433
16	0.102	0.105	0.963
17	0.088	0.106	0.829
18	−0.144	0.107	−1.354
19	0.059	0.108	0.546
20	0.037	0.108	0.339
21	0.087	0.108	0.807
22	−0.055	0.109	−0.511
23	0.010	0.109	0.088
24	−0.051	0.109	−0.472
25	0.060	0.109	0.552
26	−0.041	0.109	−0.374
27	0.096	0.109	0.881
28	−0.015	0.110	−0.139
29	0.007	0.110	0.062
30	0.022	0.110	0.201
31	−0.013	0.110	−0.122
32	−0.123	0.110	−1.122
33	0.028	0.111	0.256
34	0.013	0.111	0.122
35	0.109	0.111	0.988
36	0.047	0.111	0.423
37	0.054	0.112	0.481
38	0.018	0.112	0.159

BOX–PIERCE TEST CHISQUARE STATISTIC WITH 20 DEGREES OF FREEDOM = 64.142

FIGURE 12.6a   *Sample Autocorrelation Function for Time Series #3*

possibly some lower order autoregressive terms. In addition to the previously discussed dummy-variable multiple regression model, six Box-Jenkins seasonal models were considered. We summarize these models below. For each model we give the model equation, the operators in the general multiplicative seasonal model included in the model, and the type of differencing used.

I.   Dummy Variable Multiple Regression Model

Model 1: $y^{*}{}_{t} = \beta_0 + \beta_1 t + \beta_{S2} x_{S2,t} + \cdots + \beta_{S12} x_{S12,t} + \epsilon_t$

II.   Seasonal Box-Jenkins Models

Model 2: $y^{*}{}_{t} = \delta + y^{*}{}_{t-12} - \theta_{1,12}\epsilon_{t-12} + \epsilon_t$
 Operator: Seasonal moving-average of order 1
 Differencing: $z_t = y^{*}{}_t - y^{*}{}_{t-12}$

```
        VALUE    S.E.   T VALUE   -1.0              0.0              1.0
                                  +---------------------------------------+
   1    0.193   0.080    2.415   I                 ******               I
   2   -0.014   0.080   -0.169   I                 *                    I
   3   -0.256   0.080   -3.198   I           *******                    I
   4   -0.062   0.080   -0.777   I                **                    I
   5   -0.176   0.080   -2.199   I             *****                    I
   6    0.119   0.080    1.481   I                 ****                 I
   7    0.042   0.080    0.528   I                 **                   I
   8    0.002   0.080    0.029   I                 *                    I
   9    0.056   0.080    0.701   I                 **                   I
  10    0.081   0.080    1.007   I                 ***                  I
  11   -0.035   0.080   -0.437   I                 *                    I
  12   -0.325   0.080   -4.057   I          *********                   I
  13    0.190   0.080    2.373   I                 *****                I
  14    0.126   0.080    1.572   I                 ****                 I
  15   -0.036   0.080   -0.449   I                 *                    I
  16    0.018   0.080    0.221   I                 *                    I
  17    0.009   0.080    0.116   I                 *                    I
  18   -0.038   0.080   -0.471   I                 *                    I
  19    0.227   0.080    2.830   I                 ******               I
  20    0.040   0.080    0.496   I                 **                   I
  21    0.010   0.080    0.122   I                 *                    I
  22    0.018   0.080    0.220   I                 *                    I
  23   -0.035   0.080   -0.436   I                 *                    I
  24   -0.121   0.080   -1.505   I              ****                    I
  25    0.175   0.080    2.192   I                 *****                I
  26   -0.009   0.080   -0.116   I                 *                    I
  27    0.038   0.080    0.474   I                 *                    I
  28    0.026   0.080    0.324   I                 *                    I
  29   -0.082   0.080   -1.029   I               ***                    I
  30   -0.002   0.080   -0.023   I                 *                    I
  31    0.105   0.080    1.306   I                 ***                  I
  32   -0.119   0.080   -1.485   I               ***                    I
  33    0.077   0.080    0.956   I                 **                   I
  34   -0.024   0.080   -0.299   I                 *                    I
  35    0.002   0.080    0.021   I                 *                    I
  36   -0.053   0.080   -0.665   I                **                    I
  37    0.162   0.080    2.027   I                 *****                I
  38    0.019   0.080    0.238   I                 *                    I
                                  +---------------------------------------+
```

FIGURE 12.6b *Sample Partial Autocorrelation Function for Time Series #3*

Model 3: $y^*_t = \delta + \phi_1 y^*_{t-1} + y^*_{t-12} - \phi_1 y^*_{t-13} - \theta_{1,12}\epsilon_{t-12} + \epsilon_t$
 Operators: Seasonal moving-average of order 1
 Nonseasonal autoregressive of order 1
 Differencing: $z_t = y^*_t - y^*_{t-12}$

Model 4: $y^*_t = \delta + \phi_1 y^*_{t-1} + \phi_2 y^*_{t-2} + y^*_{t-12} - \phi_1 y^*_{t-13}$
 $- \phi_2 y^*_{t-14} - \theta_{1,12}\epsilon_{t-12} + \epsilon_t$
 Operators: Seasonal moving-average of order 1
 Nonseasonal autoregressive of order 2
 Differencing: $z_t = y^*_t - y^*_{t-12}$

Model 5: $y^*_t = \delta + \phi_1 y^*_{t-1} + \phi_2 y^*_{t-2} + \phi_3 y^*_{t-3} + y^*_{t-12} - \phi_1 y^*_{t-13}$
 $- \phi_2 y^*_{t-14} - \phi_3 y^*_{t-15} - \theta_{1,12}\epsilon_{t-12} + \epsilon_t$
 Operators: Seasonal moving average of order 1
 Nonseasonal autoregressive of order 3
 Differencing: $z_t = y^*_t - y^*_{t-12}$

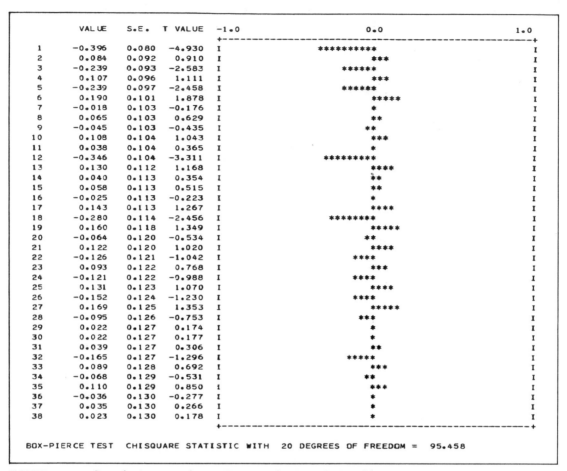

FIGURE 12.7a   *Sample Autocorrelation Function for Time Series #4*

Model 6: $y*_t = \delta + \phi_1 y*_{t-1} + \phi_2 y*_{t-2} + \phi_3 y*_{t-3} + y*_{t-12} - \phi_1 y*_{t-13}$
$- \phi_2 y*_{t-14} - \phi_3 y*_{t-15} - \theta_1 \epsilon_{t-1} - \theta_{1,12} \epsilon_{t-12} + \theta_1 \theta_{1,12} \epsilon_{t-13}$
$+ \epsilon_t$

Operators: Seasonal moving-average of order 1
Nonseasonal autoregressive of order 3
Nonseasonal moving-average of order 1

Differencing: $z_t = y*_t - y*_{t-12}$

Model 7: $y*_t = \delta + y*_{t-1} + y*_{t-12} - y*_{t-13} - \theta_1 \epsilon_{t-1}$
$- \theta_{1,12} \epsilon_{t-12} + \theta_1 \theta_{1,12} \epsilon_{t-13} + \epsilon_t$

Operators: Seasonal moving-average of order 1
Nonseasonal moving-average of order 1

Differencing: $z_t = y*_t - y*_{t-1} - y*_{t-12} + y*_{t-13}$

We will illustrate the derivation of the model equations for these models by

```
          VALUE   S.E.   T VALUE   -1.0                            0.0                                    1.0
                                   +-------------------------------------------------------------------+
    1     -0.396  0.080  -4.930    I                     **********                                     I
    2     -0.087  0.080  -1.079    I                        ***                                        I
    3     -0.284  0.080  -3.532    I                     ********                                       I
    4     -0.125  0.080  -1.561    I                       ****                                         I
    5     -0.348  0.080  -4.335    I                     *********                                      I
    6     -0.172  0.080  -2.138    I                      *****                                         I
    7     -0.102  0.080  -1.264    I                       ***                                          I
    8     -0.125  0.080  -1.561    I                       ****                                         I
    9     -0.088  0.080  -1.099    I                       ***                                          I
   10      0.032  0.080   0.394    I                          *                                         I
   11      0.234  0.080   2.911    I                          *******                                   I
   12     -0.287  0.080  -3.568    I                     ********                                        I
   13     -0.130  0.080  -1.617    I                       ****                                          I
   14      0.032  0.080   0.398    I                          *                                          I
   15     -0.030  0.080  -0.378    I                          *                                          I
   16     -0.021  0.080  -0.266    I                          *                                          I
   17      0.028  0.080   0.347    I                          *                                          I
   18     -0.208  0.080  -2.592    I                      ******                                         I
   19      0.018  0.080   0.221    I                          *                                          I
   20      0.029  0.080   0.359    I                          *                                          I
   21      0.025  0.080   0.308    I                          *                                          I
   22      0.083  0.080   1.031    I                          ***                                        I
   23      0.099  0.080   1.235    I                          ***                                        I
   24     -0.173  0.080  -2.150    I                      *****                                          I
   25      0.039  0.080   0.480    I                          **                                         I
   26     -0.024  0.080  -0.304    I                          *                                          I
   27      0.028  0.080   0.345    I                          *                                          I
   28      0.101  0.080   1.261    I                          ***                                        I
   29     -0.001  0.080  -0.014    I                          *                                          I
   30     -0.067  0.080  -0.834    I                         **                                          I
   31      0.127  0.080   1.580    I                          ****                                       I
   32     -0.086  0.080  -1.071    I                         ***                                         I
   33     -0.006  0.080  -0.069    I                          *                                          I
   34     -0.031  0.080  -0.385    I                          *                                          I
   35      0.028  0.080   0.348    I                          *                                          I
   36     -0.163  0.080  -2.026    I                      *****                                          I
   37      0.001  0.080   0.016    I                          *                                          I
   38      0.052  0.080   0.644    I                          **                                         I
                                   +-------------------------------------------------------------------+
```

FIGURE 12.7b   *Sample Partial Autocorrelation Function for Time Series #4*

considering Model 5. Since this model includes a seasonal moving-average operator of order 1 and a nonseasonal autoregressive operator of order 3, the general multiplicative seasonal model is

$$\phi_3(B)z_t = \delta + \theta_1(B^{12})\epsilon_t$$

or

$$(1 - \phi_1 B - \phi_2 B^2 - \phi_3 B^3)z_t = \delta + (1 - \theta_{1,12}B^{12})\epsilon_t$$

Since $z_t = y^*_t - y^*_{t-12}$ for this model, we have

$$(1 - \phi_1 B - \phi_2 B^2 - \phi_3 B^3)(y^*_t - y^*_{t-12}) = \delta + (1 - \theta_{1,12}B^{12})\epsilon_t$$

Expansion of this expression yields the following.

$$y^*_t - \phi_1 B y^*_t - \phi_2 B^2 y^*_t - \phi_3 B^3 y^*_t$$
$$- y^*_{t-12} + \phi_1 B y^*_{t-12} + \phi_2 B^2 y^*_{t-12} + \phi_3 B^3 y^*_{t-12}$$
$$= \delta + \epsilon_t - \theta_{1,12} B^{12} \epsilon_t$$

Application of the backshift operator gives us the following.

$$y^*_t - \phi_1 y^*_{t-1} - \phi_2 y^*_{t-2} - \phi_3 y^*_{t-3}$$
$$- y^*_{t-12} + \phi_1 y^*_{t-13} + \phi_2 y^*_{t-14} + \phi_3 y^*_{t-15}$$
$$= \delta + \epsilon_t - \theta_{1,12} \epsilon_{t-12}$$

Finally, this is equivalent to

$$y^*_t = \delta + \phi_1 y^*_{t-1} + \phi_2 y^*_{t-2} + \phi_3 y^*_{t-3}$$
$$+ y^*_{t-12} - \phi_1 y^*_{t-13} - \phi_2 y^*_{t-14}$$
$$- \phi_3 y^*_{t-15} - \theta_{1,12} \epsilon_{t-12} + \epsilon_t$$

12–4 ESTIMATION AND DIAGNOSTIC CHECKING

In order to choose among the several candidate models, it was necessary to estimate the parameters of each of them and to examine their properties. Usually such a comparison will lead one to choose a model or will at least provide direction to the search for other models that would warrant consideration. Summary results for Models 1–7 are given in Table 12.2. This table presents the estimates of the parameters in each model. Also, t-statistics for each of the estimates are given in parentheses below the estimates. The table also gives the Box-Pierce Chi-Square statistic (with 20 degrees of freedom) for each of the fitted models and indicates lags at which the residuals possess significant autocorrelations. Additionally, the value of s is presented for each of the fitted models. Because of the steady trend in the data, a constant term was included in the models for estimation purposes. The dummy-variable multiple regression model has been discussed in Section 12–2. The other models are discussed and compared below.

Model 2: In the identification stage, it was decided that a seasonal moving-average term would probably be needed in the model. When the very simple model including only this operator was fit to the log room averages, reasonably good results were obtained. For example, s was reduced to .0206, less than the value of .0217 for the dummy-variable multiple regression model. However, the autocorrelation function of the residuals from this model has significant autocorrelations at four lags and has a 20 degrees of freedom Box-Pierce Chi-Square statistic of $\chi^2 = 51.38$,

TABLE 12.2   *Comparison of Models 1–7*

	Model						
	1	2	3	4	5	6	7
Number of regular differences	0	0	0	0	0	0	1
Number of seasonal differences	0	1	1	1	1	1	1
Number of parameters	12	1	2	3	4	5	2
$\hat{\phi}_1$	—	—	.2977 (3.83)	.2779 (3.43)	.2922 (3.67)	.3518 (1.06)	—
$\hat{\phi}_2$	—	—	—	.1132 (1.40)	.1674 (2.04)	.1389 (1.18)	—
$\hat{\phi}_3$	—	—	—	—	−.2408 (−3.02)	−.2438 (−2.92)	—
$\hat{\theta}_1$	—	—	—	—	—	.0792 (0.23)	.8289 (18.5)
$\hat{\theta}_{1,12}$	—	.5509 (7.84)	.5962 (8.47)	.6552 (10.27)	.5917 (8.17)	.5896 (8.11)	.5889 (8.55)
$\hat{\delta}$	—	.0330	.0232	.0201	.0258	.0249	.0003
Box-Pierce χ^2(20 D.O.F.)	91.22	51.38	38.84	32.50	26.72	29.70	51.72
Significant autocorrelations	Lags 1, 2, 11, 12, 13	Lags 1, 5, 13, 14	Lags 3, 5, 18	Lags 3, 5, 18	Lag 10	Lag 10	Lags 1, 3, 4, 5, 15
s	.0217	.0206	.0197	.0197	.0192	.0193	.0213

a clear indication of an inadequate model. As in the dummy-variable multiple regression model, the patterns in the autocorrelation and partial autocorrelation functions of the residuals seem to indicate an autoregressive process.

Models 3, 4, and 5: These models include the seasonal moving-average operator of Model 2 as well as autoregressive operators of orders 1, 2, and 3 respectively. Model 3 showed significant improvement over Model 2 with s decreasing to .0197 and with $\chi^2 = 38.84$ and only three significant autocorrelations in the autocorrelation function of the residuals. Moving to Model 4 provided only modest improvement with $\chi^2 = 32.50$. Addition of the third autoregressive term in Model 5 further reduced s to .0192 and χ^2 to 26.7. Adding more terms to the model might result in a smaller value for s, but in the interests of parsimony and because such improvement would probably be marginal at best, Model 5 was chosen for use in forecasting. In fact, on these grounds it would be easy to argue for use of Model 3 instead of the more complicated Model 5.

Figure 12.8 shows TSERIES computer output for fitting Model 5 to the log room averages. Estimation was terminated after seven iterations when it was determined that further reduction in the sum of squares could not be obtained. All of

```
TERMS IN THE ASSUMED STATIONARY (DIFFERENCED) MODEL
   3 REGULAR AUTOREGRESSIVE TERM(S)
   0 SEASONAL AUTOREGRESSIVE TERM(S)
   0 REGULAR MOVING AVERAGE TERM(S)
   1 SEASONAL MOVING AVERAGE TERM(S)


BEGINING ESTIMATION
ITERATION    SUM OF SQUARES      PARAMETER VALUES
     0      0.8109055D-01   0.100000   0.100000   0.100000   0.100000
     1      0.7091130D-01   0.126299   0.097190  -0.010000   0.208597
     2      0.6493446D-01   0.170799   0.097126  -0.120000   0.302700
     3      0.6092252D-01   0.201075   0.113114  -0.185395   0.412700
     4      0.5884861D-01   0.244349   0.138056  -0.227251   0.522700
     5      0.5847978D-01   0.282219   0.160535  -0.235949   0.597602
     6      0.5831770D-01   0.292957   0.166756  -0.244677   0.558569
     7      0.5800217D-01   0.292182   0.167383  -0.240811   0.591711

UNABLE TO REDUCE SUM OF SQUARES ANY FURTHER
LARGEST CORRECTION IS     0.3314E-01  FOR ESTIMATE NUMBER    4

ESTIMATION COMPLETED



ESTIMATED LAG COEFFICIENTS FOR THE WORKING SERIES

AUTOREGRESSIVE TERMS
      1         2         3
   0.2922    0.1674   -0.2408

MOVING AVERAGE TERMS
         12
   0.5917


                                                    95% LIMITS
  TERM#   TYPE    ORDER       ESTIMATE   STD. ERROR   T VALUE     LOWER      UPPER
 ---------------------------------------------------------------------------------
     1   REG. AR    1          0.2922      0.0796      3.6688    0.1361     0.4483
     2   REG. AR    2          0.1674      0.0820      2.0408    0.0066     0.3282
     3   REG. AR    3         -0.2408      0.0796     -3.0239   -0.3969    -0.0847
     4   SEA. MA   12          0.5917      0.0725      8.1663    0.4497     0.7338
```

FIGURE 12.8   *Excerpts from TSERIES Output—Estimation for Model 5*

the estimated parameters are significantly different from zero. In the autocorrelation function of the residuals r_{10} is the only autocorrelation that is significantly different from zero. This is probably not an indication of any problem in the model. The estimated model parameters yield the following prediction equation.

$$\hat{y}^*_t = .0258 + .2922y^*_{t-1} + .1674y^*_{t-2} - .2408y^*_{t-3} + y^*_{t-12}$$
$$- .2922y^*_{t-13} - .1674y^*_{t-14} + .2408y^*_{t-15} - .5917\epsilon_{t-12}$$

The forecasts obtained from this model are discussed in the next section.

Model 6: In a situation analogous to model building in multiple regression, an analyst might want to add more terms to a model to see if significant improvement can be obtained. This process is usually called *overfitting*. Model 6 was obtained by adding a nonseasonal moving-average operator of order 1 to Model 5. As can be seen from Table 12.2, the results of this analysis showed no significant improvement in the model. Also, the estimated moving-average parameter is not significantly different from zero ($t = 0.23$), indicating that it probably should not be in the model. One might want to try adding other terms as well. For example, similar results were obtained when a fourth-order autoregressive term was added to Model 5.

Model 7: At the identification stage it was decided that only a single seasonal difference was needed to achieve stationarity in the hotel room average data. However, it is interesting to investigate the effect of taking a regular difference in addition to the seasonal difference. As can be seen from Table 12.2, the fit obtained with Model 7 was only slightly worse than that obtained with Models 2 through 6. However, problems were indicated in the autocorrelation function of the residuals since there are several significant autocorrelations. While one could probably develop a more complex model for this differenced series, it is doubtful that significant improvement could be achieved by moving in this direction. Another result of the additional difference is that, as will be seen in the next section, prediction intervals tend to be wider at relatively large lead times.

12–5 FORECASTING

If a time series has been differenced to remove a trend, an important decision must be made before forecasting. If the sample mean of the differenced series is significantly different from zero, a constant term should be used in fitting the model. This was done for the Box-Jenkins models discussed in Section 12–4. However, inclusion of the estimated constant term in the prediction equation will imply that the past general direction and rate for the trend will continue into the future. This may not be a reasonable assumption.

Because long-range planners were predicting continued growth at the present rate, it was decided to use the constant term for forecasting the room averages. The constant term is estimated as

$$\hat{\delta} = \bar{z}(1 - \hat{\phi}_1 - \hat{\phi}_2 - \hat{\phi}_3) = .0258$$

This yields the prediction equation given in Section 12–4. Remember that this equation really gives predictions for the logs of future room averages. To obtain forecasts for actual room averages, we can simply take antilogs. That is, $\hat{y}_t = \exp \hat{y}^*_t$. Point forecasts and corresponding prediction intervals (confidence interval forecasts) obtained using Model 5 are shown in Figure 12.9. These forecasts are taken from output supplied by the TSERIES computer system. The forecasts for periods 169–180 (1977) are graphed in Figure 12.10, along with the actual observations for these periods.

The interpretation of δ is important. For Model 5, the constant term is related to the mean of the differenced series as follows.

$$\delta = \mu_z(1 - \phi_1 - \phi_2 - \phi_3)$$

```
FORECASTS IN TERMS OF THE ORIGINAL SERIES
                                95 PERCENT LIMITS
PERIOD          FORECAST         LOWER         UPPER
────────────────────────────────────────────────────
169             839.282         808.277       871.475
170             772.404         742.700       803.295
171             778.176         747.400       810.219
172             873.158         838.421       909.333
173             860.309         826.026       896.013
174             984.442         945.052      1025.472
175            1156.817        1110.529      1205.034
176            1184.384        1136.991      1233.750
177             905.888         869.634       943.655
178             905.797         869.544       943.560
179             783.736         752.368       816.411
180             890.941         855.283       928.086
181             858.496         821.849       896.777
182             794.571         760.484       830.185
183             805.146         770.474       841.378
184             904.309         865.334       945.038
185             890.905         852.499       931.040
186            1018.169         974.252      1064.065
187            1195.705        1144.131      1249.604
188            1223.750        1170.966      1278.914
189             936.085         895.707       978.282
190             936.099         895.720       978.297
191             810.065         775.124       846.583
192             920.907         881.184       962.421
```

FIGURE 12.9   *TSERIES Output of Forecasts of Monthly Hotel Room Averages
for Periods 169 to 192*

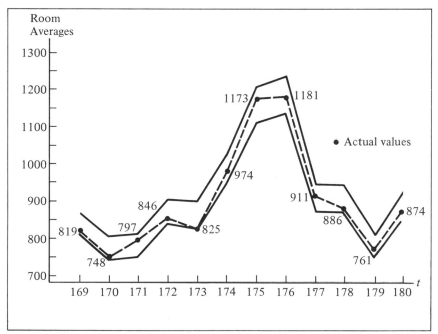

FIGURE 12.10   *95% Prediction Intervals (Confidence Interval Forecasts) and
Actual Hotel Room Averages for 1977*

The estimate of the mean of a time series that has been differenced once can be interpreted as the deterministic component of the slope of the undifferenced series. Thus, omitting δ from the prediction equation implies a mean of zero for the differenced series and that any changes in the level of the undifferenced series are stochastic with equal long-run probability of either upward or downward movement.

A positive (negative) value of $\hat{\delta}$ in the prediction equation implies positive (negative) trend in addition to any stochastic movement. For Model 5, which includes a constant term, the estimated mean of the differenced series is $\bar{z} = .0330$. Since for this model $z_t = y^*_t - y^*_{t-12}$, this implies an average increase of .0330 in each forecast over the level for the same month in the previous year. After taking antilogs, this implies a percentage growth of $100(1 - \exp\bar{z}) = 3.4$ percent for the forecast of one month over the same month in the previous year. This figure agrees closely with the inflation corrected growth rate that has been historically reported by management.

In practice it is often useful to compare the forecasts of several competing models. This is especially true if the analyst is having trouble choosing among several candidates. If the forecasts are similar the choice is not critical. If the forecasts are dissimilar, one should try to determine why they are dissimilar. This determination might help one choose the appropriate model. We will now compare the forecasts given by Model 5 with forecasts obtained from the other models we have considered.

Table 12.3 gives forecasts for each month of 1977 for each of the six Box-Jenkins seasonal models considered in Section 12–3. In each case, the constant term $\hat{\delta}$ has been included in the prediction equation (implying continued trend). Forecasts were also computed for Model 5 with the constant term $\hat{\delta}$ omitted from the model (Model 5a). The actual room averages are shown in the last column of Table 12.3.

TABLE 12.3   *Comparison of Forecasts for Seasonal Box-Jenkins Models*

Period	2	3	4	Model 5	5a	6	7	Actual
169	825	832	831	839	817	839	822	819
170	766	768	771	772	747	772	765	748
171	779	780	782	778	746	778	778	797
172	876	876	878	873	842	873	875	846
173	862	863	866	860	830	860	863	825
174	984	986	988	984	952	984	986	974
175	1156	1157	1159	1157	1120	1157	1158	1173
176	1182	1184	1186	1184	1147	1184	1186	1181
177	904	906	908	906	877	906	907	911
178	904	906	907	906	876	906	908	886
179	783	784	784	784	758	784	787	761
180	892	891	890	891	862	891	895	874

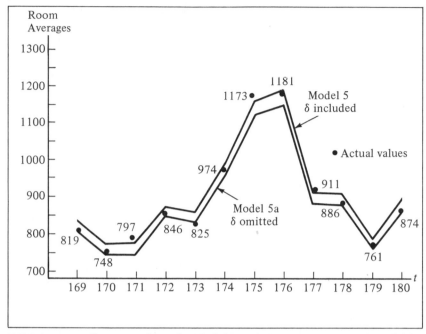

FIGURE 12.11   *Forecasts Given by Model 5 (with Constant Term) and Model 5a (without Constant Term)*

It is interesting to note that in this example the forecasts for the Box-Jenkins models assuming continued trend are all very similar over the required forecast horizon of one year. In fact, the inherent variability in the time series outweighs the differences, indicating that the choice among these models is not critical. Of course, this type of forecast behavior will not occur in general. It occurs in this case because most of the variability in the time series is due to the trend and seasonal effects, rather than to purely random components.

In Model 5a it was decided to omit the constant term in obtaining forecasts. Figure 12.11 shows the forecasts for Models 5 and 5a and the actual observations for 1977. As noted earlier, inclusion of the constant term in this model implies a continued growth of about 3.4 percent for each month over the level of the same month in the previous year. Thus Model 5a severely underpredicted the months of July and August 1977, as the previous trend did continue into 1977.

It is also interesting to compare the widths of the prediction intervals (confidence interval forecasts) obtained from several of the models. The intervals obtained at $t = 180$ and $t = 192$ using models 3, 5, and 7 are given below.

Model	Interval at $t = 180$	Interval at $t = 192$
3	[856, 928]	[882, 962]
5	[855, 928]	[881, 962]
7	[853, 939]	[877, 995]

Note that Model 7, which employed a regular difference as well as a seasonal difference, yielded an interval about 18 percent wider at $t = 180$ and about 46 percent wider at $t = 192$ than the intervals given by Models 3 and 5, which employ only a seasonal difference.

12–6   FINAL MODEL AND A WARNING

The analysts decided that Model 5 would be used to produce the needed forecasts. Before obtaining these forecasts, the model was refit using all 180 observations, yielding the following prediction equation.

$$\hat{y}^*_t = .0225 + .3100 y^*_{t-1} + .1407 y^*_{t-2} - .1412 y^*_{t-3} + y^*_{t-12}$$
$$- .3100 y^*_{t-13} - .1407 y^*_{t-14} + .1412 y^*_{t-15} - .7707 \epsilon_{t-12}$$

In the process of determining this prediction equation, several important judgments were made, the implications of which should be clearly understood. We have analyzed the natural logarithms of the room averages. This, in conjunction with the use of the constant term δ implies continued exponential rather than linear growth. The rate of growth in the present example has been small and, because it can be expected to continue, inclusion of the constant term was probably appropriate. This will not be true in all applications. Also, for some time series, use of the log transformation will actually overtransform possibly making matters worse. In such cases, a weaker transformation, such as the square root might produce good results. A paper by Chatfield and Prothero [5] and the discussion following it illustrate this problem.

PART IV SUMMARY

In Part IV you have learned how to analyze and forecast time series using the Box-Jenkins methodology. This methodology is often used when the observations in a time series are related to each other or "autocorrelated."

In Chapter 10 you learned the basic steps of the Box-Jenkins methodology. The first step in this methodology is the identification step. You have learned that in order to identify an appropriate Box-Jenkins model, it is important to produce a set of stationary time series values. These stationary time series values are often obtained by differencing the original values of the time series or by applying some other transformation to the original time series values. You then learned how to calculate the sample autocorrelation function for a set of time series values. You saw that if this function either "dies down" or "cuts off" fairly quickly, then the time series values can be assumed to be stationary. You also learned how to calculate the sample partial autocorrelation function for a set of time series values. The sample autocorrelation function and sample partial autocorrelation function are important because once a set of original time series values have been found to be or have been transformed into stationary time series values, the behavior of these functions can be used to identify the particular time series model that can be assumed to have generated the stationary time series values. In particular, you saw that whether these functions die down or cut off, and if they cut off, when they cut off, help to identify the model to be used. You also learned several rules of thumb that can be used to help classify the behavior of these functions. Once an appropriate Box-Jenkins model has been identified, the second step in this methodology involves estimation of the parameters in the model. You learned that this estimation is accomplished using computer programs and that these programs must be supplied with preliminary estimates of the parameters. The third step of the Box-Jenkins methodology is called diagnostic checking. In this step the adequacy of the tentatively identified model is tested. You learned that the overall adequacy of the model is often tested using the Box-Pierce Chi-Square statistic Q, which tests whether the autocorrelations of the residuals indicate adequacy of the model. You also saw that the adequacy of the model can be judged by considering s, which influences the length of the confidence interval forecasts obtained when using a Box-Jenkins model. If the tentatively identified model is found to be inadequate, improvements in the model must be made. Last, in Chapter 10, you learned how to carry out the fourth step in the Box-Jenkins methodology—forecasting.

In Chapter 11 you studied some of the time series models used in the Box-Jenkins methodology. First, you studied several commonly used nonseasonal models. These models are called moving-average models, autoregressive models, and mixed autoregressive–moving-average models. Then you studied Box-Jenkins models that are useful in describing and forecasting seasonal time series. In order to use these models you became familiar with the backshift operator B, the nonseasonal operator ∇, and the seasonal operator ∇_L. You learned that when dealing with seasonal time series it is often necessary to use a logarithmic transformation along with different orders of seasonal and nonseasonal differencing in order to produce a set of stationary time series values. Once a set of stationary time series values has been produced, a seasonal Box-Jenkins model must be identified. The identified model will be a special form of the general multiplicative seasonal model. You learned that this general model employs the nonseasonal autoregressive operator of order p, the nonseasonal moving-average operator of order q, the seasonal autoregressive operator of order P, and the seasonal moving-average operator of order Q. You then studied several guidelines that can be used to help identify the particular special form of the general multiplicative seasonal model to be used in representing an observed stationary time series. You should remember, however, that these guidelines do not represent a totally systematic method that can be used to identify the components of the general multiplicative seasonal model. Instead, even when using these guidelines, several iterations of the modeling process may be necessary in order to identify an appropriate seasonal model. Finally, you saw that the estimation, diagnostic checking, and forecasting procedures for seasonal Box-Jenkins models are essentially the same as those described for nonseasonal Box-Jenkins models.

In Chapter 12 you gained further insight into the Box-Jenkins modeling process by studying a rather complex case in which the Box-Jenkins methodology was used to forecast a seasonal time series concerning hotel room occupancies.

EXERCISES

1. List the basic steps involved in using the Box-Jenkins methodology.

2. Discuss the advantages of the Box-Jenkins methodology.

3. Discuss the disadvantages of the Box-Jenkins methodology.

4. Explain the difference between a stationary time series and a nonstationary time series.

5. If a set of original time series values is nonstationary, the Box-Jenkins methodology requires that a stationary time series be derived. In general, explain how this stationary time series is usually obtained.

6. Intuitively explain the meaning of ρ_K, the autocorrelation between observations separated by a lag of K time units. That is, for example, what is the relationship between observations separated by K time units when ρ_K is near 1?

7. Define the theoretical autocorrelation function and the sample autocorrelation function.

8. Describe the behavior of the theoretical autocorrelation function for a stationary time series.

9. Describe the behavior of the theoretical autocorrelation function for a nonstationary time series.

10. Define the theoretical partial autocorrelation function and the sample partial autocorrelation function.

11. Describe the behavior of the theoretical partial autocorrelation function for a stationary time series.

12. Discuss the strategy that is used in the identification of an appropriate Box-Jenkins model.

13. Suppose that more than one set of preliminary estimates for a Box-Jenkins model have been generated. How does one choose the set of preliminary estimates that should be used in computing final estimates of the model parameters?

14. What is the purpose of diagnostic checking?

15. Describe the behavior of the theoretical autocorrelation function and the theoretical partial autocorrelation function for a moving-average model of order q.

16. Describe the behavior of the theoretical autocorrelation function and the theoretical partial autocorrelation function for an autoregressive model of order p.

17. Describe the behavior of the theoretical autocorrelation function and the theoretical partial autocorrelation function for a mixed autoregressive–moving-average model of order (p,q).

18. Suppose we wish to calculate a forecast for a time series value y_{T+t}, the forecast being made at time T.
   a. If the forecast involves an error term ϵ_{T+j} and $j \leq 0$, how is ϵ_{T+j} estimated?
   b. If the forecast involves an error term ϵ_{T+j} and $j > 0$, how is ϵ_{T+j} estimated?

19. Suppose we wish to calculate a forecast for a time series value y_{T+t}, the forecast being made at time T.
   a. If the forecast involves a time series value y_{T+j} and $j \leq 0$, what value is used for y_{T+j}?

b. If the forecast involves a time series value y_{T+j} and $j > 0$, how is y_{T+j} estimated?

20. How can the estimates of the parameters in a Box-Jenkins model be updated in light of a new time series observation?

21. Explain the circumstances in which the transformation $y^*{}_t = \ln y_t$ should be used in analyzing a seasonal time series by the Box-Jenkins methodology.

22. How are spikes in the sample autocorrelation function of a stationary time series used to help identify a Box-Jenkins seasonal model which includes a seasonal moving-average operator or product of nonseasonal and seasonal moving-average operators?

23. Suppose that diagnostic checks indicate that a tentative Box-Jenkins seasonal model is inadequate. How might a new model be identified?

24. Explain what is meant by the term "most parsimonious model." Why might a forecaster use the "most parsimonious model" available?

25. Suppose that we wish to forecast a seasonal time series consisting of monthly data. If both a Box-Jenkins model and a regression model with dummy variables have mean squared errors that are nearly equal, why might the Box-Jenkins model be considered superior?

VOCABULARY LIST

Box-Jenkins methodology

Identification

Estimation

Diagnostic checking

Stationary time series

Nonstationary time series

First difference

Second difference

Autocorrelation

Theoretical autocorrelation function

Sample autocorrelation function

Partial autocorrelation

Theoretical partial autocorrelation function

Sample partial autocorrelation function

"Die down" with increasing lag K

"Cut off" after lag K

Invertibility

Preliminary estimates

Final estimates

Moving-average process

Autoregressive model

Mixed autoregressive–moving-average model

Moving-average model of order q

Autoregressive model of order p

Mixed autoregressive–moving-average model of order (p,q)

Nonseasonal autoregressive operator of order p

Nonseasonal moving-average operator of order q

Seasonal autoregressive operator of order P

Seasonal moving-average operator of order Q

Mixed autoregressive–moving-average model of order (p,P,q,Q)

Backshift operator B

QUIZ

Answer TRUE if the statement is always true. If the statement is not true, replace the underlined word(s) with a word(s) that makes the statement true.

1. The Box-Jenkins methodology is appropriately used when successive time series observations are <u>autocorrelated.</u>

2. The Box-Jenkins methodology involves several steps—<u>identification</u>, <u>estimation</u>, <u>prediction</u>, and <u>forecasting</u>.

3. The Box-Jenkins methodology would probably be most appropriately used when the data series is <u>yearly</u> rather than <u>hourly</u>.

4. The Box-Jenkins methodology <u>does</u> <u>not</u> <u>provide</u> automatic procedures that update estimates of model parameters as new data are observed.

5. If a time series is <u>stationary</u>, its values fluctuate around a constant mean, while a <u>nonstationary</u> time series has no constant mean.

6. The theoretical autocorrelation function of a stationary time series either <u>cuts off</u> after a particular lag K or <u>dies down extremely slowly</u>.

7. <u>Second differences</u> are obtained by taking differences of the <u>first differences</u>.

8. The sample autocorrelation function is a listing, or graph, of the <u>sample partial autocorrelations</u> for lags $K = 1, 2, \ldots .$

9. Each particular Box-Jenkins model is characterized by the behavior of its <u>theoretical autocorrelation function</u> and <u>theoretical partial autocorrelation function</u>.

10. The <u>estimation</u> step tests a Box-Jenkins model for adequacy.

11. A moving-average model is always stationary.

12. An autoregressive model is always stationary.

13. For a first-order moving-average model, the theoretical autocorrelation function cuts off after lag 1, while the theoretical partial autocorrelation function dies down.

14. For a first-order autoregressive model, the theoretical autocorrelation function dies down, while the theoretical partial autocorrelation function also dies down.

15. For a mixed autoregressive–moving-average model of order (p,q), the theoretical autocorrelation function cuts off after lag q, while the theoretical partial autocorrelation function cuts off after lag p.

16. The mixed autoregressive–moving-average model of order (p,q) possesses both stationarity and invertibility conditions.

17. The Box-Jenkins methodology can be used to calculate statistically correct confidence intervals for future values of a time series.

18. Regression analysis may be used to find least squares estimates of model parameters and confidence intervals for future time series values for mixed autoregressive–moving-average models.

19. Before the identification process is begun, the Box-Jenkins methodology requires that a stationary time series be derived.

20. Moving-average models possess stationarity conditions, while autoregressive models possess invertibility conditions.

21. If the sample autocorrelation function of a set of transformed time series observations dies down fairly quickly and has spikes at the seasonal lags 12, 24, 36, and 48, the time series values are probably nonstationary.

22. If the sample autocorrelation function of an observed stationary time series dies down fairly quickly, and the sample partial autocorrelation function cuts off after a lag that is nearly equal to $L, L + 1, L + 2, 2L, 2L + 1,$ or $2L + 2$, then the use of some form of the seasonal moving-average operator might be appropriate if no spikes exist at lags substantially smaller than L.

23. If use of the nonseasonal autoregressive operator is appropriate, the lag at which the sample autocorrelation function cuts off suggests the order p of the operator that should be used.

24. The use of the transformation $y^*_t = \ln y_t$ is appropriate when a seasonal time series exhibits constant variability as time advances.

25. When diagnostic checks indicate that a tentative Box-Jenkins seasonal model is inadequate, improvements in the model are often suggested by an examination of the autocorrelation function of the residuals.

REFERENCES

1. Box, G. E. P., and G. M. Jenkins, *Time Series Analysis, Forecasting and Control,* Holden-Day, Inc., San Francisco, 1970.

2. Brown, R. G., *Statistical Forecasting for Inventory Control,* McGraw-Hill Book Company, New York, 1959.

3. Brown, R. G., *Smoothing, Forecasting and Prediction of Discrete Time Series,* Prentice-Hall, Inc., Englewood Cliffs, New Jersey, 1962.

4. Brown, R. G., *Decision Rules for Inventory Management,* Holt, Rinehart and Winston, Inc., New York, 1967.

5. Chatfield, C., and D. L. Prothero, "Box-Jenkins Seasonal Forecasting: Problems in a Case Study," (with discussion), *Journal of the Royal Statistical Society,* **A136,** 1973.

6. Draper, N. R., and H. Smith, *Applied Regression Analysis,* John Wiley and Sons, Inc., New York, 1968.

7. Fuller, Wayne A., *Introduction to Statistical Time Series,* John Wiley and Sons, Inc., New York, 1976.

8. Johnson, L. A., and D. C. Montgomery, *Forecasting and Time Series Analysis,* McGraw-Hill Book Company, New York, 1976.

9. Mabert, V. A., *An Introduction to Short Term Forecasting Using the Box-Jenkins Methodology,* Publication No. 2 in the American Institute of Industrial Engineers, Inc., Monograph Series.

10. Meeker, W. Q., "TSERIES—A User-Oriented Computer Program for Time Series Analysis," *The American Statistician,* vol. 32, no. 3, 1978.

11. Miller, R. B., and D. W. Wichern, *Intermediate Business Statistics,* Holt, Rinehart and Winston, New York, 1977.

12. Nelson, C. R., *Applied Time Series Analysis for Managerial Forecasting,* Holden-Day, Inc., San Francisco, 1973.

13. Neter, John, and William Wasserman, *Applied Linear Statistical Models,* R. D. Irwin, Homewood, Illinois, 1974.

14. Wheelright, S. C., and S. Makridakis, *Forecasting Methods for Management,* John Wiley and Sons, Inc., New York, 1973.

APPENDIX A

Matrix Algebra

A *MATRIX* is a rectangular array of numbers (called elements) which is composed of rows and columns. An example of a matrix is as follows.

$$A = \begin{bmatrix} 1 & 5 & 3 & 10 \\ 12 & 6 & 7 & 4 \\ 9 & 2 & 11 & 8 \end{bmatrix}$$

Here the notation **A** is used to indicate that we are referring to a matrix rather than a number.

The *DIMENSION* of a matrix is determined by the number of rows and columns in the matrix. Since the matrix **A** has 3 rows and 4 columns, this matrix is said to have dimension 3 by 4 (commonly written 3 x 4). In general, a matrix with m rows and n columns is said to have dimension m x n. As another example, the matrix

$$X = \begin{bmatrix} 1 & 0 & 0 \\ 1 & 1 & 0 \\ 1 & 2 & 0 \\ 1 & 0 & 1 \\ 1 & 1 & 1 \\ 1 & 2 & 1 \\ 1 & 0 & 2 \\ 1 & 1 & 2 \\ 1 & 2 & 2 \end{bmatrix}$$

has dimension 9 x 3 since it has 9 rows and 3 columns.

In general, a matrix with dimension m x n can be represented as

$$\mathbf{A}_{m \times n} = \begin{bmatrix} a_{11} & a_{12} & . & . & a_{1j} & . & a_{1n} \\ a_{21} & a_{22} & . & . & . & . & a_{2n} \\ . & . & & & . & & . \\ . & . & & & . & & . \\ . & . & & & . & & . \\ . & . & & & . & & . \\ a_{i1} & a_{i2} & . & . & a_{ij} & . & a_{in} \\ . & . & & & . & & . \\ . & . & & & . & & . \\ . & . & & & . & & . \\ . & . & & & . & & . \\ a_{m1} & a_{m2} & . & . & a_{mj} & . & a_{mn} \end{bmatrix}$$

where a_{ij} is the number in the matrix in row i and column j, and the subscript m x n indicates the dimension of **A**.

A matrix that consists of one column is a *COLUMN VECTOR*. A matrix that consists of one row is a *ROW VECTOR*.

The following are examples of column vectors.

$$\mathbf{B}_{4 \times 1} = \begin{bmatrix} 5 \\ 3 \\ 4 \\ 1 \end{bmatrix} \qquad \mathbf{C}_{3 \times 1} = \begin{bmatrix} 101 \\ 73 \\ 51 \end{bmatrix}$$

The following are examples of row vectors.

$$\mathbf{E}'_{1 \times 4} = [10\ 7\ 6\ 12]$$

$$\mathbf{F}'_{1 \times 6} = [1\ 2\ 7\ 11\ 5\ 8]$$

Note that the prime mark (') is used to distinguish a row vector from a column vector.

The *TRANSPOSE* of a matrix is formed by interchanging the rows and columns of the matrix. For example, consider the following matrix.

$$\mathbf{A}_{2 \times 3} = \begin{bmatrix} 5 & 6 & 7 \\ 3 & 2 & 1 \end{bmatrix}$$

The transpose of **A**, which is denoted **A'**, is

$$A'_{3x2} = \begin{bmatrix} 5 & 3 \\ 6 & 2 \\ 7 & 1 \end{bmatrix}$$

Thus the first row of **A** is the first column of **A′** while the second row of **A** is the second column of **A′**.

Notice that the transpose of the column vector

$$E_{4x1} = \begin{bmatrix} 10 \\ 7 \\ 6 \\ 12 \end{bmatrix}$$

is the row vector $E'_{1x4} = [10\ 7\ 6\ 12]$. As a last example, consider the matrix

$$X_{9x3} = \begin{bmatrix} 1 & 0 & 0 \\ 1 & 1 & 0 \\ 1 & 2 & 0 \\ 1 & 0 & 1 \\ 1 & 1 & 1 \\ 1 & 2 & 1 \\ 1 & 0 & 2 \\ 1 & 1 & 2 \\ 1 & 2 & 2 \end{bmatrix}$$

The transpose of **X** is

$$X'_{3x9} = \begin{bmatrix} 1 & 1 & 1 & 1 & 1 & 1 & 1 & 1 & 1 \\ 0 & 1 & 2 & 0 & 1 & 2 & 0 & 1 & 2 \\ 0 & 0 & 0 & 1 & 1 & 1 & 2 & 2 & 2 \end{bmatrix}$$

Consider two matrixes, **A** and **B**, which have the same dimension. The *SUM* of **A** and **B** is a matrix obtained by adding the corresponding elements of **A** and **B**. For example, consider the following matrixes.

$$A_{2x3} = \begin{bmatrix} 1 & 4 & 2 \\ 5 & 3 & 2 \end{bmatrix} \qquad B_{2x3} = \begin{bmatrix} 7 & 0 & 4 \\ 3 & 1 & 5 \end{bmatrix}$$

The sum of the matrixes **A** and **B** is

$$C_{2x3} = A_{2x3} + B_{2x3} = \begin{bmatrix} 1+7 & 4+0 & 2+4 \\ 5+3 & 3+1 & 2+5 \end{bmatrix} = \begin{bmatrix} 8 & 4 & 6 \\ 8 & 4 & 7 \end{bmatrix}$$

In general, if $\mathbf{A}$ and $\mathbf{B}$ have the same dimension and $\mathbf{C} = \mathbf{A} + \mathbf{B}$ then

$$c_{ij} = a_{ij} + b_{ij}$$

where c_{ij} = the number in $\mathbf{C}$ in row i and column j
  a_{ij} = the number in $\mathbf{A}$ in row i and column j
  b_{ij} = the number in $\mathbf{B}$ in row i and column j

Consider two matrixes, $\mathbf{A}$ and $\mathbf{B}$, which have the same dimension. The *DIFFERENCE* of $\mathbf{A}$ and $\mathbf{B}$ is a matrix obtained by subtracting the corresponding elements of $\mathbf{A}$ and $\mathbf{B}$. For example, consider the following matrixes.

$$\mathbf{A}_{2x3} = \begin{bmatrix} 1 & 4 & 2 \\ 5 & 3 & 2 \end{bmatrix} \quad \mathbf{B}_{2x3} = \begin{bmatrix} 7 & 0 & 4 \\ 3 & 1 & 5 \end{bmatrix}$$

The difference of the matrixes $\mathbf{A}$ and $\mathbf{B}$ is

$$\mathbf{D}_{2x3} = \mathbf{A}_{2x3} - \mathbf{B}_{2x3} = \begin{bmatrix} 1-7 & 4-0 & 2-4 \\ 5-3 & 3-1 & 2-5 \end{bmatrix} = \begin{bmatrix} -6 & 4 & -2 \\ 2 & 2 & -3 \end{bmatrix}$$

In general, if $\mathbf{A}$ and $\mathbf{B}$ have the same dimension and $\mathbf{D} = \mathbf{A} - \mathbf{B}$ then

$$d_{ij} = a_{ij} - b_{ij}$$

where d_{ij} = the number in $\mathbf{D}$ in row i and column j
  a_{ij} = the number in $\mathbf{A}$ in row i and column j
  b_{ij} = the number in $\mathbf{B}$ in row i and column j

We now consider *MULTIPLICATION* of a matrix by a *NUMBER*. The product of a number λ and a matrix $\mathbf{A}$ is a matrix which is obtained by multiplying each element of $\mathbf{A}$ by the number λ. For example, consider the following matrix.

$$\mathbf{Z}_{2x3} = \begin{bmatrix} 1 & 4 & 7 \\ 3 & 2 & 3 \end{bmatrix}$$

Then

$$5\mathbf{Z}_{2x3} = 5\begin{bmatrix} 1 & 4 & 7 \\ 3 & 2 & 3 \end{bmatrix} = \begin{bmatrix} 5(1) & 5(4) & 5(7) \\ 5(3) & 5(2) & 5(3) \end{bmatrix} = \begin{bmatrix} 5 & 20 & 35 \\ 15 & 10 & 15 \end{bmatrix}$$

In general, if λ is a number, $\mathbf{A}$ is a matrix, and $\mathbf{E} = \lambda\mathbf{A}$ then

$$e_{ij} = \lambda a_{ij}$$

where e_{ij} = the number in $\mathbf{E}$ in row i and column j
  a_{ij} = the number in $\mathbf{A}$ in row i and column j

We next consider *MULTIPLICATION* of a matrix by a *MATRIX*. Consider two matrixes $\mathbf{A}$ and $\mathbf{B}$ where the number of columns in $\mathbf{A}$ is equal to the number of rows in **B**. Then the product of **A** and **B** is a matrix that is calculated so that the element in row i and column j of the product is obtained by multiplying the elements in row i of matrix **A** by the corresponding elements in column j of matrix **B** and adding the resulting products.

For example, consider the following matrixes.

$$\mathbf{A}_{2x2} = \begin{bmatrix} 4 & 3 \\ 2 & 2 \end{bmatrix} \quad \mathbf{B}_{2x2} = \begin{bmatrix} 2 & 1 \\ 3 & 5 \end{bmatrix}$$

Suppose that we wish to find the product **AB**. The number in row 1 and column 1 of the product is obtained by multiplying the elements in row 1 of **A** by the corresponding elements in column 1 of **B** and adding these products. We obtain

$$4(2) + 3(3) = 8 + 9 = 17$$

The number in row 1 and column 2 of the product is obtained by multiplying the elements in row 1 of **A** by the corresponding elements in column 2 of **B** and adding these products. We obtain

$$4(1) + 3(5) = 4 + 15 = 19$$

The number in row 2 and column 1 of the product is obtained by multiplying the elements in row 2 of **A** by the corresponding elements in column 1 of **B** and adding these products. We obtain

$$2(2) + 2(3) = 4 + 6 = 10$$

The number in row 2 and column 2 of the product is obtained by multiplying the elements in row 2 of **A** by the corresponding elements in column 2 of **B** and adding these products. We obtain

$$2(1) + 2(5) = 2 + 10 = 12$$

Thus the product **AB** is as follows.

$$\mathbf{A}_{2x2}\mathbf{B}_{2x2} = \begin{bmatrix} 4 & 3 \\ 2 & 2 \end{bmatrix} \begin{bmatrix} 2 & 1 \\ 3 & 5 \end{bmatrix} = \begin{bmatrix} 4(2) + 3(3) & 4(1) + 3(5) \\ 2(2) + 2(3) & 2(1) + 2(5) \end{bmatrix}$$
$$= \begin{bmatrix} 17 & 19 \\ 10 & 12 \end{bmatrix}$$

In general, we can multiply a matrix **A** with m rows and r columns by a matrix **B** with r rows and n columns and obtain a matrix **C** with m rows and n columns. Moreover, c_{ij}, the number in the product in row i and column j, is obtained by multiplying the elements in row i of **A** by the corresponding elements in column j of **B** and adding the resulting products. Note that the number of columns in **A** must equal the number of rows in **B** in order for this multiplication procedure to be defined.

The multiplication procedure is illustrated below.

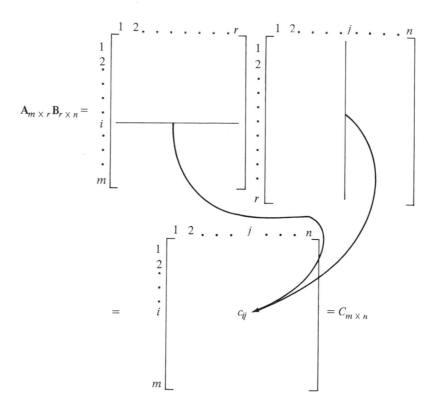

Let us now examine several more examples. Consider the following matrixes.

$$\mathbf{W}_{3x2} = \begin{bmatrix} 1 & 6 \\ 2 & 5 \\ 3 & 4 \end{bmatrix} \quad \mathbf{U}_{2x2} = \begin{bmatrix} 2 & 2 \\ 1 & 3 \end{bmatrix}$$

$$\mathbf{X}_{9x3} = \begin{bmatrix} 1 & 0 & 0 \\ 1 & 1 & 0 \\ 1 & 2 & 0 \\ 1 & 0 & 1 \\ 1 & 1 & 1 \\ 1 & 2 & 1 \\ 1 & 0 & 2 \\ 1 & 1 & 2 \\ 1 & 2 & 2 \end{bmatrix} \quad \mathbf{y}_{9x1} = \begin{bmatrix} 18 \\ 21 \\ 20 \\ 20 \\ 22 \\ 20 \\ 19 \\ 21 \\ 20 \end{bmatrix}$$

Then

$$\mathbf{W}_{3\times2}\mathbf{U}_{2\times2} = \begin{bmatrix} 1 & 6 \\ 2 & 5 \\ 3 & 4 \end{bmatrix} \begin{bmatrix} 2 & 2 \\ 1 & 3 \end{bmatrix} = \begin{bmatrix} 8 & 20 \\ 9 & 19 \\ 10 & 18 \end{bmatrix}$$

but

$$\mathbf{U}_{2\times2}\mathbf{W}_{3\times2} = \begin{bmatrix} 2 & 2 \\ 1 & 3 \end{bmatrix} \begin{bmatrix} 1 & 6 \\ 2 & 5 \\ 3 & 4 \end{bmatrix}$$

does not exist, because the number of columns in $\mathbf{U}_{2\times2}$ does not equal the number of rows in $\mathbf{W}_{3\times2}$.

$$\mathbf{X}'_{3\times9}\mathbf{X}_{9\times3} = \begin{bmatrix} 1 & 1 & 1 & 1 & 1 & 1 & 1 & 1 & 1 \\ 0 & 1 & 2 & 0 & 1 & 2 & 0 & 1 & 2 \\ 0 & 0 & 0 & 1 & 1 & 1 & 2 & 2 & 2 \end{bmatrix} \begin{bmatrix} 1 & 0 & 0 \\ 1 & 1 & 0 \\ 1 & 2 & 0 \\ 1 & 0 & 1 \\ 1 & 1 & 1 \\ 1 & 2 & 1 \\ 1 & 0 & 2 \\ 1 & 1 & 2 \\ 1 & 2 & 2 \end{bmatrix}$$

$$= \begin{bmatrix} 9 & 9 & 9 \\ 9 & 15 & 9 \\ 9 & 9 & 15 \end{bmatrix}$$

and

$$\mathbf{X}'_{3\times9}\mathbf{y}_{9\times1} = \begin{bmatrix} 1 & 1 & 1 & 1 & 1 & 1 & 1 & 1 & 1 \\ 0 & 1 & 2 & 0 & 1 & 2 & 0 & 1 & 2 \\ 0 & 0 & 0 & 1 & 1 & 1 & 2 & 2 & 2 \end{bmatrix} \begin{bmatrix} 18 \\ 21 \\ 20 \\ 20 \\ 22 \\ 20 \\ 19 \\ 21 \\ 20 \end{bmatrix} = \begin{bmatrix} 181 \\ 184 \\ 182 \end{bmatrix}$$

As a last example, consider the matrixes

$$\mathbf{A}_{2\times2} = \begin{bmatrix} 1 & 1 \\ 2 & 2 \end{bmatrix} \qquad \mathbf{B}_{2\times2} = \begin{bmatrix} 0 & 1 \\ 1 & 0 \end{bmatrix}$$

Then

$$A_{2x2}B_{2x2} = \begin{bmatrix} 1 & 1 \\ 2 & 2 \end{bmatrix} \begin{bmatrix} 0 & 1 \\ 1 & 0 \end{bmatrix} = \begin{bmatrix} 1 & 1 \\ 2 & 2 \end{bmatrix}$$

In this case **B** is said to be premultiplied by **A** or **A** is said to be postmultiplied by **B**. Now consider

$$B_{2x2}A_{2x2} = \begin{bmatrix} 0 & 1 \\ 1 & 0 \end{bmatrix} \begin{bmatrix} 1 & 1 \\ 2 & 2 \end{bmatrix} = \begin{bmatrix} 2 & 2 \\ 1 & 1 \end{bmatrix}$$

Here **A** is said to be premultiplied by **B** or **B** is said to be postmultiplied by **A**. Note that in this case **AB** is not equal to **BA**. In general, if **A** and **B** are matrixes, then **AB** $\neq$ **BA**.

We now continue by making several definitions. A matrix in which the number of rows is equal to the number of columns is called a *SQUARE MATRIX*. For example, the following matrixes are square matrixes.

$$A_{3x3} = \begin{bmatrix} 3 & 4 & 1 \\ 6 & 10 & 2 \\ 3 & 1 & 5 \end{bmatrix} \qquad B_{2x2} = \begin{bmatrix} 1 & 6 \\ 2 & 3 \end{bmatrix}$$

A square matrix in which the numbers on the main diagonal (the diagonal which runs from upper left to lower right) are 1's and all numbers off this diagonal are 0's is called an *IDENTITY MATRIX* and is denoted **I**. The following are examples of identity matrixes.

$$I_{2x2} = \begin{bmatrix} 1 & 0 \\ 0 & 1 \end{bmatrix} \qquad I_{3x3} = \begin{bmatrix} 1 & 0 & 0 \\ 0 & 1 & 0 \\ 0 & 0 & 1 \end{bmatrix}$$

Such a matrix is called an identity matrix because premultiplication or postmultiplication of a square n x n matrix **A** by the n x n identity matrix **I** leaves the matrix **A** unchanged. Consider the following example: If

$$A_{3x3} = \begin{bmatrix} 2 & 1 & 3 \\ 4 & 1 & 2 \\ 2 & 2 & 1 \end{bmatrix}$$

then

$$I_{3x3}A_{3x3} = \begin{bmatrix} 1 & 0 & 0 \\ 0 & 1 & 0 \\ 0 & 0 & 1 \end{bmatrix} \begin{bmatrix} 2 & 1 & 3 \\ 4 & 1 & 2 \\ 2 & 2 & 1 \end{bmatrix} = \begin{bmatrix} 2 & 1 & 3 \\ 4 & 1 & 2 \\ 2 & 2 & 1 \end{bmatrix}$$

and

$$\mathbf{A}_{3\times3}\mathbf{I}_{3\times3} = \begin{bmatrix} 2 & 1 & 3 \\ 4 & 1 & 2 \\ 2 & 2 & 1 \end{bmatrix} \begin{bmatrix} 1 & 0 & 0 \\ 0 & 1 & 0 \\ 0 & 0 & 1 \end{bmatrix} = \begin{bmatrix} 2 & 1 & 3 \\ 4 & 1 & 2 \\ 2 & 2 & 1 \end{bmatrix}$$

In general, if $\mathbf{A}_{m\times n}$ is not a square matrix, then

$$\mathbf{A}_{m\times n}\mathbf{I}_n = \mathbf{A}_{m\times n} \quad \text{and} \quad \mathbf{I}_m\mathbf{A}_{m\times n} = \mathbf{A}_{m\times n}$$

We now discuss two concepts known as linear independence and linear dependence. Consider the following matrix **A**.

$$\mathbf{A}_{3\times3} = \begin{bmatrix} 1 & 4 & 2 \\ 3 & 2 & 6 \\ 2 & 1 & 4 \end{bmatrix}$$

Notice that the third column in this matrix is a multiple of the first column in the matrix. In particular, the third column is simply the first column multiplied by two. That is,

$$\begin{bmatrix} 2 \\ 6 \\ 4 \end{bmatrix} = 2 \begin{bmatrix} 1 \\ 3 \\ 2 \end{bmatrix}$$

In a situation such as this, when one column in a matrix **A** is a multiple of another column in matrix **A**, the columns of **A** are said to be *LINEARLY DEPENDENT*. More generally, if one of the columns of a matrix **A** can be written as a linear combination of some of the other columns in **A**, then the columns of **A** are said to be *LINEARLY DEPENDENT*. As an example, consider the matrix

$$\mathbf{A}_{3\times4} = \begin{bmatrix} 1 & 3 & 10 & 4 \\ 5 & 2 & 5 & 1 \\ 2 & 2 & 7 & 3 \end{bmatrix}$$

In this case, column 3 is the sum of column 4, and twice column 2. That is,

$$\begin{bmatrix} 4 \\ 1 \\ 3 \end{bmatrix} + 2 \begin{bmatrix} 3 \\ 2 \\ 2 \end{bmatrix} = \begin{bmatrix} 4 \\ 1 \\ 3 \end{bmatrix} + \begin{bmatrix} 6 \\ 4 \\ 4 \end{bmatrix} = \begin{bmatrix} 10 \\ 5 \\ 7 \end{bmatrix}$$

Thus the columns of **A** are linearly dependent.

If none of the columns in a matrix **A** can be written as a linear combination of other columns in **A**, then the columns of **A** are *LINEARLY INDEPENDENT*. The maximum number of linearly independent columns in a matrix **A** is called the

RANK of the matrix. When the rank of a matrix $\mathbf{A}$ is equal to the number of columns in $\mathbf{A}$ the matrix $\mathbf{A}$ is said to be of FULL RANK.

Let us now consider a square matrix $\mathbf{A}_{nxn}$ that is of full rank. The INVERSE of the matrix $\mathbf{A}$ is another matrix, which is denoted $\mathbf{A}^{-1}$, which satisfies the condition

$$\mathbf{A}\mathbf{A}^{-1} = \mathbf{A}^{-1}\mathbf{A} = \mathbf{I}_{nxn}$$

where $\mathbf{I}_{nxn}$ is the identity matrix with dimension n x n. Again, it should be emphasized that $\mathbf{A}^{-1}$ exists if and only if $\mathbf{A}$ is a square matrix of full rank.

As an example, consider the matrix

$$\mathbf{A}_{3x3} = \begin{bmatrix} 9 & 9 & 9 \\ 9 & 15 & 9 \\ 9 & 9 & 15 \end{bmatrix}$$

The inverse of $\mathbf{A}$

$$\mathbf{A}_{3x3}^{-1} = \begin{bmatrix} 4/9 & -1/6 & -1/6 \\ -1/6 & 1/6 & 0 \\ -1/6 & 0 & 1/6 \end{bmatrix}$$

since

$$\mathbf{A}_{3x3}\mathbf{A}_{3x3}^{-1} = \begin{bmatrix} 9 & 9 & 9 \\ 9 & 15 & 9 \\ 9 & 9 & 15 \end{bmatrix} \begin{bmatrix} 4/9 & -1/6 & -1/6 \\ -1/6 & 1/6 & 0 \\ -1/6 & 0 & 1/6 \end{bmatrix}$$
$$= \begin{bmatrix} 1 & 0 & 0 \\ 0 & 1 & 0 \\ 0 & 0 & 1 \end{bmatrix}$$

and

$$\mathbf{A}_{3x3}^{-1}\mathbf{A}_{3x3} = \begin{bmatrix} 4/9 & -1/6 & -1/6 \\ -1/6 & 1/6 & 0 \\ -1/6 & 0 & 1/6 \end{bmatrix} \begin{bmatrix} 9 & 9 & 9 \\ 9 & 15 & 9 \\ 9 & 9 & 15 \end{bmatrix}$$
$$= \begin{bmatrix} 1 & 0 & 0 \\ 0 & 1 & 0 \\ 0 & 0 & 1 \end{bmatrix}$$

It can be shown that $\mathbf{A}^{-1}\mathbf{A} = \mathbf{I}_{nxn}$ if and only if $\mathbf{A}\mathbf{A}^{-1} = \mathbf{I}_{nxn}$.

Although we will not discuss them here, general formulas exist that allow the calculation of matrix inverses. Also, computer programs are often used to calculate matrix inverses.

REGRESSION CALCULATIONS USING MATRIXES

We now demonstrate how matrixes may be used in the calculation of the least squares estimates in regression analysis. The least squares estimates for the regression model

$$y_t = \beta_0 + \beta_1 x_{t1} + \beta_2 x_{t2} + \cdots + \beta_p x_{tp} + \epsilon_t$$

are given by the matrix equation $\mathbf{b} = (\mathbf{X'X})^{-1}\mathbf{X'y}$, where $\mathbf{X}$ is a matrix of the observed values of the independent variables and $\mathbf{y}$ is a column vector of the observed values of the dependent variable.

Suppose that

$$\mathbf{y} = \begin{bmatrix} 18 \\ 21 \\ 20 \\ 20 \\ 22 \\ 20 \\ 19 \\ 21 \\ 20 \end{bmatrix} \quad \text{and} \quad \mathbf{X} = \begin{bmatrix} 1 & 0 & 0 \\ 1 & 1 & 0 \\ 1 & 2 & 0 \\ 1 & 0 & 1 \\ 1 & 1 & 1 \\ 1 & 2 & 1 \\ 1 & 0 & 2 \\ 1 & 1 & 2 \\ 1 & 2 & 2 \end{bmatrix}$$

We now make the following calculations.

$$\mathbf{X'} = \begin{bmatrix} 1 & 1 & 1 & 1 & 1 & 1 & 1 & 1 & 1 \\ 0 & 1 & 2 & 0 & 1 & 2 & 0 & 1 & 2 \\ 0 & 0 & 0 & 1 & 1 & 1 & 2 & 2 & 2 \end{bmatrix}$$

$$\mathbf{X'X} = \begin{bmatrix} 1 & 1 & 1 & 1 & 1 & 1 & 1 & 1 & 1 \\ 0 & 1 & 2 & 0 & 1 & 2 & 0 & 1 & 2 \\ 0 & 0 & 0 & 1 & 1 & 1 & 2 & 2 & 2 \end{bmatrix} \begin{bmatrix} 1 & 0 & 0 \\ 1 & 1 & 0 \\ 1 & 2 & 0 \\ 1 & 0 & 1 \\ 1 & 1 & 1 \\ 1 & 2 & 1 \\ 1 & 0 & 2 \\ 1 & 1 & 2 \\ 1 & 2 & 2 \end{bmatrix}$$

$$= \begin{bmatrix} 9 & 9 & 9 \\ 9 & 15 & 9 \\ 9 & 9 & 15 \end{bmatrix}$$

Then,

$$(\mathbf{X'X})^{-1} = \begin{bmatrix} 4/9 & -1/6 & -1/6 \\ -1/6 & 1/6 & 0 \\ -1/6 & 0 & 1/6 \end{bmatrix}$$

since

$$(\mathbf{X'X})(\mathbf{X'X})^{-1} = \begin{bmatrix} 9 & 9 & 9 \\ 9 & 15 & 9 \\ 9 & 9 & 15 \end{bmatrix} \begin{bmatrix} 4/9 & -1/6 & -1/6 \\ -1/6 & 1/6 & 0 \\ -1/6 & 0 & 1/6 \end{bmatrix}$$

$$= \begin{bmatrix} 1 & 0 & 0 \\ 0 & 1 & 0 \\ 0 & 0 & 1 \end{bmatrix}$$

And

$$\mathbf{X'y} = \begin{bmatrix} 1 & 1 & 1 & 1 & 1 & 1 & 1 & 1 & 1 \\ 0 & 1 & 2 & 0 & 1 & 2 & 0 & 1 & 2 \\ 0 & 0 & 0 & 1 & 1 & 1 & 2 & 2 & 2 \end{bmatrix} \begin{bmatrix} 18 \\ 21 \\ 20 \\ 20 \\ 22 \\ 20 \\ 19 \\ 21 \\ 20 \end{bmatrix} = \begin{bmatrix} 181 \\ 184 \\ 182 \end{bmatrix}$$

Hence

$$\mathbf{b} = (\mathbf{X'X})^{-1}\mathbf{X'y} = \begin{bmatrix} 4/9 & -1/6 & -1/6 \\ -1/6 & 1/6 & 0 \\ -1/6 & 0 & 1/6 \end{bmatrix} \begin{bmatrix} 181 \\ 184 \\ 182 \end{bmatrix}$$

$$= \begin{bmatrix} 19.44 \\ 0.50 \\ 0.167 \end{bmatrix}$$

Thus the least squares estimates are given by the column vector

$$\mathbf{b} = \begin{bmatrix} 19.44 \\ 0.50 \\ 0.167 \end{bmatrix}$$

Another common matrix calculation that is made in connection with regression analysis is of the form

$$\mathbf{x_0'}(\mathbf{X'X})^{-1}\mathbf{x_0}$$

We will illustrate this type of calculation using the matrixes

$$\mathbf{x}_0 = \begin{bmatrix} 1 \\ 2 \\ 1 \end{bmatrix} \quad \text{and} \quad \mathbf{X} = \begin{bmatrix} 1 & 0 & 0 \\ 1 & 1 & 0 \\ 1 & 2 & 0 \\ 1 & 0 & 1 \\ 1 & 1 & 1 \\ 1 & 2 & 1 \\ 1 & 0 & 2 \\ 1 & 1 & 2 \\ 1 & 2 & 2 \end{bmatrix}$$

As shown above,

$$(\mathbf{X}'\mathbf{X})^{-1} = \begin{bmatrix} 4/9 & -1/6 & -1/6 \\ -1/6 & 1/6 & 0 \\ -1/6 & 0 & 1/6 \end{bmatrix}$$

Then

$$\mathbf{x}_0'(\mathbf{X}'\mathbf{X})^{-1}\mathbf{x}_0 = \begin{bmatrix} 1 & 2 & 1 \end{bmatrix} \begin{bmatrix} 4/9 & -1/6 & -1/6 \\ -1/6 & 1/6 & 0 \\ -1/6 & 0 & 1/6 \end{bmatrix} \begin{bmatrix} 1 \\ 2 \\ 1 \end{bmatrix}$$

$$= \begin{bmatrix} -1/18 & 1/6 & 0 \end{bmatrix} \begin{bmatrix} 1 \\ 2 \\ 1 \end{bmatrix} = 5/18$$

Normal Curve Areas, Critical Values of t, Percentage Points of the F-distribution, Critical Values of Chi-Square

TABLE B1   *Normal Curve Areas*

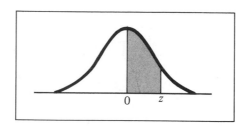

z	.00	.01	.02	.03	.04	.05	.06	.07	.08	.09
0.0	.0000	.0040	.0080	.0120	.0160	.0199	.0239	.0279	.0319	.0359
0.1	.0398	.0438	.0478	.0517	.0557	.0596	.0636	.0675	.0714	.0753
0.2	.0793	.0832	.0871	.0910	.0948	.0987	.1026	.1064	.1103	.1141
0.3	.1179	.1217	.1255	.1293	.1331	.1368	.1406	.1443	.1480	.1517
0.4	.1554	.1591	.1628	.1664	.1700	.1736	.1772	.1808	.1844	.1879
0.5	.1915	.1950	.1985	.2019	.2054	.2088	.2123	.2157	.2190	.2224
0.6	.2257	.2291	.2324	.2357	.2389	.2422	.2454	.2486	.2517	.2549
0.7	.2580	.2611	.2642	.2673	.2704	.2734	.2764	.2794	.2823	.2852
0.8	.2881	.2910	.2939	.2967	.2995	.3023	.3051	.3078	.3106	.3133
0.9	.3159	.3186	.3212	.3238	.3264	.3289	.3315	.3340	.3365	.3389
1.0	.3413	.3438	.3461	.3485	.3508	.3531	.3554	.3577	.3599	.3621
1.1	.3643	.3665	.3686	.3708	.3729	.3749	.3770	.3790	.3810	.3830
1.2	.3849	.3869	.3888	.3907	.3925	.3944	.3962	.3980	.3997	.4015
1.3	.4032	.4049	.4066	.4082	.4099	.4115	.4131	.4147	.4162	.4177
1.4	.4192	.4207	.4222	.4236	.4251	.4265	.4279	.4292	.4306	.4319
1.5	.4332	.4345	.4357	.4370	.4382	.4394	.4406	.4418	.4429	.4441
1.6	.4452	.4463	.4474	.4484	.4495	.4505	.4515	.4525	.4535	.4545
1.7	.4554	.4564	.4573	.4582	.4591	.4599	.4608	.4616	.4625	.4633
1.8	.4641	.4649	.4656	.4664	.4671	.4678	.4686	.4693	.4699	.4706
1.9	.4713	.4719	.4726	.4732	.4738	.4744	.4750	.4756	.4761	.4767
2.0	.4772	.4778	.4783	.4788	.4793	.4798	.4803	.4808	.4812	.4817
2.1	.4821	.4826	.4830	.4834	.4838	.4842	.4846	.4850	.4854	.4857
2.2	.4861	.4864	.4868	.4871	.4875	.4878	.4881	.4884	.4887	.4890
2.3	.4893	.4896	.4898	.4901	.4904	.4906	.4909	.4911	.4913	.4916
2.4	.4918	.4920	.4922	.4925	.4927	.4929	.4931	.4932	.4934	.4936
2.5	.4938	.4940	.4941	.4943	.4945	.4946	.4948	.4949	.4951	.4952
2.6	.4953	.4955	.4956	.4957	.4959	.4960	.4961	.4962	.4963	.4964
2.7	.4965	.4966	.4967	.4968	.4969	.4970	.4971	.4972	.4973	.4974
2.8	.4974	.4975	.4976	.4977	.4977	.4978	.4979	.4979	.4980	.4981
2.9	.4981	.4982	.4982	.4983	.4984	.4984	.4985	.4985	.4986	.4986
3.0	.4987	.4987	.4987	.4988	.4988	.4989	.4989	.4989	.4990	.4990

Source: A Hald, *Statistical Tables and Formulas* (New York: John Wiley & Sons, 1952), abridged from Table 1. Reproduced by permission of the publisher.

TABLE B2   *Critical Values of t*

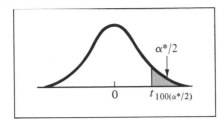

d.f.	t_{10}	t_5	$t_{2.5}$	t_1	$t_{.5}$
1	3.078	6.314	12.706	31.821	63.657
2	1.886	2.920	4.303	6.965	9.925
3	1.638	2.353	3.182	4.541	5.841
4	1.533	2.132	2.776	3.747	4.604
5	1.476	2.015	2.571	3.365	4.032
6	1.440	1.943	2.447	3.143	3.707
7	1.415	1.895	2.365	2.998	3.499
8	1.397	1.860	2.306	2.896	3.355
9	1.383	1.833	2.262	2.821	3.250
10	1.372	1.812	2.228	2.764	3.169
11	1.363	1.796	2.201	2.718	3.106
12	1.356	1.782	2.179	2.681	3.055
13	1.350	1.771	2.160	2.650	3.012
14	1.345	1.761	2.145	2.624	2.977
15	1.341	1.753	2.131	2.602	2.947
16	1.337	1.746	2.120	2.583	2.921
17	1.333	1.740	2.110	2.567	2.898
18	1.330	1.734	2.101	2.552	2.878
19	1.328	1.729	2.093	2.539	2.861
20	1.325	1.725	2.086	2.528	2.845
21	1.323	1.721	2.080	2.518	2.831
22	1.321	1.717	2.074	2.508	2.819
23	1.319	1.714	2.069	2.500	2.807
24	1.318	1.711	2.064	2.492	2.797
25	1.316	1.708	2.060	2.485	2.787
26	1.315	1.706	2.056	2.479	2.779
27	1.314	1.703	2.052	2.473	2.771
28	1.313	1.701	2.048	2.467	2.763
29	1.311	1.699	2.045	2.462	2.756
inf.	1.282	1.645	1.960	2.326	2.576

Source: From "Table of Percentage Points of the *t*-Distribution," computed by Maxine Merrington, *Biometrika 32* (1941), 300. Reproduced by permission of Biometrika Trustees.

TABLE B3   *Percentage Points of the F-Distribution*

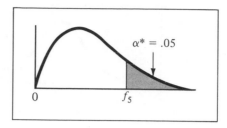

$\alpha^* = .05$

Degrees of Freedom

ν_2 \ ν_1	1	2	3	4	5	6	7	8	9
1	161.4	199.5	215.7	224.6	230.2	234.0	236.8	238.9	240.5
2	18.51	19.00	19.16	19.25	19.30	19.33	19.35	19.37	19.38
3	10.13	9.55	9.28	9.12	9.01	8.94	8.89	8.85	8.81
4	7.71	6.94	6.59	6.39	6.26	6.16	6.09	6.04	6.00
5	6.61	5.79	5.41	5.19	5.05	4.95	4.88	4.82	4.77
6	5.99	5.14	4.76	4.53	4.39	4.28	4.21	4.15	4.10
7	5.59	4.74	4.35	4.12	3.97	3.87	3.79	3.73	3.68
8	5.32	4.46	4.07	3.84	3.69	3.58	3.50	3.44	3.39
9	5.12	4.26	3.86	3.63	3.48	3.37	3.29	3.23	3.18
10	4.96	4.10	3.71	3.48	3.33	3.22	3.14	3.07	3.02
11	4.84	3.98	3.59	3.36	3.20	3.09	3.01	2.95	2.90
12	4.75	3.89	3.49	3.26	3.11	3.00	2.91	2.85	2.80
13	4.67	3.81	3.41	3.18	3.03	2.92	2.83	2.77	2.71
14	4.60	3.74	3.34	3.11	2.96	2.85	2.76	2.70	2.65
15	4.54	3.68	3.29	3.06	2.90	2.79	2.71	2.64	2.59
16	4.49	3.63	3.24	3.01	2.85	2.74	2.66	2.59	2.54
17	4.45	3.59	3.20	2.96	2.81	2.70	2.61	2.55	2.49
18	4.41	3.55	3.16	2.93	2.77	2.66	2.58	2.51	2.46
19	4.38	3.52	3.13	2.90	2.74	2.63	2.54	2.48	2.42
20	4.35	3.49	3.10	2.87	2.71	2.60	2.51	2.45	2.39
21	4.32	3.47	3.07	2.84	2.68	2.57	2.49	2.42	2.37
22	4.30	3.44	3.05	2.82	2.66	2.55	2.46	2.40	2.34
23	4.28	3.42	3.03	2.80	2.64	2.53	2.44	2.37	2.32
24	4.26	3.40	3.01	2.78	2.62	2.51	2.42	2.36	2.30
25	4.24	3.39	2.99	2.76	2.60	2.49	2.40	2.34	2.28
26	4.23	3.37	2.98	2.74	2.59	2.47	2.39	2.32	2.27
27	4.21	3.35	2.96	2.73	2.57	2.46	2.37	2.31	2.25
28	4.20	3.34	2.95	2.71	2.56	2.45	2.36	2.29	2.24
29	4.18	3.33	2.93	2.70	2.55	2.43	2.35	2.28	2.22
30	4.17	3.32	2.92	2.69	2.53	2.42	2.33	2.27	2.21
40	4.08	3.23	2.84	2.61	2.45	2.34	2.25	2.18	2.12
60	4.00	3.15	2.76	2.53	2.37	2.25	2.17	2.10	2.04
120	3.92	3.07	2.68	2.45	2.29	2.17	2.09	2.02	1.96
∞	3.84	3.00	2.60	2.37	2.21	2.10	2.01	1.94	1.88

10	12	15	20	24	30	40	60	120	∞	ν_1 / ν_2
241.9	243.9	245.9	248.0	249.1	250.1	251.1	252.2	253.3	254.3	1
19.40	19.41	19.43	19.45	19.45	19.46	19.47	19.48	19.49	19.50	2
8.79	8.74	8.70	8.66	8.64	8.62	8.59	8.57	8.55	8.53	3
5.96	5.91	5.86	5.80	5.77	5.75	5.72	5.69	5.66	5.63	4
4.74	4.68	4.62	4.56	4.53	4.50	4.46	4.43	4.40	4.36	5
4.06	4.00	3.94	3.87	3.84	3.81	3.77	3.74	3.70	3.67	6
3.64	3.57	3.51	3.44	3.41	3.38	3.34	3.30	3.27	3.23	7
3.35	3.28	3.22	3.15	3.12	3.08	3.04	3.01	2.97	2.93	8
3.14	3.07	3.01	2.94	2.90	2.86	2.83	2.79	2.75	2.71	9
2.98	2.91	2.85	2.77	2.74	2.70	2.66	2.62	2.58	2.54	10
2.85	2.79	2.72	2.65	2.61	2.57	2.53	2.49	2.45	2.40	11
2.75	2.69	2.62	2.54	2.51	2.47	2.43	2.38	2.34	2.30	12
2.67	2.60	2.53	2.46	2.42	2.38	2.34	2.30	2.25	2.21	13
2.60	2.53	2.46	2.39	2.35	2.31	2.27	2.22	2.18	2.13	14
2.54	2.48	2.40	2.33	2.29	2.25	2.20	2.16	2.11	2.07	15
2.49	2.42	2.35	2.28	2.24	2.19	2.15	2.11	2.06	2.01	16
2.45	2.38	2.31	2.23	2.19	2.15	2.10	2.06	2.01	1.96	17
2.41	2.34	2.27	2.19	2.15	2.11	2.06	2.02	1.97	1.92	18
2.38	2.31	2.23	2.16	2.11	2.07	2.03	1.98	1.93	1.88	19
2.35	2.28	2.20	2.12	2.08	2.04	1.99	1.95	1.90	1.84	20
2.32	2.25	2.18	2.10	2.05	2.01	1.96	1.92	1.87	1.81	21
2.30	2.23	2.15	2.07	2.03	1.98	1.94	1.89	1.84	1.78	22
2.27	2.20	2.13	2.05	2.01	1.96	1.91	1.86	1.81	1.76	23
2.25	2.18	2.11	2.03	1.98	1.94	1.89	1.84	1.79	1.73	24
2.24	2.16	2.09	2.01	1.96	1.92	1.87	1.82	1.77	1.71	25
2.22	2.15	2.07	1.99	1.95	1.90	1.85	1.80	1.75	1.69	26
2.20	2.13	2.06	1.97	1.93	1.88	1.84	1.79	1.73	1.67	27
2.19	2.12	2.04	1.96	1.91	1.87	1.82	1.77	1.71	1.65	28
2.18	2.10	2.03	1.94	1.90	1.85	1.81	1.75	1.70	1.64	29
2.16	2.09	2.01	1.93	1.89	1.84	1.79	1.74	1.68	1.62	30
2.08	2.00	1.92	1.84	1.79	1.74	1.69	1.64	1.58	1.51	40
1.99	1.92	1.84	1.75	1.70	1.65	1.59	1.53	1.47	1.39	60
1.91	1.83	1.75	1.66	1.61	1.55	1.50	1.43	1.35	1.25	120
1.83	1.75	1.67	1.57	1.52	1.46	1.39	1.32	1.22	1.00	∞

TABLE B4   *Critical Values of Chi-Square*

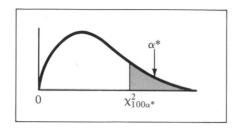

d.f.	$\chi^2(99.5)$	$\chi^2(99)$	$\chi^2(97.5)$	$\chi^2(95)$	$\chi^2(90)$
1	0.0000393	0.0001571	0.0009821	0.0039321	0.0157908
2	0.0100251	0.0201007	0.0506356	0.102587	0.210720
3	0.0717212	0.114832	0.215795	0.351846	0.584375
4	0.206990	0.297110	0.484419	0.710721	1.063623
5	0.411740	0.554300	0.831211	1.145476	1.61031
6	0.675727	0.872085	1.237347	1.63539	2.20413
7	0.989265	1.239043	1.68987	2.16735	2.83311
8	1.344419	1.646482	2.17973	2.73264	3.48954
9	1.734926	2.087912	2.70039	3.32511	4.16816
10	2.15585	2.55821	3.24697	3.94030	4.86518
11	2.60321	3.05347	3.81575	4.57481	5.57779
12	3.07382	3.57056	4.40379	5.22603	6.30380
13	3.56503	4.10691	5.00874	5.89186	7.04150
14	4.07468	4.66043	5.62872	6.57063	7.78953
15	4.60094	5.22935	6.26214	7.26094	8.54675
16	5.14224	5.81221	6.90766	7.96164	9.31223
17	5.69724	6.40776	7.56418	8.67176	10.0852
18	6.26481	7.01491	8.23075	9.39046	10.8649
19	6.84398	7.63273	8.90655	10.1170	11.6509
20	7.43386	8.26040	9.59083	10.8508	12.4426
21	8.03366	8.89720	10.28293	11.5913	13.2396
22	8.64272	9.54249	10.9823	12.3380	14.0415
23	9.26042	10.19567	11.6885	13.0905	14.8479
24	9.88623	10.8564	12.4011	13.8484	15.6587
25	10.5197	11.5240	13.1197	14.6114	16.4734
26	11.1603	12.1981	13.8439	15.3791	17.2919
27	11.8076	12.8786	14.5733	16.1513	18.1138
28	12.4613	13.5648	15.3079	16.9279	18.9392
29	13.1211	14.2565	16.0471	17.7083	19.7677
30	13.7867	14.9535	16.7908	18.4926	20.5992
40	20.7065	22.1643	24.4331	26.5093	29.0505
50	27.9907	29.7067	32.3574	34.7642	37.6886
60	35.5346	37.4848	40.4817	43.1879	46.4589
70	43.2752	45.4418	48.7576	51.7393	55.3290
80	51.1720	53.5400	57.1532	60.3915	64.2778
90	59.1963	61.7541	65.6466	69.1260	73.2912
100	67.3276	70.0648	74.2219	77.9295	82.3581

Source for Table B4: From "Tables of the Percentage Points of the χ²-Distribution," by Catherine M. Thompson, *Biometrika 32* (1941), 188–189. Reproduced by permission of Biometrika Trustees.

$\chi^2(10)$	$\chi^2(5)$	$\chi^2(2.5)$	$\chi^2(1)$	$\chi^2(.5)$	d.f.
2.70554	3.84146	5.02389	6.63490	7.87944	1
4.60517	5.99147	7.37776	9.21034	10.5966	2
6.25139	7.81473	9.34840	11.3449	12.8381	3
7.77944	9.48773	11.1433	13.2767	14.8602	4
9.23635	11.0705	12.8325	15.0863	16.7496	5
10.6446	12.5916	14.4494	16.8119	18.5476	6
12.0170	14.0671	16.0128	18.4753	20.2777	7
13.3616	15.5073	17.5346	20.0902	21.9550	8
14.6837	16.9190	19.0228	21.6660	23.5893	9
15.9871	18.3070	20.4831	23.2093	25.1882	10
17.2750	19.6751	21.9200	24.7250	26.7569	11
18.5494	21.0261	23.3367	26.2170	28.2995	12
19.8119	22.3621	24.7356	27.6883	29.8194	13
21.0642	23.6848	26.1190	29.1413	31.3193	14
22.3072	24.9958	27.4884	30.5779	32.8013	15
23.5418	26.2962	28.8454	31.9999	34.2672	16
24.7690	27.5871	30.1910	33.4087	35.7185	17
25.9894	28.8693	31.5264	34.8053	37.1564	18
27.2036	30.1435	32.8523	36.1908	38.5822	19
28.4120	31.4104	34.1696	37.5662	39.9968	20
29.6151	32.6705	35.4789	38.9321	41.4010	21
30.8133	33.9244	36.7807	40.2894	42.7956	22
32.0069	35.1725	38.0757	41.6384	44.1813	23
33.1963	36.4151	39.3641	42.9798	45.5585	24
34.3816	37.6525	40.6465	44.3141	46.9278	25
35.5631	38.8852	41.9232	45.6417	48.2899	26
36.7412	40.1133	43.1944	46.9630	49.6449	27
37.9159	41.3372	44.4607	48.2782	50.9933	28
39.0875	42.5569	45.7222	49.5879	52.3356	29
40.2560	43.7729	46.9792	50.8922	53.6720	30
51.8050	55.7585	59.3417	63.6907	66.7659	40
63.1671	67.5048	71.4202	76.1539	79.4900	50
74.3970	79.0819	83.2976	88.3794	91.9517	60
85.5271	90.5312	95.0231	100.425	104.215	70
96.5782	101.879	106.629	112.329	116.321	80
107.565	113.145	118.136	124.116	128.299	90
118.498	124.342	129.561	135.807	140.169	100

Index